LONDON MATHEMATICAL SOCIETY LECTURE

Managing Editor: Professor N.J. Hitchin, Mathematical Institute,
University of Oxford, 24–29 St Giles, Oxford OX1 3LB, United

The titles below are available from booksellers, or, in case of di

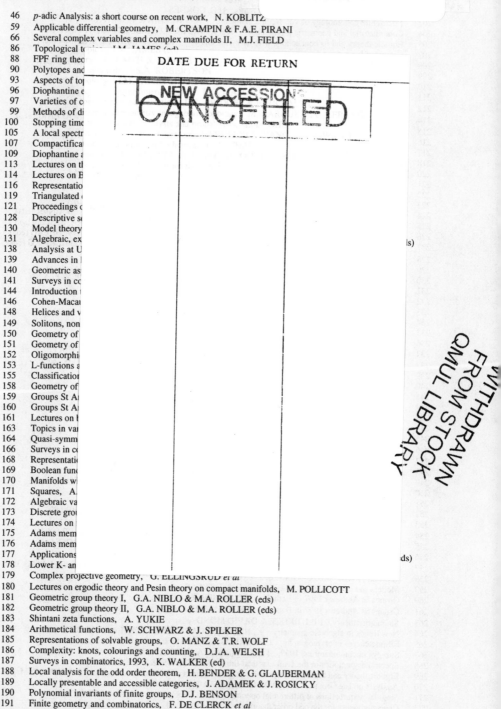

London Mathematical Society Lecture Note Series. 275

Computational and Geometric Aspects of Modern Algebra

Edited by

Michael Atkinson
University of St Andrews

Nick Gilbert
Heriot-Watt University

James Howie
Heriot-Watt University

Steve Linton
University of St Andrews

Edmund Robertson
University of St Andrews

PUBLISHED BY THE PRESS SYNDICATE OF THE UNIVERSITY OF CAMBRIDGE
The Pitt Building, Trumpington Street, Cambridge, United Kingdom

CAMBRIDGE UNIVERSITY PRESS
The Edinburgh Building, Cambridge, CB2 2RU, UK   http://www.cup.cam.ac.uk
40 West 20th Street, New York, NY 10011–4211, USA   http://www.cup.org
10 Stamford Road, Oakleigh, Melbourne 3166, Australia
Ruiz de Alarcón 13, 28014 Madrid, Spain

First published 2000

Printed in the United Kingdom at the University Press, Cambridge

A catalogue record for this book is available from the British Library

ISBN   0 521 78889 7 paperback

CONTENTS

FOREWORD

We are pleased to present this selection of articles contributed by participants at a workshop on Computational and Geometric Aspects of Modern Algebra, which took place at Heriot-Watt University, 23-31 July 1998, under the auspices of the International Centre for Mathematical Sciences (ICMS).

The workshop was generously supported by the UK Engineering and Physical Sciences Research Council, with additional financial support from the London Mathematical Society, the Edinburgh Mathematical Society, and Heriot-Watt University. Its organisation was made infinitely smoother by the various assistance of Tracey Dart and Julie Brown of ICMS, and Isobel Johnston and Fiona Paterson of Heriot-Watt.

In the preparation of this volume, we have received invaluable help and advice from Roger Astley and Tamsin van Essen of CUP, numerous anonymous referees, and of course the contributors of the articles. To all of the above we wish to record our gratitude.

M D Atkinson
N D Gilbert
J Howie
S A Linton
E F Robertson
October 1999

CGAMA 98 Participants

I Araujo (St. Andrews)
M D Atkinson (St. Andrews)
R E Arthur (St. Andrews)
L Bartholdi (Geneva)
B Baumeister (Imperial College)
N Billington (Victoria Univ)
S Billington (Warwick)
A Bis (Lodz)
B H Bowditch (Southampton)
N Brady (Cornell)
T Brady (DCU)
M R Bridson (Oxford)
C J B Brookes (Cambridge)
C M Campbell (St. Andrews)
I M Chiswell (QMW)
A Clow (Warwick)
D E Cohen (QMW)
G Conner (BYU)
J S Crisp (Southampton)
D Cruickshank (Glasgow)
R L Curtis (Geneva)
A Cutting (St. Andrews)
M Czarnecki (Lodz)
A Drapal (Prague)
A J Duncan (Newcastle)
M J Dunwoody (Southampton)
M Edjvet (Nottingham)
M Elder (Melbourne)
D B A Epstein (Warwick)
B Everitt (York)
V Felsch (Aachen)
B Fine (Fairfield)
A Fish (Warwick)
R Foord (Warwick)
N D Gilbert (Heriot-Watt)
L Grasselli (Bologna)
R I Grigorchuk (Steklov Inst.)
P Hammond (Nottingham)
J Harlander (Frankfurt)
G Havas (Queensland)

S Hermiller (Las Cruces)
A Heyworth (Bangor)
M P Hitchman (Lewis & Clark)
C Hog-Angeloni (Frankfurt)
D F Holt (Warwick)
S Holub (Prague)
J Howie (Heriot-Watt)
G Huck (Flagstaff)
A Hulpke (St. Andrews)
S P Humphries (BYU)
M R Jerrum (Edinburgh)
D L Johnson (UWI, Mona)
O Kharlampovich (McGill)
J Kortelainen (Oulu)
G Levitt (Toulouse)
S A Linton (St. Andrews)
M Lustig (Bochum)
C Maclachlan (Aberdeen)
K Madlener (Kaiserslautern)
B Makubate (Glasgow)
J Marshall (Warwick)
G Martin (Auckland)
U H M Martin (St. Andrews)
J McCammond (Texas A & M)
V Metaftsis (Univ Aegean)
W Metzler (Frankfurt)
C F Miller (Melbourne)
W D Neumann (Melbourne)
W Nickel (St. Andrews)
J-P Preaux (Marseille)
S J Pride (Glasgow)
A R Prince (Heriot-Watt)
M A Ramirez (Barcelona)
A A Ranicki (Edinburgh)
S E Rees (Newcastle)
A W Reid (Austin)
V N Remeslennikov (Omsk)
K Reynolds (Newcastle)
E F Robertson (St. Andrews)
C Roever (Oxford)

S Rosebrock (Karlsruhe)

G Rosenberger (Dortmund)

A Rosenmann (Tel Aviv)

N Ruskuc (St. Andrews)

M E Sageev (Rutgers)

M V Sapir (Vanderbilt)

Z Sela (Jerusalem)

V Yu Shavrukov (Leicester)

H Short (Marseille)

R Shwartz (Technion)

G C Smith (Bath)

A Solomon (St. Andrews)

E Souche (Marseille)

R Strebel (Fribourg)

V Sushchansky (Gliwice/Kiev)

E Swenson (BYU)

O Talelli (Athens)

R M Thomas (Leicester)

R Thomson (St. Andrews)

E Ventura (Manresa)

K Vogtmann (Cornell)

J Wang (Glasgow)

R Weidmann (Bochum)

A G Williams (Heriot-Watt)

LIE METHODS IN GROWTH OF GROUPS AND GROUPS OF FINITE WIDTH

LAURENT BARTHOLDI AND ROSTISLAV I. GRIGORCHUK

ABSTRACT. In the first, mostly expository, part of this paper, a graded Lie algebra is associated to every group G given with an N-series of subgroups. The asymptotics of the Poincaré series of this algebra give estimates on the growth of the group G. This establishes the existence of a gap between polynomial growth and growth of type $e^{\sqrt{n}}$ in the class of residually-p groups, and gives examples of finitely generated p-groups of uniformly exponential growth.

In the second part, we produce two examples of groups of finite width and describe their Lie algebras, introducing a notion of *Cayley graph* for graded Lie algebras. We compute explicitly their lower central and dimensional series, and outline a general method applicable to some other groups from the class of branch groups.

These examples produce counterexamples to a conjecture on the structure of just-infinite groups of finite width.

1. INTRODUCTION

The main goal of this paper is to present new examples of groups of finite width and to give a method of proving that some groups from the class of branch groups have finite width. This provides examples of groups of finite width with a completely new origin and answers a question asked by several mathematicians. We also give new examples of Lie algebras of finite width associated to the groups mentioned above.

The first group we study, $\mathfrak{G}$, was constructed in [Gri80] where it was shown to be an infinite torsion group; later in [Gri84] it was shown to be of intermediate growth. The second group, $\widetilde{\mathfrak{G}}$, was already considered by the second author in 1979, but was rejected at that time for not being periodic. We now know that it also has intermediate growth [BG98] and finite width.

Our interest in the finite width property comes from the theory of growth of groups. Another important area connected to this property is the theory of finite p-groups and the theory of pro-p-groups; see [Sha95b], [Sha95a, §8]

1991 *Mathematics Subject Classification.* 20F50,20F14,17B50,16P90.

The authors express their thanks to the Swiss National Science Foundation; the second author thanks the Russian Fund for Fundamental Research, research grant 01-00974 for its support.

and [KLP97] with its bibliography. More precisely, the following was discussed by many mathematicians and stated by Zel'manov in Castelvecchio in 1996 [Zel96]:

Conjecture 1.1. *Let G be a just-infinite pro-p-group of finite width. Then G is either solvable, p-adic analytic, or commensurable to a positive part of a loop group or to the Nottingham group.*

Our computations disprove this conjecture by providing a counter-example, the profinite completion of $\mathfrak{G}$ (it is a pro-p-group with $p = 2$). Note that it exhibits a behaviour specific to positive characteristic: indeed it was proved by Martinez and Zel'manov in [MZ99] that unipotence and finite width imply local nilpotence.

Before we give the definition of a group of finite width, let us recall a classical construction of Magnus [Mag40], described for instance in [HB82, Chapter VIII]. Given a group G and $\{G_n\}_{n=1}^{\infty}$ an N-series (i.e. a series of normal subgroups with $G_1 = G$, $G_{n+1} \leq G_n$ and $[G_m, G_n] \leq G_{m+n}$ for all $m, n \geq 1$), there is a canonical way of associating to G a graded Lie ring

$$(1) \qquad \mathcal{L}(G) = \bigoplus_{n=1}^{\infty} L_n,$$

where $L_n = G_n/G_{n+1}$ and the bracket operation is induced by commutation in G. Possible examples of N-series are the lower central series $\{\gamma_n(G)\}_{n=1}^{\infty}$; for an integer p, the *lower p-central series* given by $P_1(G) = G$ and $P_{n+1}(G) = P_n(G)^p[P_n(G), G]$; and, for a field $\Bbbk$, the series of $\Bbbk$-*dimension subgroups* $\{G_n\}_{n=1}^{\infty}$ defined by

$$G_n = \{g \in G | g - 1 \in \Delta^n\}, \qquad n = 1, 2, \ldots$$

where Δ is the augmentation (or fundamental) ideal of the group algebra $\Bbbk[G]$.

Tensoring the $\mathbb{Z}$-modules L_n with a suitable field $\Bbbk$, we obtain in (1) a graded Lie algebra $\mathcal{L}_{\Bbbk}(G)$. In case the N-series chosen satisfies the additional condition $G_n^p \leq G_{pn}$, and $\Bbbk$ is a field of characteristic p, the algebra $\mathcal{L}_{\Bbbk}(G)$ will then be a p-algebra or *restricted algebra*; see [Jac41] or [Jac62, Chapter V], the Frobenius operation on $\mathcal{L}_{\Bbbk}(G)$ being induced by raising to the power p in G. In this case the quotients G_n/G_{n+1} are elementary p-groups.

Many properties of a group are reflected in properties of its corresponding Lie algebra. For instance, one of the most important results obtained using the Lie method is the theorem of Zel'manov [Zel95a] asserting that if the Lie algebra $\mathcal{L}_{\mathbb{F}_p}(G)$ associated to the dimension subgroups of a finitely generated periodic residually-p group G satisfies a polynomial identity then the group G is finite ($\mathbb{F}_p$ is the prime field of characteristic p). This result gives in fact a positive solution to the Restricted Burnside Problem [VZ93, Zel95b, VZ96, Zel97]. Another example is the criterion of analyticity of pro-p-groups discovered by Lazard [Laz65].

The Lie method also applies to the theory of growth of groups, as was first observed in [Gri89]. There the second author proved that in the class of residually-p groups there is a gap between polynomial growth and growth of type $e^{\sqrt{n}}$. This result was then generalized in [LM91, Theorem D] to the class of residually-nilpotent groups, and in [CG97] the Lie method was also used to prove that certain one-relator groups with exponential-growth Lie algebra $\mathcal{L}_{\mathbb{k}}(G)$ have uniformly exponential growth. If a group G is finitely generated, then its Lie algebra $\mathcal{L}_{\mathbb{k}}(G) = \bigoplus L_n \otimes \mathbb{k}$ is also finitely generated, and the growth of $\mathcal{L}_{\mathbb{k}}(G)$ is by definition the growth of the sequence $\{b_n = \dim(L_n \otimes \mathbb{k})\}_{n=1}^{\infty}$.

The investigation of the growth of graded algebras related to groups has its own interest and is related to other topics. One of the first results in this direction is the Golod-Shafarevich inequality [GS64] which plays an important role in group, number and field theories. The idea of Golod and Shafarevich was used by Lazard in the proof of the aforementioned criterion of analyticity (he even used the notation 'gosha' for the growth of the algebras). Vershik and Kaimanovich observed the relation between the growth of gosha, amenability, and asymptotic behaviour of random walks (see Section 4 below).

For our purposes it will be sufficient to consider only the fields $\mathbb{Q}$ and $\mathbb{F}_p$. Let G_n be the corresponding series of dimension subgroups, which is also an N-series, and let $\mathcal{L}_{\mathbb{k}}(G)$ be the associated Lie algebra. If $\mathcal{L}_{\mathbb{k}}(G)$ is of polynomial growth of degree $d \geq 0$, then the growth of G is at least $e^{n^{1-1/(d+2)}}$, and if $\mathcal{L}_{\mathbb{k}}(G)$ is of exponential growth, then G is of uniformly exponential growth.

If $\mathbb{k} = \mathbb{Q}$ and G is residually-nilpotent and $b_n = 0$ for some n, then G is nilpotent; indeed G_n must be finite for that n, whence $\gamma_n(G)$ is finite too, and since $\bigcap_{k \geq 1} \gamma_k(G) = 1$ this implies that $\gamma_N(G) = 1$ for some N. It follows that G has polynomial growth [Mil68]. In fact polynomial growth is equivalent to virtual nilpotence [Gro81a].

If $\mathbb{k} = \mathbb{F}_p$ and G is a residually-p group and $b_n = 0$ for some n, then G is a linear group over a field, by Lazard's theorem [Laz65] and therefore has either polynomial or exponential growth, by the Tits alternative [Tit72].

Finally, if $b_n \geq 1$ for all n then, independent of $\mathbb{k}$, the growth of G is at least $e^{\sqrt{n}}$. Keeping in mind that polynomial growth $b_n \sim n^d$ of $\mathcal{L}_{\mathbb{k}}(G)$ implies a lower bound $e^{n^{1-1/(d+2)}}$ for the growth of G, we conclude that examples of groups with growth exactly $e^{\sqrt{n}}$ are to be found amongst the class of groups for which the sequence $\{b_n\}_{n=1}^{\infty}$ is uniformly bounded, or at least bounded in average. This key observation leads to the notion of *groups of finite width*. We present two versions of the definition:

Definition 1.2. Let G be a group and $\mathbb{k} \in \{\mathbb{Q}, \mathbb{F}_p\}$ a field. If $\mathbb{k} = \mathbb{Q}$, assume G is residually-nilpotent; if $\mathbb{k} = \mathbb{F}_p$, assume G is residually-p.

1. G has *finite C-width* if there is a constant K with $[\gamma_n(G) : \gamma_{n+1}(G)] \leq K$ for all n.

2. *G* has *finite D-width with respect to* $\Bbbk$ if there is a constant *K* with
$b_n \leq K$ for all *n*, where $\{b_n\}_{n=1}^{\infty}$ is the growth of $\mathcal{L}_{\Bbbk}(G)$ constructed
from the dimension subgroups.

A third notion can be defined, that of *finite averaged width*; see [Gri89]
or [KLP97, Definition I.1.ii]. From our point of view *D*-width is more natural;
but the first notion is more commonly used in the theory of finite *p*-groups
and pro-*p*-groups, see for instance [KLP97, Definition I.1.i]. The examples
we will produce are of finite width according to both definitions. That one
of our groups has finite width was conjectured in [Gri89]; it was proven that
the numbers b_n are bounded in average. Rozhkov then confirmed this conjecture in [Roz96a] by computing explicitly the b_n; but the proof had gaps,
one of which was filled in [Roz96b]. We fix another gap in the "Technical
Lemma 4.3.2" of [Roz96b] while simplifying and clarifying Rozhkov's proof,
and also outline a general method, connected to ideas of Kaloujnine [Kal46].

We recall in the next section known notions on algebras associated to
groups, and construct in Section 3 a torsion group of uniformly exponential
growth. Section 5 describes a class of groups acting on rooted trees, and the
next two sections detail for two specific examples the indices of the lower
central and dimensional series. More specifically, we compute in Theorem 6.6
and 7.6 the indices of these series for the group $\mathfrak{G}$ and an overgroup $\widetilde{\mathfrak{G}}$. We
also obtain in the process the structure of the Lie algebras $L(\mathfrak{G})$ (associated
to the lower central series) and $\mathcal{L}_{\mathbb{F}_2}(\mathfrak{G})$ (associated to the dimension series)
in Theorem 6.7, and that of $L(\widetilde{\mathfrak{G}})$ and $\mathcal{L}_{\mathbb{F}_2}(\widetilde{\mathfrak{G}})$ in Theorem 7.7. They are
described using Cayley graphs of Lie algebras, see Subsection 6.1.

Throughout this paper groups shall act on the left. We use the notational
conventions $[x, y] = xyx^{-1}y^{-1}$ and $x^y = yxy^{-1}$.

Both authors wish to thank Aner Shalev and Efim Zelmanov for their
interest and generous contribution through discussions.

2. GROWTH OF GROUPS AND ASSOCIATED GRADED ALGEBRAS

Let *G* be a group, $\{\gamma_n(G)\}_{n=1}^{\infty}$ the lower central series of *G*, $\Bbbk \in \{\mathbb{Q}, \mathbb{F}_p\}$
a prime field, $\Delta = \ker(\varepsilon) < \Bbbk[G]$ the augmentation ideal, where $\varepsilon(\sum k_i g_i) =$
$\sum k_i$ is the augmentation map $\Bbbk[G] \to \Bbbk$, and $\{G_n\}_{n=1}^{\infty}$ the series of dimension
subgroups of *G* [Zas40, Jen41]. Recall that

$$G_n = \{g \in G \,|\, g - 1 \in \Delta^n\}.$$

The restrictions we impose on $\Bbbk$ are not important, as G_n depends only on
the characteristic of $\Bbbk$. We suppose throughout that *G* is residually-nilpotent
if $\Bbbk = \mathbb{Q}$ and is residually-*p* if $\Bbbk = \mathbb{F}_p$.

If $\Bbbk = \mathbb{Q}$, then G_n is the isolator of $\gamma_n(G)$, as was proved in [Jen55] (see
also [Pman77, Theorem 11.1.10] or [Pas79, Theorem IV.1.5]); i.e.

$$G_n = \sqrt{\gamma_n(G)} = \{g \in G \,|\, g^\ell \in \gamma_n(G) \text{ for an } \ell \in \mathbb{N}\}.$$

Note that in [Pman77] these results are stated for finite p-groups. They nevertheless hold in the more general setting of residually-nilpotent or residually-p groups.

If $\Bbbk = \mathbb{F}_p$, then $\gamma_n(G) \leq G_n \leq \sqrt{\gamma_n(G)}$, and the G_n can be defined in several different ways, for instance by the relation

$$G_n = \prod_{i \cdot p^j \geq n} \gamma_i^{p^j}(G)$$

due to Lazard [Laz53], or recursively as

$$(2) \qquad G_n = [G, G_{n-1}]G_{\lceil n/p \rceil}^p,$$

where $\lceil n/p \rceil$ is the least integer greater than or equal to n/p. In characteristic p, the series $\{G_n\}_{n=1}^{\infty}$ is called the lower p-central, Brauer, Jennings, Lazard or Zassenhaus series of G. The quotients G_n/G_{n+1} are elementary abelian p-groups and define the fastest-decreasing central series with this property [Jen55].

Let

$$\mathcal{A}(G) = \mathcal{A}_{\Bbbk}(G) = \bigoplus_{n=0}^{\infty} \Delta^{n+1}/\Delta^n$$

be the associative graded algebra with product induced linearly from the group product (see [Pman77, Pas79] for more details).

If $\Bbbk = \mathbb{Q}$, consider the following graded Lie algebras over $\Bbbk$:

$$\mathcal{L}(G) = \bigoplus_{n=1}^{\infty} \left(G_n/G_{n+1} \otimes_{\mathbb{Z}} \mathbb{Q} \right), \qquad L(G) = \bigoplus_{n=1}^{\infty} \left(\gamma_n(G)/\gamma_{n+1}(G) \otimes_{\mathbb{Z}} \mathbb{Q} \right).$$

If $\Bbbk = \mathbb{F}_p$, consider the restricted Lie $\mathbb{F}_p$-algebra

$$\mathcal{L}_p(G) = \bigoplus_{n=1}^{\infty} \left(G_n/G_{n+1} \right).$$

Then Quillen's Theorem [Qui68] asserts that $\mathcal{A}(G)$ is the universal enveloping algebra of $\mathcal{L}(G)$ in characteristic 0 and is the universal p-enveloping algebra of $\mathcal{L}_p(G)$ in positive characteristic.

Let us introduce the following numbers:

$$a_n(G) = \dim_{\Bbbk}(\Delta^n/\Delta^{n+1}), \qquad b_n(G) = \operatorname{rank}(G_n/G_{n+1}).$$

Here by the rank of the G-module M we mean the torsion-free rank $\dim_{\mathbb{Q}}(M \otimes \mathbb{Q})$ in characteristic 0 and the p-group rank $\dim_{\mathbb{F}_p}(M \otimes \mathbb{F}_p)$, equal to the minimal number of generators, in positive characteristic. Note that in zero-characteristic $b_n = \operatorname{rank}(\gamma_n(G)/\gamma_{n+1}(G))$, because the natural map

$$\gamma_n(G)/\gamma_{n+1}(G) \to G_n/G_{n+1}$$

has finite kernel and cokernel.

The following result is due to Jennings. The case $\mathbb{k} = \mathbb{F}_p$ appears in [Jen41] and the case $\mathbb{k} = \mathbb{Q}$ appears in [Jen55]; but see also [Pman77, Theorem 3.3.6 and 3.4.10].

$$(3) \qquad \sum_{n=0}^{\infty} a_n(G)t^n = \begin{cases} \prod_{n=1}^{\infty} \left(\frac{1-t^{pn}}{1-t^n}\right)^{b_n(G)} & \text{if } \mathbb{k} = \mathbb{F}_p, \\ \prod_{n=1}^{\infty} \left(\frac{1}{1-t^n}\right)^{b_n(G)} & \text{if } \mathbb{k} = \mathbb{Q}. \end{cases}$$

The series $\sum_{n=0}^{\infty} a_n(G)t^n$ is the Hilbert-Poincaré series of the graded algebra $\mathcal{A}(G)$. The equation (3) expresses this series in terms of the numbers $b_n(G)$; the relation between the sequences $\{a_n(G)\}_{n=0}^{\infty}$ and $\{b_n(G)\}_{n=1}^{\infty}$ is quite complicated. We shall be interested in asymptotic growth of series, in the following sense:

Definition 2.1. Let f and g be two functions $\mathbb{R}_+ \to \mathbb{R}_+$. We write $f \precsim g$ if there is a constant $C > 0$ such that $f(x) \leq C + Cg(Cx + C)$ for all $x \in \mathbb{R}_+$, and write $f \sim g$ if $f \precsim g$ and $g \precsim f$.

A series $\{a_n\}_{n=0}^{\infty}$ defines a function $f : \mathbb{R}_+ \to \mathbb{R}_+$ by $f(x) = a_{\lfloor x \rfloor}$, and for two series $a = \{a_n\}$ and $b = \{b_n\}$ we write $a \precsim b$ and $a \sim b$ when the same relations hold for their associated functions.

The main facts are presented in the following statement:

Proposition 2.2. *Let $\{a_n\}$ and $\{b_n\}$ be connected by the one of the relations (3). Then*

1. *$\{b_n\}$ grows exponentially if and only if $\{a_n\}$ does, and we have*

$$\limsup_{n \to \infty} \frac{\ln a_n}{n} = \limsup_{n \to \infty} \frac{\ln b_n}{n}.$$

2. *If $b_n \sim n^d$ then $a_n \sim e^{n^{(d+1)/(d+2)}}$.*

Proof. We first suppose $\mathbb{k} = \mathbb{Q}$, and prove Part 1 following [Ber83]. Let $A = \limsup (\ln a_n)/n$ and $B = \limsup (\ln b_n)/n$. Clearly $A \geq B$ as $a_n \geq b_n$ for all n; we now prove that $A \leq B$. Define

$$f(z) = \prod_{n=1}^{\infty} (1 - e^{-nz})^{-b_n},$$

viewed as a complex analytic function in the half-plane $\Re(z) > B$. We have $|1 - e^{-nz}|^{-1} \leq (1 - e^{-n\Re z})^{-1}$, from which $|f(z)| \leq f(\Re z)$. Now applying the Cauchy residue formula,

$$a_n = \frac{1}{2\pi} \int_{-\pi}^{\pi} f(u + iv)e^{n(u+iv)}dv \leq \frac{1}{2\pi} \int_{-\pi}^{\pi} |f(u + iv)|e^{nu}dv \leq e^{nu}f(u)$$

for all $u > B$, so

$$A = \limsup_{n \to \infty} \frac{\ln a_n}{n} \leq \limsup_{u > B, n \to \infty} \left(u + \frac{\ln f(u)}{n}\right) = B.$$

For $\mathbb{k} = \mathbb{F}_p$, Part 1 holds *a fortiori*.

Part 2 for $\mathbb{k} = \mathbb{Q}$ is a consequence of a result by Meinardus ([Mei54]; see also [And76, Theorem 6.2]). More precisely, when $b_n = n^d$, his result implies that

$$a_n \approx \frac{e^{\zeta'(-d)}}{\sqrt{2\pi(d+2)n}} \left(\frac{(d+1)!\zeta(d+2)}{n} \right)^{\frac{1-2\zeta(-d)}{2+4d}} e^{n\frac{d+2}{d+1}\left(\frac{(d+1)!\zeta(d+2)}{n} \right)^{\frac{1}{1+2d}}},$$

where '$\approx$' means that the quotient tends to 1 as $n \to \infty$, and ζ is the Riemann zeta function.

We sketch the proof for $\mathbb{k} = \mathbb{Q}$ below: we suppose that $b_n \sim n^d$, so $A = B = 0$ by Part 1, and compute

$$\frac{d}{du} \ln f(u) = \sum_{n=1}^{\infty} -b_n \frac{-ne^{-nu}}{1 - e^{-nu}} \sim \frac{1}{u^{d+2}} \sum_{n=1}^{\infty} \frac{(nu)^{d+1}}{e^{nu} - 1} u$$

$$\sim \frac{1}{u^{d+2}} \int_0^{\infty} \frac{w^{d+1}}{e^w - 1} dw = \frac{C}{u^{d+2}}.$$

Thus $\ln f(u) \sim C/u^{d+1}$, and the inequality

$$\log a_n \leq nu + \log f(u) \sim nu + C/u^{d+1}$$

is tight by the saddle-point principle when the right-hand side is minimized. This is done by choosing $u = n^{-1/(d+2)}$, whence as claimed $\log a_n \sim n^{1-1/(d+2)}$.

Finally, we show that (3) yields the same asymptotics when $\mathbb{k} = \mathbb{F}_p$ as when $\mathbb{k} = \mathbb{Q}$. Clearly

$$\prod_{n=1}^{\infty}(1 + t^n)^{b_n} \leq \prod_{n=1}^{\infty}(1 + t^n + \cdots + t^{(p-1)n})^{b_n} \leq \prod_{n=1}^{\infty}(1 + t^n + \dots)^{b_n}$$

for all $p \geq 2$, where for two power series $\sum e_t^n$ and $\sum f_n t^n$ the inequality $\sum e_t^n \leq \sum f_n t^n$ means that $e_n \leq f_n$ for all n. It thus suffices to consider the case $p = 2$. For this purpose define

$$g(z) = \prod_{n=1}^{\infty}(1 + e^{-nz})^{b_n},$$

and compare the series developments of $\log(f)$ and $\log(g)$ in e^{-z}. From $-\log(1 - z) = \sum_{n \geq 1} \frac{z^n}{n}$ it follows that

$$\log f(z) = \sum_{n \geq 1} f_n e^{-nz}, \quad f_n = \sum_{d|n} \frac{1}{d},$$

$$\log g(z) = \sum_{n \geq 1} g_n e^{-nz}, \quad g_n = \sum_{d|n} \frac{(-1)^{d+1}}{d},$$

so both series have the same odd-degree coefficients, and thus $\log f \sim \log g$. Their exponentials then have the same asymptotics; more precisely, $f_n \leq g_{2n-1}$ for all n, so $e^z \log f(2z) \leq \log g(z)$ termwise, and $f(2z) \leq g(z)$. $\qquad\square$

2.1. Growth of Groups. Let G be a finitely generated group with a fixed semigroup system S of generators (i.e. such that every element $g \in G$ can be expressed a product $g = s_1 \ldots s_n$ for some $s_i \in S$). Let $\gamma_G^S(n)$ be the growth function of (G, S); recall that it is

$$\gamma_G^S(n) = \#\{g \in G | \, |g| \le n\},$$

where $|g|$ is the minimal number of generators required to express g as a product.

The following observations are well-known:

Lemma 2.3. *Let G be a group and consider two finite generating sets S and T. Then $\gamma_G^S \sim \gamma_G^T$, with $\sim$ given in Definition 2.1.*

It is then meaningful to consider the *growth* γ_G of G, which is the $\sim$-equivalence class containing its growth functions γ_G^S.

Lemma 2.4. *Let G be a finitely generated group, $H < G$ a finitely generated subgroup and K a quotient of G. Then $\gamma_H \precsim \gamma_G$ and $\gamma_K \precsim \gamma_G$.*

Proof. Let S be a finite generating set for H; choose a generating set $T \supset S$ for G. Apply Definition 2.1 with $C = 1$ to obtain $\gamma_H^S \precsim \gamma_G^T$. Clearly $\gamma_K^T(n) \le \gamma_G^T(n)$ for all n. ∎

Lemma 2.5 ([Gri89]). *For any field $\Bbbk$ and any group G with generating set S the inequalities $a_n(G) \le \gamma_G^S(n)$ hold for all $n \ge 0$.*

Proof. Fix a generating set S. The identities

$$xy-1 = (x-1)+(y-1)+(x-1)(y-1), \qquad x^{-1}-1 = -(x-1)-(x-1)(x^{-1}-1)$$

show that

$$xy - 1 \equiv (x - 1) + (y - 1), \qquad x^{-1} - 1 \equiv -(x - 1) \quad \mod \Delta^2,$$

so Δ^n is generated over $\Bbbk$ by Δ^{n+1} and elements of the form

$$x_0(s_1 - 1)x_1(s_2 - 1) \ldots (s_n - 1)x_n,$$

for all $s_i \in S$ and $x_i \in \Bbbk[G]$. Now $x_i \equiv \varepsilon(x_i) \in \Bbbk$ modulo Δ, so Δ^n/Δ^{n+1} is spanned by the

$$(s_1 - 1)(s_2 - 1) \ldots (s_n - 1), \qquad s_i \in S.$$

All these elements are in the vector subspace S_n of $\Bbbk[G]$ spanned by products of at most n generators, and by definition S_n is of dimension $\gamma_G^S(n)$. ∎

Corollary 2.6. $\{a_n(G)\}_{n=0}^\infty \precsim \gamma_G$.

Combining Proposition 2.2 and Lemma 2.5, we obtain as

Corollary 2.7. *If there exist $C > 0$ and $d \ge 0$ such that $b_n \ge Cn^d$ for all n, then $\gamma_G(n) \succsim e^{1-1/(d+2)}$. In particular, if $b_n \ne 0$ for all n, then $\gamma_G(n) \succsim e^{\sqrt{n}}$.*

We shall say a group G is of *subradical growth* if $\gamma_G \precnsim e^{\sqrt{n}}$.

Theorem 2.8 ([Gri89]). *Let G be a finitely generated residually-p group. If G is of subradical growth then G is virtually nilpotent and $\gamma_G(n) \sim n^d$ for some $d \in \mathbb{N}$.*

Proof. By the previous corollary, $b_n(G) = 0$ for some n. Consider the p-completion $\widehat{G}$ of G. As Lie algebras, $\mathcal{L}_{\mathbb{F}_p}(G)$ and $\mathcal{L}_{\mathbb{F}_p}(\widehat{G})$ coincide, so $b_n(\widehat{G}) = 0$. By Lazard's criterion $\widehat{G}$ is an analytic pro-p-group [Laz65] and thus is linear over a field. Since G is residually-p it embeds in $\widehat{G}$ so is also linear. By the Tits alternative [Tit72] either G contains a free group on two generators (contradicting the assumption on the growth of G) or G is virtually solvable. By the results of Milnor and Wolf every virtually solvable group is either of exponential growth or is virtually nilpotent [Mil68, Wol68]. The asymptotic growth is invariant under taking finite-index subgroups, and the growth of a nilpotent group is polynomial of degree $\sum_{k \geq 1} k b_k$, as was shown by Guivarc'h and Bass [Gui70, Gui73, Bas72]. $\qquad \square$

In the class of residually-p groups, Theorem 2.8 improves Gromov's result [Gro81a] that a finitely generated group G having polynomial growth is virtually nilpotent, in that the assumption is weakened from 'polynomial growth' to 'subradical growth'. Lubotzky and Mann have shown the same result for residually nilpotent groups of subradical growth. It is not known whether subradical growth does imply virtual nilpotence, and whether there exist groups of precisely radical growth. Certainly the right place to look for such examples is among groups of finite width, or groups satisfying some tight condition on the growth of their b_n.

Therefore new examples of groups of finite width are of special interest. Below we shall give two examples of such groups and outline a method of constructing new examples; but first a consequence of 2.8 is

Theorem 2.9. *The growth $\gamma_{\mathfrak{G}}$ of the group $\mathfrak{G}$ satisfies*

$$e^{\sqrt{n}} \precsim \gamma_{\mathfrak{G}}(n) \precsim e^{n^{1/(1 - \log_2 \eta)}},$$

where η is the real root of $X^3 + X^2 + X - 2$.

Proof. If $\mathfrak{G}$ were nilpotent it would be finite, as it is finitely generated and torsion; since it is infinite 2.8 yields the left inequality.

The right inequality was proven by the first author in [Bar98], using purely combinatorial techniques. $\qquad \square$

Note that the estimate from below can be obtained directly as in [Gri84], by showing that for an appropriate S the growth function γ_G^S satisfies

$$\gamma_G^S(4n) \geq \gamma_G^S(n)^2.$$

The second author conjectured in 1984 that the left inequality is in fact an equality, but Leonov recently announced that this is not the case [Leo98].

For our second example $\tilde{\mathfrak{G}}$ it is only known that

$$e^{\sqrt{n}} \precsim \gamma_{\tilde{\mathfrak{G}}} \precsim e^n,$$

as is shown in [BG98].

Lemma 2.5 can also be used to study uniformly exponential growth, as was observed in [CG97]. Let

$$\omega_G^S = \lim_{n \to \infty} \sqrt[n]{\gamma_G^S(n)}$$

be the base of exponential growth of G with respect to the generating set S and let $\omega_G = \inf_S \omega_G^S$, the infimum being taken over all finite generating sets.

Definition 2.10. The group G has *uniformly exponential growth* if $\omega_G > 1$.

(See [Gro81b] for the original definition and motivations, and [GH97] for more details on this notion.) For instance, the free groups of rank ≥ 2, and more generally, the non-elementary hyperbolic groups have uniformly exponential growth [Kou98]. It is currently not known whether there exists a group of exponential but not uniformly exponential growth.

Corollary 2.11. *If for some* $\Bbbk \in \{\mathbb{Q}, \mathbb{F}_p\}$ *the algebra* $\mathcal{A}_{\Bbbk}(G)$ *has exponential growth then* G *has uniformly exponential growth. (We do not need here the assumption that* G *is residually-p or residually nilpotent.)*

In the next section we will combine this idea with the Golod-Shafarevich construction to produce examples of finitely generated residually finite p-groups of uniformly exponential growth.

3. Torsion Groups of Uniformly Exponential Growth

As a reference to the Golod-Shafarevich construction we recommend the original paper [GS64], one of the books [Her94, Koc70], or [HB82, § VIII.12].

Consider the free associative algebra A over the field $\mathbb{F}_p$ on the generators $x_1, \ldots, x_d$ for some $d \geq 2$. The algebra A is graded: $A = \bigoplus_{n=0}^{\infty} A_n$ where A_n is spanned by the monomials of degree n, with $A_0 = \mathbb{F}_p 1$. Elements of the subspace A_n are called *homogeneous of degree n*.

Consider an ideal $\mathcal{I}$ in A generated by r_1 homogeneous elements of degree 1, r_2 of degree 2, etc. (We make this homogeneity assumption for simplicity; it is not necessary, as was indicated in [Koc70].) Let $B = A/\mathcal{I}$. Then B is also a graded algebra: $B = \bigoplus_{n=0}^{\infty} B_n$ and if $H_B(t) = \sum_{n=0}^{\infty} d_n t^n$ be the Hilbert-Poincaré series of B, i.e. $d_n = \dim_{\mathbb{F}_p} B_n$, then the Golod-Shafarevich inequality

$$(4) \qquad\qquad H_B(t)(1 - dt + H_R(t)) \geq 1$$

holds; here $H_R(t) = \sum_{n=1}^{\infty} r_n t^n$, and for the comparison of two power series the same agreement holds as in the previous section.

Suppose that for some $\xi \in (0,1)$ the series $H_R(t)$ converges at ξ and $1 - d\xi + H_R(\xi) \leq 0$. Then the series $H_B(t)$ cannot converge at $t = \xi$, so the coefficients d_n of $H_B(t)$ grow exponentially and

$$\limsup_{n \to \infty} \sqrt[n]{d_n} \geq \frac{1}{\xi}.$$

Golod proves in [Gol64] that $\mathcal{I}$ can be chosen in such a way that the ideal $\mathcal{D} = \bigoplus_{n=1}^{\infty} B_n$ will be a p-nilalgebra (i.e. for all $y \in \mathcal{D}$ there is an $n \in \mathbb{N}$ such that $y^{p^n} = 0$).

The construction of the relators goes as follows: enumerate first as $\{y_k\}_{k=1}^{\infty}$ all elements of the algebra A (this is possible since A is countable). Start with $\mathcal{I}_0 = 0$; then if y_k is not a nilelement of $A/\mathcal{I}_{k-1}$ take $\ell_k \geq 3$ sufficiently large so that the least degree of monomials in $y_k^{p^{\ell_k}}$ is larger than all degrees of monomials in $\mathcal{I}_{k-1}$. Construct $\mathcal{I}_k$ by adding to $\mathcal{I}_{k-1}$ all homogeneous parts of the polynomial $y_k^{p^{\ell_k}}$. Let finally $\mathcal{I} = \bigcup_{n=0}^{\infty} \mathcal{I}_n$.

The numbers r_k will then all be 0 or 1 with $r_k = 0$ for $k < p^3$, so taking $\xi = 3/4$ we have

$$1 - d\xi + H_R(\xi) \leq 1 - 2\xi + \frac{\xi^{2^3}}{1 - \xi} < 0$$

and $B = A/\mathcal{I}$ is of exponential growth at least $(4/3)^n$. Let $\overline{x}_1, \ldots, \overline{x}_d$ be the images of $x_1, \ldots, x_d$ in B, and let G be the group generated by the elements $s_i = 1 + \overline{x}_i$; they are invertible because the $\overline{x}_i$ are p-nilelements and B is of characteristic p. The vector subspace of B spanned by G is B itself, so B is a quotient of the group algebra $\mathbb{F}_p[G]$.

Theorem 3.1. *G is a finitely generated residually finite p-group of uniformly exponential growth.*

Proof. That G is a p-group was observed by Golod and follows from the fact that $\mathcal{D} = \bigoplus_{n=1}^{\infty} B_n$ is a p-nilalgebra. Let π be the natural map $\mathbb{F}_p[G] \to B$. Then $\mathcal{D}$ is generated by $\pi(\Delta)$ and more generally $\bigoplus_{n=N}^{\infty} B_n = \pi(\Delta^N)$, so by Lemmata 2.5 and 2.4 there is a $\xi < 1$ such that the estimate

$$\frac{1}{\xi^n} \leq \dim_{\mathbb{F}_p} B_n \leq a_n(G) \leq \gamma_G^T(n), \qquad n = 1, 2, \ldots$$

holds for any system T of generators of G. $\qquad\square$

4. GROWTH OF ALGEBRAS AND AMENABILITY

As was mentioned in the introduction, there is an interesting question (due to Vershik) on the relation between the amenability of a group and the growth of related algebras. Let us formulate our version of this question:

Problem 4.1. 1. *Let G be amenable. Does $b_n(G)$ grow subexponentially for any field $\Bbbk$?*

2. *Suppose G is residually nilpotent (or residually-p) and $b_n(G)$ grows subexponentially for the field $\mathbb{Q}$ (or $\mathbb{F}_p$). Is then the group G amenable?*

There is a chance that for at least one of these questions the answer is affirmative.

For solvable groups (which are amenable) the associated algebras have subexponential growth, as follows from computations by Berezniĭ [Ber83] (based on computations for free solvable algebras given by Egorychev [Ego84]) of the growth of the $b_n(G)$ of free solvable groups, and the functoriality of the growth (see also Petrogradskiĭ [Pet93], where some mistakes from [Ber83] are corrected).

On the other hand there is some similarity between the asymptotics of random walks on solvable groups and the growth of $b_n(G)$ [Kaĭ80] which gives a hope that subexponential growth of algebras implies (under the residuality hypothesis) subexponential decay of the probability of returning to the origin for symmetric random walks on a group. Then Kesten's criterion [Kes59] can be invoked to imply the amenability of G.

5. GROUPS ACTING ON ROOTED TREES

We now consider examples of groups whose lower central series and dimension series we can compute explicitly. Let Σ be a finite alphabet, and Σ^* the set of finite sequences over Σ. This set has a natural rooted tree structure: the vertices are finite sequences, and the edges are all the $(\sigma, \sigma s)$ for $\sigma \in \Sigma^*$ and $s \in \Sigma$; the root vertex is $\emptyset$, the empty sequence. By $\mathrm{Aut}(\Sigma^*)$ we mean the bijections of Σ^* that preserve the tree structure, i.e. preserve length and prefixes. We write $\sigma \Sigma^*$ for the subtree of Σ^* below vertex σ: it is isomorphic to Σ^* but rooted at σ.

Let G be a finitely generated subgroup of $\mathrm{Aut}(\Sigma^*)$ acting transitively on Σ^n for all n (such an action will be called *spherically transitive.*) We denote by $\mathrm{st}_G(\sigma)$ the stabilizer of the vertex σ in G, and by $\mathrm{st}_G(n)$ the stabilizer of all vertices of length n. An arbitrary element $g \in \mathrm{st}_G(n)$ can be identified with a tuple $(g_\sigma)_{|\sigma|=n}$ of tree automorphisms; we write this monomorphism

$$\phi_n : \mathrm{st}_G(n) \hookrightarrow \prod_{\sigma \in \Sigma^n} \mathrm{Aut}(\sigma \Sigma^*).$$

We define the *vertex group* or *rigid stabilizer* $\mathrm{rist}_G(\sigma)$ of the vertex σ by

$$\mathrm{rist}_G(\sigma) = \{g \in G \mid g\tau = \tau \ \forall \tau \in \Sigma^* \setminus \sigma\Sigma^*\},$$

and the nth *rigid stabilizer* as the group generated by the length-n vertex groups: $\mathrm{rist}_G(n) = \langle \mathrm{rist}_G(\sigma) : |\sigma| = n \rangle$. Since G acts transitively on Σ^n the vertex groups of vertices at level n are all conjugate. Therefore $\mathrm{rist}_G(n)$ is a direct product of $|\Sigma|^n$ copies of $\mathrm{rist}_G(\sigma)$ for a σ of length n.

Definition 5.1. A finitely generated group G is called a *branch group* if

1. G acts faithfully on Σ^* and transitively on Σ^n for all $n \geq 0$;
2. $[G : \operatorname{rist}_G(n)]$ is finite for all $n \geq 0$.

5.1. The Modules V_n. Let G be a group acting on a regular rooted tree Σ^*, where Σ contains p elements for some prime p; for ease of notation suppose $\Sigma = \mathbb{F}_p$. Assume moreover that at each vertex G acts as a power of the cyclic permutation $\epsilon = (0, 1, \ldots, p-1)$ of Σ. Let $V_n = \mathbb{F}_p[G/\operatorname{st}_G(0^n)]$; it is a vector space of dimension p^n, as G acts transitively on Σ^n, and has a natural G-module structure coming from the action of G on $G/\operatorname{st}_G(0^n)$. Identify $G/\operatorname{st}_G(0^n)$ with the set Σ^n of vertices at level n, and also with the set of monomials over $\{X_1, \ldots, X_n\}$ of degree $< p$ in each variable, by

$$\sigma = \sigma_1 \ldots \sigma_n \leftrightarrow X_1^{\sigma_1} \ldots X_n^{\sigma_n}.$$

Under this identification, we can write

$$
\begin{aligned}
V_n &= \mathbb{F}_p[X_1, \ldots, X_n]/(X_1^p - 1, \ldots, X_n^p - 1) \\
&= \mathbb{F}_p[X_1]/(X_1^p - 1) \otimes \cdots \otimes \mathbb{F}_p[X_n]/(X_n^p - 1).
\end{aligned}
$$

We write $g\sigma$ the action of $g \in G$ on $\sigma \in V_n$, and denote by $[g, \sigma] = \sigma - g\sigma$ the "Lie action" of G on V_n. For $r \in \{0, \ldots, p^n - 1\}$ we write $r = r_n \ldots r_1$ in base p, and define

$$v_n^r = (1 - X_1)^{r_1} \ldots (1 - X_n)^{r_n} \in V_n,$$
$$V_n^r = \langle v_n^r, \ldots, v_n^{p^n - 1} \rangle.$$

We extend the last definition to $V_n^r = 0$ when $r \geq p^n$. There is a natural projective sequence

$$\cdots \to V_n \to V_{n-1} \to \cdots \to V_0 = \mathbb{F}_p$$

of G-modules, and at each step n a sequence of V_n-submodules

$$V_n^{p^n} = 0 \subset \cdots \subset V_n^{p^n - p^{n-1}} \subset \cdots \subset V_n^1 \subset V_n^0 = V_n$$

each having codimension 1 in the next. Moreover $V_{n-1}^{p^{n-1} - i}$ and $V_n^{p^n - i}$ are naturally isomorphic under multiplication by $(1 - X_n)^{p-1}$; thus $V_n^{p^n - p^{n-1}}$ is isomorphic to $V_{n-1}^0 = V_{n-1}$ as a G-module.

Lemma 5.2. 1. *The inclusion $[G, V_n^r] \subset V_n^{r+1}$ holds for all n and all r.*

2. *If G contains for all $m \leq n$ an element g_m such that*

$$g_m(0^m) = 0^{m-1}1, \qquad g_m(\sigma x) = \sigma' x \quad \forall \sigma \in \Sigma^{m-1} \setminus \{0^{m-1}\}, x \in \Sigma$$

(where in the second condition σ' is an arbitrary function of σ), then $[G, V_n^r] = V_n^{r+1}$ for all n and all r.

A G-module V having the property $\dim V^{(n)}/V^{(n+1)} = 1$ for all n, where the $V^{(n)}$ are defined inductively by $V^{(0)} = V$ and $V^{(n+1)} = [G, V^{(n)}]$ is called *uniserial*. This notion was introduced by Leedham-Green [LG94]; see also [DdSMS91, page 111].

Note that every element of G can be described by a colouring $\{g_\sigma\}_{\sigma\in\Sigma^*}$ of the vertices of Σ^* by elements of the cyclic group $C_p = \langle\epsilon\rangle$. The condition in the lemma amounts to the existence, for all m, of an element g_m whose colouring is ϵ at the vertex 0^m, and is 1 on all other vertices of the m-th level as well as on all vertices 0^i, for $i < m$. Note also that this implies that the action is spherically transitive.

Proof. We proceed by induction on (n, r) in lexicographic order. For $n = 0$ the claim holds trivially; suppose thus $n \geq 1$. In order to prove $[G, V_n^r] \subset V_n^{r+1}$, it suffices to check that for all $g \in G$ we have $[g, v_n^r] \in V_n^{r+1}$, as the V_n^r form an ascending tower of subspaces. During the proof we will consider V_{n-1} as a subspace of V_n; beware though that it is not a submodule. We shall write '$*$' for the action of G on $V_{n-1} \subset V_n$, and '$\cdot$' for that of G on V_n.

Observe that if $v \in V_{n-1}$ then $g \cdot (vX_n^i) = (g \cdot v)X_n^i$. Thus $g \cdot v - g * v$ is always divisible by $1 - X_n$ because if $g \cdot v = \sum_{s=0}^{p-1} \Psi_s X_n^s$ for some $\Psi_s \in V_{n-1}$ then $g * v = \sum_{s=0}^{p-1} \Psi_s$ and

$$(5) \qquad g \cdot v - g * v = (1 - X_n) \sum_{s=1}^{p-1} -\Psi_s(1 + X_n + \cdots + X_n^{s-1}).$$

Write $r = r_n \ldots r_1$ in base p. For some Φ and Ψ_s in V_{n-1}, we may write

$$v_n^r = \Phi(1 - X_n)^{r_n}, \qquad g \cdot v_n^r = \sum_{s=0}^{p-1} \Psi_s X_n^s(1 - X_n)^{r_n}.$$

Then by induction

$$[g, v_n^r] = \underbrace{\underbrace{\left(\Phi - \sum_{s=0}^{p-1} \Psi_s\right)(1 - X_n)^{r_n}}_{\in V_{n-1}^{(r+1)\bmod p^{n-1}}} + \underbrace{\sum_{s=0}^{p-1} \Psi_s(1 - X_n^s)(1 - X_n)^{r_n}}_{\in V_n^{(r_n+1)p^{n-1}}\subseteq V_n^{r+1}}}_{\in V_n^{r+1}},$$

as in the second summand $(1 - X_n^s)(1 - X^n)^{r_n}$ is divisible by $(1 - X_n)^{r_n+1}$. This proves the first claim of the lemma.

Next, we prove $[G, V_n^r] \supset V_n^{r+1}$ by showing that $v_n^{r+1} \in [G, V_n^r]$. As above, write $r = r_n \ldots r_1$ in base p. If $(r_1, \ldots, r_{n-1}) \neq (p - 1, \ldots, p - 1)$, we have $v_n^{r+1} = v_{n-1}^{r+1\bmod p^{n-1}}(1-X_n)^{r_n}$, and by induction $v_{n-1}^{r+1\bmod p^{n-1}} = \sum_s \alpha_s[g_s, v_{n-1}^{i_s}]$ for some $\alpha_s \in \mathbb{F}_p$, $g_s \in G$ and $i_s \geq r \bmod p^{n-1}$. Then

$$v_n^{r+1} = \underbrace{\sum_s \alpha_s\left[g_s, v_n^{i_s+r_np^{n-1}}\right]}_{\in [G,V_n^r]} + \underbrace{\sum_s \alpha_s\left(g_s \cdot v_n^{i_s+r_np^{n-1}} - (g_s * v_{n-1}^{i_s})(1 - X_n)^{r_n}\right)}_{\in V_n^{(r_n+1)p^{n-1}}\subseteq V_n^{r+2}\subseteq [G,V_n^{r+1}]\subseteq [G,V_n^r]}$$

where the last inclusions hold by (5) and induction. Finally, if $r = (r_n + 1)p^{n-1} - 1$, note that

$$v_n^r = (1 + X_1 + \cdots + X_1^{p-1}) \cdots (1 + X_{n-1} + \cdots + X_{n-1}^{p-1})(1 - X_n)^{r_n}$$
$$= (1 - X_n)^{r_n} + P(1 - X_n)^{r_n},$$

where $P = \sum_{\sigma \in \Sigma^{n-1} \setminus \{0^{n-1}\}} X_1^{\sigma_1} \cdots X_n^{\sigma_n}$ is invariant under g_n; thus

$$v_n^{r+1} = (1 - X_n)^{r_n} - X_n(1 - X_n)^{r_n} = [g_n, v_n^r] \in [G, V_n^r].$$

$\square$

The strategy we follow to compute the lower central series or dimension series of G in the examples of Sections 6 and 7 is the following:

- We recognize some $\gamma_m(G)$ or G_m as a subgroup of G simply obtained from rigid stabilizers in G.
- We identify a quotient $\gamma_m(G)/N$ or G_m/N with a direct sum of copies of the module V_n defined above, for an appropriate subgroup N.
- We show that N is a further term of the lower central or dimensional series, allowing the process to repeat.

Then the exact terms of the lower central or dimension series are obtained by pulling back the appropriate V_n^r through the identification.

6. THE GROUP $\mathfrak{G}$

Let $\Sigma = \mathbb{F}_2$, the field on two elements. For $x \in \mathbb{F}_2$ set $\bar{x} = 1 - x$, and define the automorphisms a, b, c, d of Σ^* as follows:

$$a(x\sigma) = \bar{x}\sigma,$$

$$b(0x\sigma) = 0\bar{x}\sigma, \qquad b(1\sigma) = 1c(\sigma),$$
$$c(0x\sigma) = 0\bar{x}\sigma, \qquad c(1\sigma) = 1d(\sigma),$$
$$d(0x\sigma) = 0x\sigma, \qquad d(1\sigma) = 1b(\sigma).$$

Thus for instance b acts on the subtree $0\Sigma^*$ as c, while c acts on it as d, etc. Note that all generators are of order 2 and $\{1, b, c, d\}$ forms a Klein group. Set $\mathfrak{G} = \langle a, b, c, d \rangle$. For ease of notation, we shall identify elements of $\mathrm{st}_{\mathfrak{G}}(n)$ with their image under ϕ_n by writing $\phi_n(g) = (g_1, \ldots, g_{2^n})_n$ (omitting the subscript n if it is obvious from context); for instance we will write $b = (a, c)$, $c = (a, d)$ and $d = (1, b)$. Set $x = [a, b]$, and set

$$K = \langle x \rangle^{\mathfrak{G}} = \langle x, (x, 1), (1, x) \rangle.$$

Note that $(x, 1) = [b, d^a]$ and $(1, x) = [b^a, d]$. Also, K is a subgroup of finite index (actually index 16) in $\mathfrak{G}$, and contains $K \times K$ as a subgroup of finite index (actually index 4); for more details see [Har98] or [BG98]. Set also $T = \langle x^2 \rangle^{\mathfrak{G}} = K^2$, and for any $Q \le K$ define $Q_m = Q \times \cdots \times Q$ (2^m copies). Clearly $Q_m \le \mathrm{st}_{\mathfrak{G}}(m)$ and acts on each subtree starting on level m by the corresponding factor. For $m \ge 1$ set $N_m = K_m \cdot T_{m-1}$.

For $m \geq 2$, we have $\mathrm{rist}_{\mathfrak{G}}(m) = K_{m-2}$, so $\mathfrak{G}$ is a branch group.

Lemma 6.1. *The mapping*

$$\alpha \oplus \beta : N_m/N_{m+1} \longrightarrow V_m \oplus V_{m-1}$$

is an isomorphism for all m, where the V_m are the modules defined in Subsection 5.1, α maps $(1, \ldots, 1, x, 1, \ldots, 1) \in K_m$ to the monomial in V_m corresponding to the vertex at the x's position, and β maps $(1, \ldots, 1, x^2, 1, \ldots, 1) \in T_{m-1}$ to the corresponding monomial in V_{m-1}.

Proof. We first suppose $m = 1$. Then $N_1/N_2 = \langle x^2, (1, x), (x, 1) \rangle / N_2$; it is easy to check that $x^4 = (x^2, x^2)$ modulo K_2, so all generators of N_1/N_2 are of order 2. Further, $[x^2, (1, x)] \in K_2$ and $[x^2, (x, 1)] \in K_2$, so the quotient N_1/N_2 is the elementary abelian group 2^3, and $\alpha \oplus \beta$ is an isomorphism in that case.

For $m > 1$ it suffices to note that both sides of the isomorphism are direct sums of 2^{m-1} terms on each of which the lemma for $m = 1$ can be applied. $\square$

Lemma 6.2. *The following equalities hold in $\mathfrak{G}$:*

$$[x, a] = x^2, \qquad [x, b] = x^2,$$
$$[x, c] = x(1, x^{-1})x, \qquad [x, d] = (1, x),$$
$$[x^2, a] = x^4 = ((U, V)x^2, (V, U)x^2), \qquad [x^2, b] = x^4,$$
$$[x^2, c] = ((U, V)x^2, (1, x)), \qquad [x^2, d] = (1, (U, 1)x^2),$$

where $U = (1, x^{-1})x$ and $V = (x^{-1}, 1)x^{-1}$ are in K.

Proof. Direct computation; see also [Roz96b], where different notations are used. $\square$

Lemma 6.3. *If $Q \not\geq N_{m+1}$ contains $g = (x, \ldots, x) \in K_m$, then $[Q, \mathfrak{G}] \geq N_{m+1}$.*

Proof. Let $b_m \in \{b, c, d\}$ be such that it acts like b on $1^m \Sigma^*$. Then

$$h = [g, b_m] = (1, \ldots, 1, [x, b])_m = (1, \ldots, 1, x^2)_m \in T_m.$$

Conjugating h by elements of g yields all cyclic permutations of the above vector, so as $[G, \mathfrak{G}]$ is normal in G it contains T_m. Likewise, let d_m act like d on $1^m \Sigma^*$. Then

$$[g, d_m] = (1, \ldots, 1, [x, a], [x, d])_m = (1, \ldots, 1, x^2, (1, x))_m;$$

using $T_m \leq [Q, \mathfrak{G}]$, we obtain $(1, \ldots, 1, (1, x))_m = (1, \ldots, 1, x)_{m+1} \in [Q, \mathfrak{G}]$, so by the same conjugation argument $[Q, \mathfrak{G}] \geq K_{m+1}$. $\square$

Theorem 6.4. *For all $m \geq 1$ we have:*

1. $\gamma_{2^m+1}(\mathfrak{G}) = N_m$.
2. $\gamma_{2^m+1+r}(\mathfrak{G}) = N_{m+1}\alpha^{-1}(V_m^r)\beta^{-1}(V_{m-1}^r)$ *for* $r = 0, \ldots, 2^m$.
3.

$$\mathrm{rank}(\gamma_n(\mathfrak{G})/\gamma_{n+1}(\mathfrak{G})) = \begin{cases} 3 & \text{if } n = 1, \\ 2 & \text{if } n = 2, \\ 2 & \text{if } n = 2^m + 1 + r, \text{ with } 0 \leq r < 2^{m-1}, \\ 1 & \text{if } n = 2^m + 1 + r, \text{ with } 2^{m-1} \leq r \leq 2^m. \end{cases}$$

Proof. First compute $\gamma_2(\mathfrak{G}) = \mathfrak{G}' = \langle[a,d], K\rangle$; it is of index 8 in $\mathfrak{G}$, with quotient generated by $\{a, b, c\}$. Compute also $\gamma_3(\mathfrak{G}) = \langle x^2 = [x,a], (1,x) = [x,d]\rangle^{\mathfrak{G}} = N_1$ of index 2 in $\gamma_2(\mathfrak{G})$, with quotient generated by $\{x^2, (1,x)\}$. This gives the basis of an induction on $m \geq 1$ and $0 < r \leq 2^m$.

Assume that $\gamma_{2^m+1}(\mathfrak{G}) = N_m$. Note that the hypothesis of Lemma 5.2 is satisfied; indeed g_m can even be chosen among the conjugates of b, c or d. Consider the sequence of quotients $Q_r = N_{m+1}\gamma_{2^m+1+r}(\mathfrak{G})/N_{m+1}$ for $r \geq 0$. Lemmata 6.1 and 5.2 tell us that $Q_r = \alpha^{-1}(V_m^r) \oplus \beta^{-1}(V_{m-1}^r)$; in particular $Q_r \ni (x, \ldots, x) = \alpha^{-1}(v_m^{2^m-1})$ for all $r < 2^m$, and then Lemma 6.3 tells us that $\gamma_{2^m+1+r}(\mathfrak{G}) \geq N_{m+1}$ for $r \leq 2^m$. When $r = 2^m$ we have $\gamma_{2^{m+1}+1}(\mathfrak{G}) = N_{m+1}$ and the induction can continue. $\square$

Lemma 6.5. *For all $m \geq 1$ and $r \in \{0, \ldots, 2^m - 1\}$ we have:*

$$(\alpha^{-1}V_m^r)^2 = \beta^{-1}(V_m^r) \leq N_{m+1};$$
$$(\beta^{-1}V_{m-1}^r)^2 = \beta^{-1}(V_m^{r+2^{m-1}}) \leq N_{m+1}.$$

Proof. Write $\alpha^{-1}(v_m^r) = (x^{i_1}, \ldots, x^{i_{2^m}})$ or $\beta^{-1}(v_m^r) = (x^{2i_1}, \ldots, x^{2i_{2^m}})$ for some $i_* \in \{0, 1\}$. Then these claims follow immediately, using Lemma 6.2, from

$$(\alpha^{-1}v_m^r)^2 = (x^{i_1}, \ldots, x^{i_{2^m}})^2 = (x^{2i_1}, \ldots, x^{2i_{2^m}}) = \beta^{-1}(v_m^r),$$
$$(\beta^{-1}v_{m-1}^r)^2 = (x^{2i_1}, \ldots, x^{2i_{2^{m-1}}})^2 = (x^{4i_1}, \ldots, x^{4i_{2^{m-1}}})$$
$$\equiv (x^{2i_1}, x^{2i_1}, \ldots, x^{2i_{2^{m-1}}}, x^{2i_{2^{m-1}}}) = \beta^{-1}(v_m^{r+2^{m-1}}) \mod N_{m+1}.$$

$\square$

Theorem 6.6. *For all $m \geq 1$ we have:*

1. $\mathfrak{G}_{2^m+1} = N_m$.
2.

$$\mathfrak{G}_{2^m+1+r} = \begin{cases} N_{m+1}\alpha^{-1}(V_m^r)\beta^{-1}(V_{m-1}^{r/2}) & \text{if } 0 \leq r \leq 2^m \text{ is even}, \\ N_{m+1}\alpha^{-1}(V_m^r)\beta^{-1}(V_{m-1}^{(r-1)/2}) & \text{if } 0 \leq r \leq 2^m \text{ is odd}. \end{cases}$$

3.

$$\mathrm{rank}(\mathfrak{G}_i/\mathfrak{G}_{i+1}) = \begin{cases} 3 & \text{if } i = 1, \\ 2 & \text{if } i > 1 \text{ is even}, \\ 1 & \text{if } i > 1 \text{ is odd}. \end{cases}$$

Proof. First compute $\mathfrak{G}_2 = \gamma_2(\mathfrak{G})$ and $\mathfrak{G}_3 = \gamma_3(\mathfrak{G}) = N_1$. This gives the basis of an induction on $m \geq 1$ and $0 \leq r \leq 2^m$. Assume $\mathfrak{G}_{2^m+1} = N_m$. Consider the sequence of quotients $Q_{m,r} = N_{m+1}\mathfrak{G}_{2^m+1+r}/N_{m+1}$ for $r \geq 0$. We have $Q_{m,r} = [\mathfrak{G}, Q_{m,r-1}]Q^2_{m-1,\lfloor r/2\rfloor}$ by (2). Lemmata 6.1, 6.5 and 5.2 tell us that $Q_r = \alpha^{-1}(V^r_m) \oplus \beta^{-1}(V^{\lfloor r/2\rfloor}_{m-1})$; in particular $Q_r \ni (x,\dots,x) = \alpha^{-1}(v^{2^m-1}_m)$ for all $r < 2^m$, and then Lemma 6.3 tells us that $\mathfrak{G}_{2^m+1+r} \geq N_{m+1}$ for $r \leq 2^m$. When $r = 2^m$ we have $\mathfrak{G}_{2^{m+1}+1} = N_{m+1}$ and the induction can continue. $\square$

6.1. Cayley graphs of Lie algebras.

We introduce the notion of *Cayley graph* for graded Lie algebras. Let $L = \bigoplus^\infty_{n=1} L_n$ be a graded Lie algebra generated by a finite set S of degree one elements. Fix a basis $(\ell_{n,1}, \dots, \ell_{n,\dim L_n})$ of L_n for every n, and give each L_n an orthogonal scalar product $\langle \ell_{n,i}|\ell_{n,j}\rangle = \delta_{i,j}$. The *Cayley graph* of L is defined as follows: its vertices are the $(i,j) \in \mathbb{N}^2$ with $i \geq 1$ and $1 \leq j \leq \dim L_n$. For every $s \in S$ and $i,j,k \in \mathbb{N}$ there is an edge from (i,j) to $(i+1,k)$ labeled by s and with weight $\langle [\ell_{i,j}, s]|\ell_{i+1,k}\rangle$. By convention edges of weight 0 are not represented. Additionally, if L is a p-algebra, there is an unlabeled edge of length $(p-1)i$ from (i,j) to (pi,k) with weight $\langle \ell^p_{i,j}|\ell_{pi,k}\rangle$.

Clearly, the Cayley graph of a Lie algebra L determines the structure of L. It is a connected graph, because S is a generating set. The geometric growth of the graph is the same as the growth of the algebra.

As a simple example, consider the quaternion group $Q = \{\pm 1, \pm i, \pm j, \pm k\}$ generated by $\{i,j\}$, and its dimension series $Q_1 = Q$, $Q_2 = \{\pm 1\}$ and $Q_3 = 1$ over the field $\mathbb{F}_2$. Then the Cayley graph of $\mathcal{L}(Q)$ is

We now describe the Cayley graphs of L and $\mathcal{L}_{\mathbb{F}_2}$ associated respectively to the lower central and dimensional series of $\mathfrak{G}$. Fix $S = \{a,b,c,d\}$ as a generating set for $\mathfrak{G}$, and extend it to $\overline{S} = \{a,b,c,d,\{^b_c\},\{^b_d\},\{^c_d\}\}$. Define the transformation σ on $\overline{S}^*$ by

$$\sigma(a) = a\{^b_c\}a, \quad \sigma(b) = d, \sigma(c) = b, \sigma(d) = b,$$

naturally extended to subsets. (For any fixed $g \in G$, one may obtain all elements $h \in \mathrm{st}_{\mathfrak{G}}(1)$ with $\phi(h) = (g,*)$ by computing $\sigma(g)$ and making all possible choices of a letter from the braced symbols. This explains the definition of $\overline{S}$.)

Theorem 6.7. *The Cayley graph of $L(\mathfrak{G})$ is as follows:*

where $x_m^r = \alpha^{-1}(v_m^r)$ and $z_m^r = \beta^{-1}(v_m^r)$. The edge $(x_m^{2^m-1}, x_{m+1}^0)$ is labelled by $\sigma^m\{{}^c_d\}$, the edge $(x_m^{2^m-1}, z_m^0)$ is labelled by $\sigma^m\{{}^b_d\}$, and the paths from x_m^0 to $x_m^{2^m-1}$ and from z_m^0 to $z_m^{2^m-1}$ are labelled by $\sigma^{m-1}(a)$.

The Cayley graph of $\mathcal{L}_{\mathbb{F}_2}(\mathfrak{G})$ is as follows:

with the same rule for labellings as for $L(\mathfrak{G})$; and power maps from x_m^r to z_m^r.

Note that as we are in characteristic 2 the non-zero weights can only be 1 and thus are not indicated.

7. The Group $\widetilde{\mathfrak{G}}$

We describe here the lower central and dimension series for a group $\widetilde{\mathfrak{G}}$ containing the previous section's group $\mathfrak{G}$ as a subgroup. More details about $\widetilde{\mathfrak{G}}$ can be found in [BG98].

As in Section 6 set $\Sigma = \mathbb{F}_2$, and define automorphisms $\tilde{b}$, $\tilde{c}$ and $\tilde{d}$ of Σ^* by

$$\tilde{b}(0x\sigma) = 0\overline{x}\sigma, \qquad \tilde{b}(1\sigma) = 1\tilde{c}(\sigma),$$

$$\tilde{c}(0\sigma) = 0\sigma, \qquad \tilde{c}(1\sigma) = 1\tilde{d}(\sigma),$$

$$\tilde{d}(0\sigma) = 0\sigma, \qquad \tilde{d}(1\sigma) = 1\tilde{b}(\sigma).$$

Note that all generators are of order 2 and $\{\tilde{b}, \tilde{c}, \tilde{d}\}$ generate the elementary abelian group 2^3. Set $\widetilde{\mathfrak{G}} = \langle a, \tilde{b}, \tilde{c}, \tilde{d} \rangle$. Clearly, $\mathfrak{G} = \langle a, b = \tilde{b}\tilde{c}, c = \tilde{c}\tilde{d}, d = \tilde{d}\tilde{b} \rangle$ is a subgroup of $\widetilde{\mathfrak{G}}$. Its index is infinite, because $\mathfrak{G}$ is a torsion group while $w = a\tilde{b}\tilde{c}\tilde{d}$ has infinite order, because $w^2 = (w^a, w)$. Set $x = [a, \tilde{b}]$, $y = [a, \tilde{d}]$, and

$$\tilde{K} = \langle x, y \rangle^{\widetilde{\mathfrak{G}}}.$$

Then $\tilde{K}$ is a subgroup of finite index (actually index 32) in $\mathfrak{G}$, and contains $\tilde{K} \times \tilde{K}$ as a subgroup of finite index (actually index 8). Set also $\tilde{T} = \langle x^2 \rangle^{\mathfrak{G}} = \tilde{K}^2$, and for any $Q \leq \tilde{K}$ define $Q_m = Q \times \cdots \times Q$ (2^m copies). For $m \geq 1$ set $\tilde{N}_m = \tilde{K}_m \cdot \tilde{T}_{m-1}$.

For $m \geq 2$, we have $\text{rist}_{\tilde{\mathfrak{G}}}(m) = \tilde{K}_{m-2}$, so $\tilde{\mathfrak{G}}$ is a branch group.

Lemma 7.1. *The mapping*

$$\alpha \oplus \beta \oplus \gamma : \tilde{N}_m / \tilde{N}_{m+1} \longrightarrow V_m \oplus V_m \oplus V_{m-1}$$

is an isomorphism for all m, where the V_m are the modules defined in Subsection 5.1, α maps $(1, \ldots, x, \ldots, 1) \in \tilde{K}_m$ to the monomial in V_m corresponding to the vertex in x's position, and β maps $(1, \ldots, y, \ldots, 1) \in \tilde{K}_m$ to the corresponding vertex in V_m, and γ maps $(1, \ldots, x^2, \ldots, 1) \in \tilde{T}_{m-1}$ to the corresponding monomial in V_{m-1}.

Proof. We first suppose $m = 1$. Then

$$\tilde{N}_1 / \tilde{N}_2 = \langle x^2, (1, x), (x, 1), (1, y), (y, 1) \rangle / \tilde{N}_2;$$

it is easy to check that $x^4 = 1$, so all generators of $\tilde{N}_1 / \tilde{N}_2$ are of order 2. Further, all commutators of generators belong to $\tilde{K}_2$, so the quotient $\tilde{N}_1 / \tilde{N}_2$ is the elementary abelian group 2^5, and $\alpha \oplus \beta \oplus \gamma$ is an isomorphism in that case.

For $m > 1$ it suffices to note that both sides of the isomorphism are direct sums of 2^{m-1} terms on each of which the lemma for $m = 1$ can be applied. $\square$

Lemma 7.2. *The following equalities hold in $\tilde{\mathfrak{G}}$:*

$$[x, a] = x^2, \qquad [x, \tilde{b}] = x^2$$
$$[x, \tilde{c}] = (1, y), \qquad [x, \tilde{d}] = (1, x),$$
$$[x^2, a] = 1, \qquad [x^2, \tilde{b}] = 1,$$
$$[x^2, \tilde{c}] = 1, \qquad [x^2, \tilde{d}] = (1, x(x, 1)x),$$
$$[y, a] = 1, \qquad [y, \tilde{b}] = (x^{-1}, 1),$$
$$[y, \tilde{c}] = 1, \qquad [y, \tilde{d}] = 1.$$

Proof. Direct computation. $\square$

Lemma 7.3. *If $Q \gneq \tilde{N}_{m+1}$ contains $g = (x, \ldots, x) \in \tilde{K}_m$, then $[Q, \tilde{\mathfrak{G}}] \geq \tilde{N}_{m+1}$.*

Proof. Let $\tilde{b}_m \in \{\tilde{b}, \tilde{c}, \tilde{d}\}$ be such that it acts like $\tilde{b}$ on $1^m \Sigma^*$. Then

$$[g, \tilde{b}_m] = (1, \ldots, 1, [x, \tilde{b}])_m = (1, \ldots, 1, x^2)_m \in \tilde{T}_m,$$

so by a conjugation argument $[Q, \tilde{\mathfrak{G}}] \geq \tilde{T}_m$. Likewise, let $\tilde{c}_m$ and $\tilde{d}_m$ act like c and d on $1^m \Sigma^*$. Then

$$[g, \tilde{c}_m] = (1, \ldots, 1, [x, a], [x, \tilde{c}])_m = (1, \ldots, 1, x^2, (1, y))_m,$$
$$[g, \tilde{d}_m] = (1, \ldots, 1, (1, x))_m.$$

Using $\tilde{T}_m \leq [Q, \tilde{\mathfrak{G}}]$, we obtain $(1, \ldots, 1, (1, y))_m = (1, \ldots, 1, y)_{m+1} \in [Q, \tilde{\mathfrak{G}}]$, so again by a conjugation argument $[Q, \tilde{\mathfrak{G}}] \geq \tilde{K}_{m+1}$. $\square$

Theorem 7.4. *For all $m \geq 1$ we have:*

1. $\gamma_{2^m+1}(\widetilde{\mathfrak{G}}) = \tilde{N}_m$.
2. $\gamma_{2^m+1+r}(\widetilde{\mathfrak{G}}) = \tilde{N}_{m+1}\alpha^{-1}(V_m^r)\beta^{-1}(V_m^r)\gamma^{-1}(V_{m-1}^r)$ *for* $r = 0, \ldots, 2^m$.
3.

$$\operatorname{rank}(\gamma_n(\widetilde{\mathfrak{G}})/\gamma_{n+1}(\widetilde{\mathfrak{G}})) = \begin{cases} 4 & \text{if } n = 1, \\ 3 & \text{if } n = 2, \\ 3 & \text{if } n = 2^m + 1 + r, \text{ with } 0 \leq r < 2^{m-1}, \\ 2 & \text{if } n = 2^m + 1 + r, \text{ with } 2^{m-1} \leq r \leq 2^m. \end{cases}$$

Proof. First compute $\gamma_2(\widetilde{\mathfrak{G}}) = \widetilde{\mathfrak{G}}' = \langle [a, \tilde{c}], \tilde{K} \rangle$, of index 16 in $\widetilde{\mathfrak{G}}$, and $\gamma_3(\widetilde{\mathfrak{G}}) = \langle x^2, (1, x), (1, y) \rangle^{\widetilde{\mathfrak{G}}} = \tilde{N}_1$, with $x^2 = [x, a]$, $(1, x) = [x, \tilde{d}]$ and $(1, y) = [x, \tilde{c}]$. This gives the basis of an induction on $m \geq 1$ and $0 \leq r \leq 2^m$.

Assume that $\gamma_{2^m+1}(\widetilde{\mathfrak{G}}) = \tilde{N}_m$. Note that the hypothesis of Lemma 5.2 is satisfied for $\widetilde{\mathfrak{G}}$, as it holds for $\mathfrak{G} < \widetilde{\mathfrak{G}}$. Consider the sequence of quotients $Q_r = \tilde{N}_{m+1}\gamma_{2^m+1+r}(\widetilde{\mathfrak{G}})/\tilde{N}_{m+1}$ for $r \geq 0$. Lemmata 7.1 and 5.2 tell us that $Q_r = \alpha^{-1}(V_m^r) \oplus \beta^{-1}(V_m^r) \oplus \gamma^{-1}(V_{m-1}^r)$; in particular $Q_r \ni (x, \ldots, x) = \alpha^{-1}(v_m^{2^m-1})$ for all $r < 2^m$, and then Lemma 7.3 tells us that $\gamma_{2^m+1+r}(\widetilde{\mathfrak{G}}) \geq \tilde{N}_{m+1}$ for $r \leq 2^m$. When $r = 2^m$ we have $\gamma_{2^m+1+1}(\widetilde{\mathfrak{G}}) = \tilde{N}_{m+1}$ and the induction can continue. $\qquad\square$

Lemma 7.5. *For all $m \geq 1$ and $r \in \{0, \ldots, 2^m - 1\}$ we have:*

$$(\alpha^{-1}V_m^r)^2 = \gamma^{-1}(V_m^r) \leq \tilde{N}_{m+1};$$
$$(\beta^{-1}V_m^r)^2 = 1 \leq \tilde{N}_{m+1};$$
$$(\gamma^{-1}V_{m-1}^r)^2 = \gamma^{-1}(V_m^{r+2^{m-1}}) \leq \tilde{N}_{m+1}.$$

Proof. Write $\alpha^{-1}(v_m^r) = (x^{i_1}, \ldots, x^{i_{2^m}})$, $\beta^{-1}(v_m^r) = (y^{i_1}, \ldots, y^{i_{2^m}})$ or $\gamma^{-1}(v_m^r) = (x^{2i_1}, \ldots, x^{2i_{2^m}})$ for some $i_* \in \{0, 1\}$. Then these claims follow immediately, using Lemma 7.2, from

$$(\alpha^{-1}v_m^r)^2 = (x^{i_1}, \ldots, x^{i_{2^m}})^2 = (x^{2i_1}, \ldots, x^{2i_{2^m}}) = \gamma^{-1}(v_m^r),$$
$$(\beta^{-1}v_m^r)^2 = (y^{i_1}, \ldots, y^{i_{2^m}})^2 = (y^{2i_1}, \ldots, y^{2i_{2^m}}) = (1, \ldots, 1),$$
$$(\gamma^{-1}v_{m-1}^r)^2 = (x^{2i_1}, \ldots, x^{2i_{2^{m-1}}})^2 = (x^{4i_1}, \ldots, x^{4i_{2^{m-1}}})$$
$$\equiv (x^{2i_1}, x^{2i_1}, \ldots, x^{2i_{2^{m-1}}}, x^{2i_{2^{m-1}}}) = \gamma^{-1}(v_m^{r+2^{m-1}}) \quad \text{mod } \tilde{N}_{m+1}.$$

$\qquad\square$

Theorem 7.6. *For all $m \geq 1$ we have:*

1. $\widetilde{\mathfrak{G}}_{2^m+1} = \tilde{N}_m$.
2.

$$\widetilde{\mathfrak{G}}_{2^m+1+r} = \begin{cases} \tilde{N}_{m+1}\alpha^{-1}(V_m^r)\beta^{-1}(V_m^r)\gamma^{-1}(V_{m-1}^{r/2}) & \text{if } 0 \leq r \leq 2^m \text{ is even}, \\ \tilde{N}_{m+1}\alpha^{-1}(V_m^r)\beta^{-1}(V_m^r)\gamma^{-1}(V_{m-1}^{(r-1)/2}) & \text{if } 0 \leq r \leq 2^m \text{ is odd}. \end{cases}$$

3.
$$\operatorname{rank}(\widetilde{\mathfrak{G}}_i/\widetilde{\mathfrak{G}}_{i+1}) = \begin{cases} 4 & \text{if } i = 1, \\ 3 & \text{if } i > 1 \text{ is even}, \\ 2 & \text{if } i > 1 \text{ is odd}. \end{cases}$$

Proof. First compute $\widetilde{\mathfrak{G}}_2 = \gamma_2(\widetilde{\mathfrak{G}})$ and $\widetilde{\mathfrak{G}}_3 = \gamma_3(\widetilde{\mathfrak{G}}) = \tilde{N}_1$. This gives the basis of an induction on $m \geq 1$ and $0 \leq r \leq 2^m$. Assume $\mathfrak{G}_{2^m+1} = \tilde{N}_m$. Consider the sequence of quotients $Q_{m,r} = \tilde{N}_{m+1}\widetilde{\mathfrak{G}}_{2^m+1+r}/\tilde{N}_{m+1}$ for $r \geq 0$. We have $Q_{m,r} = [\widetilde{\mathfrak{G}}, Q_{m,r-1}]Q^2_{m-1,\lfloor r/2 \rfloor}$ by (2). Lemmata 7.1, 7.5 and 5.2 tell us that $Q_r = \alpha^{-1}(V_m^r) \oplus \beta^{-1}(V_m^r) \oplus \gamma^{-1}(V_{m-1}^{\lfloor r/2 \rfloor})$; in particular $Q_r \ni (x, \ldots, x) = \alpha^{-1}(v_m^{2^m-1})$ for all $r < 2^m$, and then Lemma 7.3 tells us that $\mathfrak{G}_{2^m+1+r} \geq \tilde{N}_{m+1}$ for $r \leq 2^m$. When $r = 2^m$ we have $\mathfrak{G}_{2^m+1+1} = \tilde{N}_{m+1}$ and the induction can continue. $\qquad\square$

7.1. The Lie Algebra Structures.

We describe here the Cayley graphs of L and $\mathcal{L}_{\mathbb{F}_p}$ associated respectively to the lower central and dimension series of $\widetilde{\mathfrak{G}}$. Consider $\tilde{S} = \{a, \tilde{b}, \tilde{c}, \tilde{d}\}$ and define the transformation $\tilde{\sigma}$ on $\tilde{S}^*$ by

$$\tilde{\sigma}(a) = a\tilde{b}a, \quad \tilde{\sigma}(\tilde{b}) = \tilde{d}, \quad \tilde{\sigma}(\tilde{c}) = \tilde{b}, \quad \tilde{\sigma}(\tilde{d}) = \tilde{b}.$$

Theorem 7.7. *The Cayley graph of $L(\widetilde{\mathfrak{G}})$ is as follows:*

where $x_m^r = \alpha^{-1}(v_m^r)$, $y_m^r = \beta^{-1}(v_m^r)$ and $z_m^r = \gamma^{-1}(v_m^r)$. The edge $(x_m^{2^m-1}, x_{m+1}^0)$ is labelled by $\tilde{\sigma}^m(\tilde{d})$, the edges $(x_m^{2^m-1}, y_{m+1}^0)$ and $(x_m^{2^m-1}, z_m^0)$ are labelled by $\tilde{\sigma}^m(\tilde{b})$, the edges $(x_m^{2^m-1}, z_m^0)$ and $(y_m^{2^m-1}, x_{m+1}^0)$ are labelled by $\tilde{\sigma}^m(\tilde{c})$, and the paths from x_m^0 to $x_m^{2^m-1}$, from y_m^0 to $y_m^{2^m-1}$ and from z_m^0 to $z_m^{2^m-1}$ are labelled by $\tilde{\sigma}^{m-1}(a)$.

The Cayley graph of $\mathcal{L}_{\mathbb{F}_2}(\widetilde{\mathfrak{G}})$ is as follows:

with the same labellings as for $L(\widetilde{\mathfrak{G}})$; and power maps from x_m^r to z_m^r.

8. OTHER FRACTAL GROUPS

The technique involved in the proof of the results of the last three sections show that for a group G acting on a tree Σ^* by powers of the cyclic permutation $\epsilon = (0, 1, \ldots, p-1)$ at each vertex, G has finite width when the following conditions are satisfied:

1. the corresponding action on a sequence $\{V_n\}_{n=0}^{\infty}$ of G-modules as defined in Subsection 5.1 has the *bounded corank property*, i.e. there is a constant C such that
   $$\dim V_n^r/[G, V_n^r] \leq C$$
   for all $n \geq 0$ and $0 \leq r \leq p^n - 1$.
2. There is a descending sequence $\{N_m\}_{m=1}^{\infty}$ of normal subgroups of G satisfying the condition that for all m the quotients N_m/N_{m+1} are isomorphic to some direct sum $\bigoplus_{i=1}^K V_{m+\delta_i}$ for fixed K and δ_i.

Let us mention that the p-groups G_ω, for arbitrary $p \geq 2$ and $\omega \in \{0, \ldots, p\}^{\mathbb{N}}$ constructed in [Gri84, Gri85] all satisfy Condition 1. Also, the group $\langle a, t \rangle < \text{Aut}(\Sigma_p^*)$, $p \geq 3$, where a acts as ϵ on the root vertex and trivially elsewhere and t is defined recursively by $t = (a, 1, \ldots, 1, t)$, satisfies Condition 1. We believe that this last group also satifies Condition 2, as do all G_ω for periodic sequences ω. Note that $\mathfrak{G}$ is a particular case of G_ω when $p = 2$ and $\omega = 012012\ldots$. Therefore they all 'should' have finite width.

Meanwhile, the Gupta-Sidki groups constructed in [GS83] do not satisfy Condition 1. As was proved recently by the first author, the growth of the Lie algebra $\mathcal{L}_{\mathbb{F}_p}(G)$ coincides with the spherical growth of the Schreier graph of G relatively to $\text{st}_G(e)$, where e is an infinite geodesic path in the tree Σ^*. For our groups $\mathfrak{G}$ and $\widetilde{\mathfrak{G}}$ the spherical growth is bounded and this is why these groups have bounded width. For the Gupta-Sidki groups, the spherical growth of the Schreier graph is unbounded (it grows approximately as $\sqrt{n}$), and therefore these groups do not have the finite width property. It also follows from these considerations that their growth is at least $e^{n^{1-1/(1/2+2)}} = e^{n^{3/5}}$.

9. PROFINITE GROUPS OF FINITE WIDTH

Finally we wish to explain how our results in the previous sections lead to counterexamples to Conjecture 1.1 stated in the introduction. Let $\widehat{G}$ be the profinite completion of $G = \mathfrak{G}$ or $\widetilde{\mathfrak{G}}$.

Theorem 9.1. *The group $\widehat{G}$ is a just-infinite pro-2-group of finite width which does not belong to the list of Conjecture 1.1 (which consists of solvable groups, p-adic analytic groups, and groups commensurable to positive parts of loop groups or to the Nottingham group).*

Its proof relies on the following notion:

Definition 9.2. Let $G < \operatorname{Aut}(\Sigma^*)$ be a group acting on a rooted tree. G has the *congruence subgroup property* if for any finite-index subgroup H of G there is an n such that $\operatorname{st}_G(n) < H < G$.

Proof. G has the congruence property. This is well known for $\mathfrak{G}$ (see for instance [Gri98]); while for $\tilde{\mathfrak{G}}$ the subgroup $\tilde{K}$ contains $\operatorname{st}_{\tilde{\mathfrak{G}}}(4)$ and enjoys the property that every subgroup of finite index in $\tilde{\mathfrak{G}}$ contains $\tilde{K}_m = \tilde{K} \times \cdots \times \tilde{K}$ for some m; see [BG98].

The profinite completion of G with respect to its subgroups $\operatorname{st}_G(n)$ is therefore a pro-2-group and coincides with the closure of G in $\operatorname{Aut}(\{0,1\}^*)$. The closure of a branch group is again a branch group, as is observed in [Gri98].

The criterion of just-infiniteness for profinite branch groups is the same as the one for discrete branch groups given in [BG98]; it is that $\overline{K}/\overline{K}'$ (respectively $\overline{\tilde{K}}/\overline{\tilde{K}}'$) be finite, where $\overline{K}$ and $\overline{\tilde{K}}$ are the closures of K and $\tilde{K}$. Now $|\overline{K}/\overline{K}'| \le |K/K'| < \infty$, the last inequality following from a computation in [BG98]. The same inequalities hold for $\tilde{\mathfrak{G}}$, and this proves the just-infiniteness of $\widehat{G}$.

The group $\widehat{G}$ has finite width for both versions of Definition 1.2. This is clear for D-width, because the discrete and pro-p Lie algebras $\mathcal{L}(G)$ and $\mathcal{L}(\widehat{G})$ are isomorphic. The finiteness of C-width follows from the inequalities

$$\left| \gamma_n(\widehat{G})/\gamma_{n+1}(\widehat{G}) \right| \le |\gamma_n(G)/\gamma_{n+1}(G)| < \infty,$$

which again are consequences of the congruence property of G.

Finally, $\widehat{G}$ does not belong to the list of groups given in Conjecture 1.1: it is neither solvable, because G isn't, nor p-adic analytic, by Lazard's criterion [Laz65] (its Lie algebra $\mathcal{L}(\widehat{G}) = \mathcal{L}(G)$ would have a zero component in some dimension). The other groups in the list of Conjecture 1.1 are *hereditarily just-infinite* groups, that is, groups every open subgroup of which is just-infinite [KLP97, page 5]. Profinite just-infinite branch groups are never just-infinite, as is shown in [Gri98]. □

REFERENCES

[And76]   George E. Andrews, *The theory of partitions*, Addison-Wesley Publishing Co., Reading, Mass.-London-Amsterdam, 1976, Encyclopedia of Mathematics and its Applications, Vol. 2.

[Bar98]   Laurent Bartholdi, *The growth of Grigorchuk's torsion group*, Internat. Math. Res. Notices **20** (1998), 1349–1356.

[BG98]   Laurent Bartholdi and Rostislav I. Grigorchuk, *On the parabolic subgroups and growth of certain fractal groups*, to appear, 1998.

[Bas72]   Hyman Bass, *The degree of polynomial growth of finitely generated nilpotent groups*, Proc. London Math. Soc. (3) **25** (1972), 603–614.

[Ber83]   Alexander E. Bereznii, *Discrete subexponential groups*, Zap. Nauchn. Sem. Leningrad. Otdel. Mat. Inst. Steklov. (LOMI) **123** (1983), 155–166 (Russian).

[CG97] Tullio G. Ceccherini-Silberstein and Rostislav I. Grigorchuk, *Amenability and growth of one-relator groups*, Enseign. Math. (2) **43** (1997), no. 3–4, 337–354.

[DdSMS91] John D. Dixon, Marcus P. F. du Sautoy, Avinoam Mann, and Dan Segal, *Analytic pro-p-groups*, Cambridge University Press, Cambridge, 1991.

[Ego84] Georgiĭ P. Egorychev, *Integral representation and the computation of combinatorial sums*, American Mathematical Society, Providence, R.I., 1984, Translated from the Russian by H. H. McFadden, Translation edited by Lev J. Leifman.

[Gol64] Evgueniĭ S. Golod, *On nil-algebras and finitely approximable p-groups*, Izv. Akad. Nauk SSSR Ser. Mat. **28** (1964), 273–276.

[GS64] Evgueniĭ S. Golod and Igor R. Shafarevich, *On the class field tower*, Izv. Akad. Nauk SSSR Ser. Mat. **28** (1964), no. 2, 261–272 (Russian).

[Gri80] Rostislav I. Grigorchuk, *On Burnside's problem on periodic groups*, Funktsional. Anal. i Prilozhen. **14** (1980), no. 1, 53–54, English translation: Functional Anal. Appl. **14** (1980), 41–43.

[Gri84] Rostislav I. Grigorchuk, *Degrees of growth of finitely generated groups and the theory of invariant means*, Izv. Akad. Nauk SSSR Ser. Mat. **48** (1984), no. 5, 939–985, English translation: Math. USSR-Izv. **25** (1985), no. 2, 259–300.

[Gri85] Rostislav I. Grigorchuk, *Degrees of growth of p-groups and torsion-free groups*, Mat. Sb. (N.S.) **126(168)** (1985), no. 2, 194–214, 286.

[Gri89] Rostislav I. Grigorchuk, *On the Hilbert-Poincaré series of graded algebras that are associated with groups*, Mat. Sb. (N.S.) **180** (1989), no. 2, 207–225, 304, English translation: Math. USSR-Sb. **66** (1990), no. 1, 211–229.

[Gri98] Rostislav I. Grigorchuk, *Just infinite branched groups*, 1998, to appear in Horizons in Profinite Groups (Dan Segal ed.), Birkhaüser, Basel.

[GH97] Rostislav I. Grigorchuk and Pierre de la Harpe, *On problems related to growth, entropy, and spectrum in group theory*, J. Dynam. Control Systems **3** (1997), no. 1, 51–89.

[Gro81a] Mikhael Gromov, *Groups of polynomial growth and expanding maps*, Inst. Hautes Études Sci. Publ. Math. (1981), no. 53, 53–73.

[Gro81b] Mikhael Gromov, *Structures métriques pour les variétés riemanniennes*, CEDIC, Paris, 1981, Edited by J. Lafontaine and P. Pansu.

[Gui70] Yves Guivarc'h, *Groupes de Lie à croissance polynomiale*, C. R. Acad. Sci. Paris Sér. A-B **271** (1970), A237–A239.

[Gui73] Yves Guivarc'h, *Croissance polynomiale et périodes des fonctions harmoniques*, Bull. Soc. Math. France **101** (1973), 333–379.

[GS83] Narain Gupta and Said Sidki, *On the Burnside problem for periodic groups*, Math. Z. **182** (1983), 385–388.

[Har98] Pierre de la Harpe, *Topics in geometric group theory*, University of Chicago Press, 1998, to appear; preprint at http://www.unige.ch/math/biblio/preprint/pp98.html.

[Her94] Israel N. Herstein, *Noncommutative rings*, Carus Mathematical Monographs, vol. 15, Mathematical Association of America, Washington, DC, 1994, Reprint of the 1968 original, With an afterword by Lance W. Small.

[HB82] Bertram Huppert and Norman Blackburn, *Finite groups II*, Springer-Verlag, Berlin, 1982, AMD, 44.

[Jac41] Nathan Jacobson, *Restricted Lie algebras of characteristic p*, Trans. Amer. Math. Soc. **50** (1941), 15–25.

[Jac62] Nathan Jacobson, *Lie algebras*, Interscience Publishers (a division of John Wiley & Sons), New York-London, 1962, Interscience Tracts in Pure and Applied Mathematics, No. 10.

[Jen41]   Stephen A. Jennings, *The structure of the group ring of a p-group over a modular field*, Trans. Amer. Math. Soc. **50** (1941), 175–185.

[Jen55]   Stephen A. Jennings, *The group ring of a class of infinite nilpotent groups*, Canad. J. Math. **7** (1955), 169–187.

[Kaĭ80]   Vadim A. Kaĭmanovič, *The spectral measure of transition operator and harmonic functions connected with random walks on discrete groups*, Zap. Nauchn. Sem. Leningrad. Otdel. Mat. Inst. Steklov. (LOMI) **97** (1980), 102–109, 228–229, 236, Problems of the theory of probability distributions, VI.

[Kal46]   Lev A. Kaloujnine, *Sur les p-groupes de Sylow du groupe symétrique du degré p^m. (Suite centrale ascendante et descendante.)*, C. R. Acad. Sci. Paris Sér. I Math. **223** (1946), 703–705.

[Kes59]   Harry Kesten, *Symmetric random walks on groups*, Trans. Amer. Math. Soc. **92** (1959), 336–354.

[KLP97]   Gundel Klaas, Charles R. Leedham-Green, and Wilhelm Plesken, *Linear pro-p-groups of finite width*, Lecture Notes in Mathematics, vol. 1674, Springer-Verlag, Berlin, 1997.

[Koc70]   Helmut Koch, *Galoissche Theorie der p-Erweiterungen*, Springer-Verlag, Berlin, 1970, Mit einem Geleitwort von I. R. Šafarevič.

[Kou98]   Malik Koubi, *Croissance uniforme dans les groupes hyperboliques*, to appear, 1998.

[Laz53]   Michel Lazard, *Sur les groupes nilpotents et les anneaux de lie*, Ann. École Norm. Sup. (3) **71** (1953), 101–190.

[Laz65]   Michel Lazard, *Groupes analytiques p-adiques*, Inst. Hautes Études Sci. Publ. Math. **26** (1965), 389–603.

[Leo98]   Youriĭ G. Leonov, *On lower estimation of growth for some torsion groups*, to appear, 1998.

[LG94]   Charles R. Leedham-Green, *The structure of finite p-group's*, J. London Math. Soc. (2) **50** (1994), no. 1, 49–67.

[LM91]   Alexander Lubotzky and Avinoam Mann, *On groups of polynomial subgroup growth*, Invent. Math. **104** (1991), no. 3, 521–533.

[Mag40]   Wilhelm Magnus, *Über Gruppen und zugeordnete Liesche Ringe*, J. Reine Angew. Math. **182** (1940), 142–149.

[Mei54]   Günter Meinardus, *Asymptotische Aussagen über Partitionen*, Math. Z. **59** (1954), 388–398.

[Mil68]   John W. Milnor, *Growth of finitely generated solvable groups*, J. Differential Geom. **2** (1968), 447–449.

[MZ99]   Consuelo Martinez and Efim I. Zel'manov, *Nil algebras and unipotent groups of finite width*, To appear in Advances in Math., 1999.

[Pas79]   Inder Bir S. Passi, *Group rings and their augmentation ideals*, Lecture Notes in Mathematics, vol. 715, Springer, Berlin, 1979.

[Pman77]   Donald S. Passman, *The algebraic structure of group rings*, Wiley-Interscience [John Wiley & Sons], New York, 1977, Pure and Applied Mathematics.

[Pet93]   Victor M. Petrogradskiĭ, *Some type of intermediate growth in Lie algebras*, Uspekhi Mat. Nauk **48** (1993), no. 5(293), 181–182.

[Qui68]   Daniel G. Quillen, *On the associated graded ring of a group ring*, J. Algebra **10** (1968), 411–418.

[Roz96a]   Alexander V. Rozhkov, *Lower central series of a group of tree automorphisms*, Mat. Zametki **60** (1996), no. 2, 225–237, 319.

[Roz96b]   Alexander V. Rozhkov, *Conditions of finiteness in groups of automorphisms of trees*, Habilitation thesis, Chelyabinsk, 1996.

[Sha95a]   Aner Shalev, *Finite p-groups*, Finite and locally finite groups (Istanbul, 1994), Kluwer Acad. Publ., Dordrecht, 1995, pp. 401–450.

[Sha95b]   Aner Shalev, *Some problems and results in the theory of pro-p groups*, Groups '93 Galway/St. Andrews, Vol. 2, London Math. Soc. Lecture Note Ser., vol. 212, Cambridge Univ. Press, Cambridge, 1995, pp. 528–542.

[Tit72]   Jacques Tits, *Free subgroups in linear groups*, J. Algebra **20** (1972), 250–270.

[VZ93]   Michael Vaughan-Lee and Efim I. Zel'manov, *Upper bounds in the restricted Burnside problem*, J. Algebra **162** (1993), no. 1, 107–145.

[VZ96]   Michael Vaughan-Lee and Efim I. Zel'manov, *Upper bounds in the restricted Burnside problem. II*, Internat. J. Algebra Comput. **6** (1996), no. 6, 735–744.

[Wol68]   Joseph A. Wolf, *Growth of finitely generated solvable groups and curvature of Riemanniann manifolds*, J. Differential Geom. **2** (1968), 421–446.

[Zas40]   Hans Zassenhaus, *Ein Verfahren, jeder endlichen p-Gruppe einen Lie-Ring mit der Characteristik p zuzuordnen*, Abh. Math. Sem. Univ. Hamburg **13** (1940), 200–207.

[Zel95a]   Efim I. Zel'manov, *Lie ring methods in the theory of nilpotent groups*, Groups '93 Galway/St. Andrews, Vol. 2, London Math. Soc. Lecture Note Ser., vol. 212, Cambridge Univ. Press, Cambridge, 1995, pp. 567–585.

[Zel95b]   Efim I. Zel'manov, *More on Burnside's problem*, Combinatorial and geometric group theory (Edinburgh, 1993), London Math. Soc. Lecture Note Ser., vol. 204, Cambridge Univ. Press, Cambridge, 1995, pp. 314–321.

[Zel96]   Efim I. Zel'manov, *Talk at the ESF conference on algebra and discrete mathematics "Group Theory: from Finite to Infinite"*, Castelvecchio Pascoli, 13–18 July 1996.

[Zel97]   Efim I. Zel'manov, *On the restricted Burnside problem*, Fields Medallists' lectures, World Sci. Ser. 20th Century Math., vol. 5, World Sci. Publishing, River Edge, NJ, 1997, pp. 623–632.

Laurent Bartholdi
Section de Mathématiques
Université de Genève
CP 240, 1211 Genève 24
Switzerland
Laurent.Bartholdi@math.unige.ch

Rostislav Grigorchuk
Steklov Mathematical Institute
Gubkina Street 8
Moscow 117966
Russia
grigorch@mi.ras.ru

TRANSLATION NUMBERS OF GROUPS ACTING ON QUASICONVEX SPACES

GREGORY R. CONNER

1. INTRODUCTION

Suppose M is a compact Riemannian manifold of nonpositive sectional curvature and $\widetilde{M}$ is its universal cover, then every element of the covering isometry group fixes at least one geodesic line in $\widetilde{M}$ and acts as a translation of each line it fixes. It turns out that the displacement of these translations, which is called the *translation number*, $\tau(g)$, of the covering isometry g, is independent of which line is chosen and that

$$\tau(g) = \limsup_{n \to \infty} \frac{\|g^n\|}{n}.$$

This notion can be extended to a larger class of groups. Let G be a group which is equipped with a metric, d, which is invariant under left multiplication by elements of G (i.e., $d(xy, xz) = d(y, z) \,\forall\, x, y, z \in G$.) We then say that d is a *left-invariant group metric* (or a *metric* for brevity) on G. Let $\| \, \| : G \longrightarrow \mathbb{Z}$ be defined by $\|x\| = d(x, 1_G)$. Examples of such metrics are a finitely generated group equipped with a word metric or the pull-back metric on a group of isometries of a metric space. For x in G, let

$$\tau(x) = \lim_{n \to \infty} \frac{\|x^n\|}{n}.$$

This limit exists (see [GS]) and is called the *translation number* of x.

This paper will discuss the properties of translation numbers of group actions on metric spaces in general and on "nonpositively curved spaces" in particular. The notion of translation number is quite a useful one since it has both algebraic and geometric aspects. It allows us to assign a "length" to each element of a group which is endowed with a group invariant metric (such as a word metric or a pull-back metric). Numerous authors have used the notion of translation numbers in their work, including Busemann [Bu], the author [C1, C2], Gersten/Short [GS], Gromov [Gr] and Gromoll/Wolf [GW]. In [C1] we define a group to be *translation discrete* if it supports a group invariant metric for which the translation numbers of the nontorsion elements are bounded away from zero and prove that solvable translation discrete groups of finite cohomological dimension are finite extensions of $\mathbb{Z}^n$. Gromov [Gr] has shown that translation numbers corresponding to a word metric in a word hyperbolic group are rational with bounded denominator (and thus are discrete). The same is shown to be true for finitely generated nilpotent groups in [C2].

Of late there has been particularly keen interest in actions on "nonpositively curved" and "negatively curved" spaces. There are several notions of nonpositive curvature in use today. Chief among these is the condition CAT(0). We will call a geodesic metric space *convex* if the distance between the midpoints of two sides of any geodesic triangle is no more than half the length of the third side. This notion is more general than CAT(0) and is originally due to Busemann [Bu] for G-spaces (in his notation convex would mean straight and nonpositively curved), but we will use it in this more general setting.

The standard model for combinatorial negative curvature is the notion of a *Gromov–hyperbolic* space. This is a space in which there is a global constant δ such that any point on a side of a geodesic triangle is at distance at most δ from the union of the other two sides. This notion is a quasiisometry invariant and so does not "see" the fine structure of the metric.

We propose a common generalization which shares a number of the good points of both hyperbolicity and convexity. Our generalization has neither the strong negative curvature condition of hyperbolicity nor the strong local requirements of convexity. As we shall see in Theorem 3.3, this class of spaces behaves nicely with respect to translation numbers in a way similar to that of a convex space, but allows small local changes in the metric to stay in the class. We define a geodesic metric space to be δ-*quasiconvex* if the distance between the midpoints of any two sides of any geodesic triangle is no more than half the length of the remaining side plus δ.

We prove the following results:

Theorem 3.2. A finitely generated group has a quasiconvex Cayley graph if and only if it is word hyperbolic.

Theorem 4.1. Let the group G act properly discontinuously cocompactly by isometries on a proper δ-quasiconvex metric space M. Then G is translation discrete if and only if G does not contain an essential Baumslag-Solitar quotient as a subgroup.

Recently, Poleksic has announced that cocompact groups of isometries of quasiconvex spaces contain no Baumslag-Solitar quotients. One can combine that result with the previous result to get that cocompact groups of isometries of quasiconvex spaces are translation discrete.

Corollary 4.3. Let the group G act properly discontinuously cocompactly on a proper δ-quasiconvex metric space. Then any finitely generated residually finite normal subgroup of G is translation discrete.

2. PRELIMINARIES

2.1. Notation. If M is a metric space, i a nonnegative real number and m a point in M, then $B_i(m)$ will denote the closed ball of radius i about the point m. We say that a metric space is *proper* if its closed balls are compact, and that it is *geodesic* if given any two points in the space there is a rectifiable curve joining the two points whose length is the distance between the two points. We say that a group *acts discretely* on a topological space if the orbit of each point is a discrete closed set.

Suppose M is a metric space with a basepoint, m_0, and G is a group of isometries of M. Then we can *induce the metric from M to G* by defining a left-invariant (semi)metric, d_G, on G by defining the distance between each pair of elements g and h in G to be $d_G(h, g) = d_M(h(m_0), g(m_0))$. It is easily verified that d_G is, indeed, a left-invariant (semi)metric. If m_0 is not a fixed point of any element of G then d is a metric. We call the (semi)norm $\|g\| = d(g, 1_G)$ the *induced norm*.

We note that a discrete group of fixed-point-free isometries of a proper metric space M acts freely, properly discontinuously.

2.2. Displacement.

Definition 2.1 (Displacement Function). Suppose the group G acts by isometries on the metric space M. For each $g \in G$ we define the *displacement function* of g to be the function $d_g : M \longrightarrow \mathbb{R}^*$ given by $d_g(m) = d(gm, m)$ and define $d_0(g) = \inf_{m \in M} d_g(m)$. An isometry, g, of M is called *semisimple* if there is an $m \in M$ such that $d_0(g) = d(m, gm)$. For all $m \in M$ define $r_0 : M \longrightarrow \mathbb{R}^*$ by $r_0(m) = \inf_{g \in G \setminus \{1\}} d_g(m)$ and define $\lambda_0(M) = \inf_{m \in M} r_0(m)$.

It is evident that $d_g(m) = d_{hgh^{-1}}(hm) \; \forall \; h, g \in G, \; m \in M$, that d_g is a Lipschitz-(2) continuous function, and that d_0 is a class function (i.e. its value depends only on the conjugacy class of the element).

By definition, $r_0(m) = r_0(gm) \; \forall \, g \in G$. The fact that r_0 is a Lipschitz-(2) continuous function follows immediately from the fact that d_g is Lipschitz-(2) continuous.

We leave most of the proof of the following two lemmas to the interested reader.

Lemma 2.2. *If the group G acts cocompactly by isometries on the proper metric space M, then every element of G is semisimple. In particular if D is a compact subset of M such that $M = G \cdot D$, then for each $g \in G$,*

$$d_0(g) = \min_{h \in G,\, m \in D} d_{hgh^{-1}}(m).$$

Lemma 2.3. *If the group G acts discretely cocompactly by isometries on the proper metric space M, then*

1. *For each $m \in M$ there is a $g \in G$ such that $r_0(m) = d_g(m)$.*
2. *There exists $g \in G, m \in M$ such that $d_0(g) = d(m, gm) = \lambda_0(G)$.*
3. *There exists $\epsilon > 0$ such that $g \in G$ has a fixed-point if and only if $d_0(g) < \epsilon$.*

Proof. We omit the easy proofs of the first two parts of the statement and prove the third part. Let D be a compact set so that $G \cdot D = M$. Suppose $\forall\, \epsilon > 0$ there exists a fixed-point-free $g \in G$ such that $d_0(g) < \epsilon$. Then there is a sequence of points (x_i) in M and a sequence of fixed-point-free group elements (g_i) in $G \setminus \{1\}$ so that $d(g_i x_i, x_i) < 1/8^i$. By the definition of D, for each x_i there is an $h_i \in G$ so that $h_i x_i \in D$. Since D is compact we may replace (x_i) by a terminal subsequence so that $(h_i x_i)$ converges to a point x of D, and $d(h_i x_i, x) < 1/8^i$. Now we have

$$
\begin{aligned}
d(h_i g_i h_i^{-1} x, x) &\leq\ d(h_i g_i h_i^{-1} x, h_i g_i h_i^{-1}(h_i x_i)) + d(h_i g_i h_i^{-1}(h_i x_i), h_i x_i) + \\
&\qquad d(h_i x_i, x) \\
&=\ 2d(x, h_i x_i) + d(h_i g_i x_i, h_i x_i) \\
&=\ 2d(x, h_i x_i) + d(g_i x_i, x_i) \\
&\leq\ 3 \cdot 1/8^i \\
&<\ 1/2^i.
\end{aligned}
$$

Since G acts discretely, this implies that there is a constant C such that $i > C \Rightarrow h_i g_i h_i^{-1} x = x$. This shows that $i > C \Rightarrow g_i$ has a fixed-point, which contradicts the choice of (g_i).

Lemma 2.4. *Let G act properly discontinuously cocompactly by isometries on a proper metric space M, then $\forall\, \epsilon > 0$ there are only finitely many conjugacy classes K with $d_0(K) < \epsilon$. In particular, 0 is an isolated point of $d_0(G)$.*

Proof. Let D be a compact set such that $G \cdot D = M$. Fix $m_0 \in M$. Choose R such that $B_R(m_0) \supseteq D$. Let $B = B_{R+\epsilon}(m_0)$. Suppose K is a conjugacy class of G with $d_0(K) < \epsilon$. Now choose $k \in K$ so that there is an $x \in D$ so that

$d_k(x) = d_0(k) < \epsilon$. Then $k \cdot x \in k \cdot B \cap B \neq \emptyset$. Since $\{g \mid g \cdot B \cap B\} \neq \emptyset\}$ is finite there can be only finitely many such conjugacy classes K.

2.3. Translation Numbers.

Definition 2.5 (Translation number). Suppose G is a group equipped with a norm $\| \ \|$. We define the *translation number*, $\tau(g)$ of an element g in G by

$$\tau(g) = \lim_{n \to \infty} \frac{\|g^n\|}{n}.$$

It is shown in [GS] that this limit exists. A group is called *translation discrete* if it supports a metric such that the translation numbers of its nontorsion elements are bounded away from zero.

Observation 2.5.1. Suppose M is a metric space, G a group of isometries of M equipped with an induced metric, then

1. τ is invariant under change of basepoint.
2. $\tau(g) \leq d_g(m)$ for all $g \in G$ and $m \in M$.
3. $\tau(g) \leq d_0(g) \leq \|g\|$ for every g in G.

The following is a result of Gersten and Short ([GS]).

Theorem 2.6. *If G is a group endowed with a left-invariant metric then for every pair of elements x and y of G we have*

1. $0 \leq \tau(x) = \lim_{n \to \infty} \frac{\|x^n\|}{n} \leq \|x\|$.
2. $\tau(x) = \tau(y^{-1}xy)$.
3. $\tau(x^n) = |n| \cdot \tau(x) \, \forall n \in \mathbb{Z}$.
4. $\tau(x) = \tau(x^{-1})$.
5. *If x and y commute then $\tau(xy) \leq \tau(x) + \tau(y)$.*

2.4. Baumslag-Solitar quotients.

Definition 2.7 (Baumslag-Solitar groups). We call the class of groups

$$B_{n,m} = \langle a, b \mid a^{-1}b^m a = b^n \rangle$$

the *Baumslag-Solitar groups*. We call the group H an *essential Baumslag-Solitar quotient* if H is a quotient group of $B_{n,m}$, $m > n > 0$ such that the image of b in H has infinite order.

Observation 2.7.1. It is evident that a group contains an essential Baumslag-Solitar quotient as a subgroup if and only if the group contains an element of infinite order having two different conjugate powers. One can apply Theorem 2.6 to see that such an element must have zero translation number in any metric. It follows that a translation discrete group cannot contain an essential Baumslag-Solitar quotient as a subgroup. In [GS] it is shown that a biautomatic group equipped with a word metric contains no nontorsion element with zero translation number, thus such a group cannot contain an essential Baumslag-Solitar quotient as a subgroup.

3. QUASICONVEXITY

Definition 3.1 (δ-Midpoint Convex). We call a geodesic triangle in a geodesic metric space δ-*midpoint convex* if the distance between the midpoints of any two sides of the triangle is at most δ more than half the length of the remaining side.

Similarly we call a function, f, on a subinterval I of the real numbers a δ-*midpoint convex* function, if for any two points a and b in I,

$$f\left(\frac{a+b}{2}\right) \leq \frac{a+b}{2} + \delta.$$

We say f is δ-*quasiconvex* if whenever $a \leq t \leq b \in I$ then

$$f(t) \leq (t-a)\frac{f(b)-f(a)}{b-a} + f(a) + \delta.$$

We will call a geodesic metric space δ-*quasiconvex* if every geodesic triangle in it is δ-midpoint convex. A 0-quasiconvex space is called *convex*.

Troyanov shows in [GH] that $\mathbb{R}^2$ with the ℓ_P metric with $P > 2$ is convex but not CAT(0). It is an easy exercise to show that convex spaces are contractible (retract along unique geodesics) and that quasiconvex spaces are combable (i.e., they are CW-Lipschitz contractible).

One can show that a continuous δ-midpoint convex function on a subinterval of $\mathbb{R}$ is β-quasiconvex for some β. Thus the distance function for geodesics in a quasiconvex metric space is quasiconvex. This means that if we are given any two geodesics, and we reparameterize them both to be constant speed on the unit interval then the distance between corresponding points will be a quasiconvex function of the parameter. More precisely, if we choose γ, η of lengths l_1 and l_2 respectively, then the distance $d(\gamma(t \cdot l_1), \eta(t \cdot l_2))$ will be a quasiconvex function of t on $[0, 1]$.

It is now evident that Gromov–hyperbolic metric spaces are quasiconvex since δ-thin triangles are γ-midpoint convex for some γ, and every geodesic triangle is "almost" a tripod.

Theorem 3.2. *A finitely generated group has a quasiconvex Cayley graph if and only if it is word hyperbolic.*

Proof. By the preceding paragraph, a word hyperbolic group has a quasiconvex Cayley graph. On the other hand, a δ-quasiconvex Cayley graph has the property that any two geodesic paths which have the same initial point and end at distance at one apart must stay within $\delta + 1$ of each other along their entire length. It follows from [ECHLPT, Theorem 3.2.1 and Lemma 2.5.5] that a finitely generated group with a quasiconvex Cayley graph is bi-automatic with a geodesic language. In [Pa], Papasoglu shows that a group with a regular language satisfying the above thinness property (a strongly geodesically automatic group in his notation) is word hyperbolic.

Note 3.2.1. The preceding theorem shows that the property of having a quasi-convex Cayley graph is a quasiisometry invariant of finitely generated groups, leading one to ask if the property of being a "quasiconvex group" (i.e. being a finitely generated cocompact group of isometries of a quasiconvex space) might also be a quasiisometry invariant. One might also ask if there exist convex groups which are not CAT(0) groups or if there exist quasiconvex groups which are neither CAT(0) nor word hyperbolic.

The following is the main result of this section.

Theorem 3.3. *Let G be a group of isometries of a δ-quasiconvex space M, and equip G with an induced metric. Then for all $\gamma \in G$,*

$$d_0(\gamma) \geq \tau(\gamma) \geq d_0(\gamma) - \delta.$$

Proof. Let $m \in M$. Let η be a geodesic segment joining m and γm. Let ω be a geodesic segment joining m and $\gamma^2 m$. Let s, t be the midpoints of η and $\gamma\eta$ respectively. Since γ acts as an isometry of M it follows that $t = \gamma(s)$.

Applying the fact that the triangle $(\eta, \gamma\eta, \omega)$ is δ-midpoint convex, we have $d_0(\gamma) \leq d(s, t) \leq \delta + 1/2 \cdot d(m, \gamma^2(m))$ and thus $d(m, \gamma^2(m)) \geq 2(d_0(\gamma) - \delta)$. Since m was arbitrary, we deduce that $d_0(\gamma^2) \geq 2(d_0(\gamma) - \delta)$. Since γ was also arbitrary, we may use induction to show that $d_0(\gamma^{2^n}) \geq 2^n(d_0(\gamma) - \delta)$, and hence $\|\gamma^{2^n}\| \geq 2^n(d_0(\gamma) - \delta)$. So,

$$\tau(\gamma) \;=\; \lim_{n \to \infty} \frac{\|\gamma^n\|}{n} = \lim_{n \to \infty} \frac{\|\gamma^{2^n}\|}{2^n} \geq \lim_{n \to \infty} \frac{2^n(d_0(\gamma) - \delta)}{2^n} = d_0(\gamma) - \delta.$$

The following classical result, whose short proof we include for the sake of completeness, can be deduced from Busemann, [Bu, 4.11,6.1].

Corollary 3.4. *Let G act as a cocompact group of isometries of the convex space M, then for every $g \in G$ either $\tau(g) = 0$, in which case g fixes some point of M, or $\tau(g) > 0$ and there exists an infinite geodesic which g leaves invariant and translates by an amount $\tau(g)$.*

Proof. Let $g \in G$. By Lemma 2.2, g is semisimple. Choose $m \in M$ such that $d_g(m) = d_0(g)$. By Theorem 3.3, $\tau(g) = d_0(g)$. If $d_0(g) = 0$, then $d_g(m) = d(gm, m) = 0$ and thus $gm = m$.

Now suppose $d_0(g) > 0$. Let T be the geodesic segment joining m and gm and let $S = \langle g \rangle \cdot T$. Now,

$$d_{g^n}(m) \leq |n| \cdot length(T) = |n| \cdot d_g(m) = |n| \cdot d_0(g) = |n| \cdot \tau(g) = \tau(g^n) \leq d_{g^n}(m).$$

Using a convexity argument, one can show that S is an infinite geodesic and that every point of S is translated by g an amount $\tau(g)$.

Definition 3.5 (Infinite Quasigeodesic). Suppose M is a metric space. Then an *infinite* (λ, ϵ)-*quasigeodesic* in M is a (λ, ϵ)-quasi-isometric embedding of $(-\infty, \infty)$ into M.

The following is the result analogous to Corollary 3.4 in a quasiconvex setting

Corollary 3.6. *Let G act as a cocompact group of isometries of the quasiconvex space M, then for every $g \in G$ such that $\tau(g) > 0$ there exists an infinite quasigeodesic, S_g, which g leaves invariant and translates by an amount $d_0(g)$.*

Proof. Let $g \in G$. By Lemma 2.2, g is semisimple. Choose $m \in M$ such that $d_g(m) = d_0(g)$. By Theorem 3.3, $d_0(g) \leq \tau(g) + \delta$, where δ is the quasiconvexity constant for M. Let T_g be the geodesic segment joining m and gm and let $S_g = \langle g \rangle \cdot T$. Clearly one can parameterize S_g as a (nonembedded) line by $L_g : \mathbb{R} \to S_g$ where L_g maps the interval $[n \cdot d_0(g), (n+1) \cdot d_0(g)]$ to the geodesic segment $g^n T_g$. Now, $|n| \cdot \tau(g) = \tau(g^n) \leq d_0(g^n) \leq d_{g^n}(m) \leq |n| \cdot length(T_g) = |n| \cdot d_g(m) = |n| \cdot d_0(g) \leq |n| \cdot (\tau(g) + \delta)$, and thus L is a quasigeodesic embedding of $\mathbb{R}$ into M. Now, if x is in the image of L_g, choose n such that $x \in g^n T_g$, so that $d_0(g) \leq d(x, g(x)) \leq d(x, L_g(n+1)) + d(L_g(n+1), g(x)) = d(x, L_g(n+1)) + d(L_g(n), x) = length(g^n(T_g)) = d_0(g)$. ∎

We now state two well-known results which follow immediately from Theorem 3.3 and Corollary 3.4.

Corollary 3.7. *Let G act as a discrete cocompact group of isometries of the convex space M. Then $g \in G$ has finite order if and only if g fixes a point of M.*

Corollary 3.8. *If G is a group of isometries acting properly discontinuously and cocompactly on a convex metric space M then G is translation discrete. Furthermore, if G acts freely on M then G is torsion–free and $G \backslash M$ is a $K(G, 1)$*

Proof. Applying Theorem 3.3 we have that $\tau = d_0$. By Lemma 2.3, there is an $\epsilon > 0$ so that for every $g \in G, d_0(g) < \epsilon$ only if g fixes a point. By

Corollary 3.7, we see that an element of G has translation number less than ϵ if and only if it has finite order, and thus G is translation discrete. If, in addition, the action of G is fixed–point–free we note that G acts as covering isometries of M and that Corollary 3.7 implies that G is torsion-free. Then $G\backslash M$ has fundamental group G and has as its universal cover the contractible space M. Thus $G\backslash M$ is a $K(G,1)$.

4. MAIN RESULTS

Theorem 4.1. *Suppose the group G acts properly discontinuously cocompactly by isometries on a proper δ-quasiconvex metric space M. Then G is translation discrete if and only if it does not contain an essential Baumslag-Solitar quotient as a subgroup.*

Proof. By Observation 2.7.1 we see that if G is translation discrete then G does not contain an essential Baumslag-Solitar quotient.

On the other hand, suppose G contains no essential Baumslag-Solitar quotients. If every non-torsion element of G has nonzero translation number, then G is translation discrete by Lemma 2.4 and Theorem 3.3 . By Theorem 2.6 and Theorem 3.3 we need only show that every non-torsion element g of G has a power g^n so that $d_0(g^n) > \delta$. Let $g \in G$ and consider the set $\{d_0(g^n) \,|\, n \in \mathbb{N}\}$. If each element of this set is less than δ we see that two different positive powers of g must be conjugate since there are only finitely many conjugacy classes K with $d_0(K) < \delta$. Considering Observation 2.7.1, we see that since G contains no essential Baumslag-Solitar quotients as subgroups, g must be a torsion element.

The next result follows from the above and Observation 2.7.1.

Corollary 4.2. *A biautomatic group which acts properly discontinuously cocompactly by isometries on a proper δ-quasiconvex metric space is translation discrete.*

Corollary 4.3. *If the group G acts properly discontinuously cocompactly by isometries on a proper δ-quasiconvex metric space, then any finitely generated residually finite normal subgroup of G is translation discrete.*

Proof. Let H be a finitely generated residually finite normal subgroup of G. By Theorem 2.6 and Theorem 3.3 we need only show that every non-torsion element h of H has a power h^n so that $d_0(h^n) > \delta$. Since H is finitely generated residually finite we choose a characteristic subgroup H' of H of finite index so that H' misses the finitely many non-trivial conjugacy classes

K with $d_0(K) \leq \delta$. Since H' has finite index in H, every non-torsion element, h, of H has a power, say h^n, which lies in $H' \backslash \{1\}$ and thus $d_0(h^n) > \delta$.

5. APPLICATIONS

The purpose of this section is to give the reader a flavor of how the results in this paper can be applied.

Note 5.0.1. In [C1] it is shown that a finite cohomological dimension solvable subgroup of a translation discrete group is a finite extension of $\mathbb{Z}^n$ for some n.

Theorem 5.1. *If G is a discrete linear group acting freely as a cocompact, properly discontinuous group of isometries of a quasiconvex space, then G is translation discrete and any solvable subgroup of G is a finite extension of $\mathbb{Z}^n$ for some n.*

Proof. Since G is a linear group it is residually finite and thus is translation discrete by Corollary 4.3. Since G is a discrete linear group it has finite cohomological dimension. The result now follows from the above note.

The following classical result follows immediately from [C1] and Corollary 3.8

Corollary 5.2 (Gromoll–Wolf Theorem). *Let M be a compact Riemannian manifold of nonpositive sectional curvature. Then any solvable subgroup of $\pi_1(M)$ is Abelian-by-finite. In particular, $\pi_1(M)$ is a finite extension of $\mathbb{Z}^n$ where $n \leq \dim(M)$.*

REFERENCES

[Bu]   H. Busemann, *Spaces with Non-positive Curvature*, Acta Math **80**, 259-310 (1960).

[C1]   G. Conner, *Discreteness Properties of Translation Numbers in Solvable Groups*, J. Group Theory, (to appear).

[C2]   G. Conner, *Properties of Translation Numbers in Nilpotent Groups*, Communications in Algebra, 26(4),1069-1080(1998).

[ECHLPT] D. B. A. Epstein, J. W. Cannon, D. F. Holt, S. V. F. Levy, M. S. Paterson, and W. P. Thurston, *Word Processing in Groups*, Jones and Bartlett Publishers, Inc. (1992)

[GS]   S. M. Gersten and H. B. Short, *Rational Subgroups of Biautomatic Groups*, Annals of Math. **134** (1991), 125-158.

[GH]   M. Troyanov, *Espaces á courbure négative et groupes hyperboliques*, in: *Sur les groupes hyperboliques d'apres Mikhael Gromov*, edited by E. Ghys and P. de la Harpe, Progress in Maths Vol. 83, Birkhäuser (1990).

[GW]   D. Gromoll and J. A. Wolf, *Some Relations Between the Metric Structure and the Algebraic Structure of the Fundamental Group in Manifolds of Nonpositive Curvature*, Bul. Amer. Math. Soc. 4 (1971), 545-552.

[Gr]   M. Gromov, *Hyperbolic Manifolds, Groups and Actions*, Riemann Surfaces and Related Topics, Proceedings of the 1978 Stony Brook Conference, Princeton University Press (1980).

[Pa]   P. Papasoglu, *Strongly geodesically automatic groups are hyperbolic.*, Invent. Math. 121 (1995), no. 2, 323-334.

Gregory R. Conner
Math Department
Brigham Young Univ.
Provo, UT. 84602
USA
conner@math.byu.edu

ON A TERM REWRITING SYSTEM CONTROLLED BY SEQUENCES OF INTEGERS

ALEŠ DRÁPAL

ABSTRACT. A convergent term-rewriting system is developed for free algebras $A(\cdot, \triangleright)$ that satisfy the equality $x \cdot (y \cdot z) = (x \triangleright y) \cdot (x \cdot z)$.

We shall consider algebraic systems $A(\cdot, \triangleright)$, in which the binary operations $\cdot$ and $\triangleright$ satisfy the identity

$$x \cdot (y \cdot z) = (x \triangleright y) \cdot (x \cdot z)$$

for all $x, y, z \in A$. Our main result is the proof that the free algebra satisfying the above identity allows a definition of a convergent term-rewriting system that chooses a unique representative in every block of the congruence that is induced by this identity on the absolutely free algebra of terms $W = W(\cdot, \triangleright)$. The terms are constructed over a set of variables X.

Each rewriting rule is associated with a finite sequence of integers that has certain properties. There are infinitely many such sequences, and hence our system contains infinitely many rewriting rules. To prove that the rewriting system is convergent, we just need to show that it is locally confluent and finitely terminating [6][7]. The quasiorder needed to prove the latter fact is presented in Section 3 in a form that can make it useful also in other settings.

In other sections we are concerned with the specific properties of our identity. The identity seems to be of some importance in conjugacy closed loops [9][10][12], and this paper can be regarded, to some extent, as a preparation for the study of free conjugacy closed loops. A quasigroup (or a loop) $Q(\cdot)$ is called *left conjugacy closed* if all its left translations $L_a : x \mapsto ax$ are closed under conjugation. It is easy to see that a loop $Q(\cdot)$ is left conjugacy closed if and only if $b(a(b \backslash x)) = ((ba)/b)x$ holds for all $a, b, x \in Q$ (where $b \backslash x = L_b^{-1}(x)$ is the left division and $x/b = R_b^{-1}(x)$ is the right division). Setting $x = bc$ we obtain

$$b(ac) = ((ba)/b)(bc),$$

and this takes us to our equation, with $b \triangleright a = (ba)/b$. We shall mention in Section 5 how an approach using two operations might lead to the solution of the word problem in left conjugacy closed loops. In this context interesting connections to the braid group and the left distributive law seem to start to appear.

Partially supported by Grant Agency of Charles University, grant number 715.

The left distributive law

$$x \cdot (y \cdot z) = (x \cdot y) \cdot (x \cdot z)$$

has received much attention in the last decade, and quite a lot is known about free left distributive systems (some references can be found in Section 5). The identity that is studied in this paper resembles the left distributive identity closely enough to justify a few remarks comparing the equational theories of both identities.

If a term t_1 can be transformed to a term t_2 by successive applications of the left distributive law, then there always exists a term t_3 such that both t_1 and t_2 can be transformed to t_3 just by successive applications of the left distributive expansion (expansion means here an application of the law that increases the size of the term). This fact is used to construct for each free left distributive system a well behaved linear order.

The introduction of the operation ▷ kills this sort of confluency. On the other hand, it also restricts the positions where distribution can be iterated, and this fact allows us to keep track of the history of applications of the law with far less sophisticated means (when compared to left distributivity). Loosely speaking, one can just describe the movement along the right edge of a term, and use a certain simple criterion to decide when a series of law applications is to be considered as a contraction. The purpose of this paper is to show that a collection of all these compound contractions (which can be effectively enumerated) constitutes a convergent rewriting system, and thus yields a normal form.

The normal forms known for free left distributive systems [4] [15] are of a different nature – their construction can be described as a sort of controlled expansion. It is not clear if a contractive normal form is possible in the case of left distributivity.

1. ACTIONS BY SEQUENCES OF INTEGERS

We shall work with (finite) sequences of integers and (finite) sequences of terms. A sequence will be expressed by a juxtaposition of its elements, and we shall use • to denote concatenation of two sequences. The empty sequence will be always denoted by ε, and the length of a sequence, say a sequence a, will be denoted by $|a|$. A continuous subsequence of a sequence a is called a *factor* of this sequence.

A sequence of integers is said to be *reduced*, if

(1) it does not contain 0, and
(2) whenever it contains a factor $i_1 \ldots i_k$ with $i_k = -i_1$, then there exists j, $1 < j < k$, with $\big||i_j| - |i_1|\big| = 1$.

The above definition will look more natural after we introduce actions of integer sequences on sequences of terms. It will be observed that two integers with their absolute values differing by more than one yield commuting actions, and hence an action of an integer sequence can always be represented by an action of a reduced sequence.

Note that a reduced sequence never contains, by (2), a factor $+i - i$, and that (1) can be, strictly speaking, deduced from (2) as well. On the other hand, by our definition, the empty sequence ε is reduced. Thus every factor of a reduced sequence is reduced as well.

An action $a(t)$ of an integer sequence $a = i \bullet b$, i an integer, on a term sequence t is set to equal $i(b(t))$. However, the action of an integer, say i, on t, will not be always defined. Hence $a(t)$ can be defined, in general, only for some t; and we say that a is *compatible* with t in such a case.

Put $\varepsilon(t) = t = 0(t)$; for any sequence $t = t_0 \ldots t_n$ of terms from W. Assume now $i > 0$. If $i > n$, then neither i nor $-i$ is compatible with t. If $i \leq n$ holds, then $i(t)$ is always defined and equals

$$t_0 \ \ldots \ t_{i-2}(t_{i-1} \triangleright t_i)t_{i-1}t_{i+1} \ldots t_n,$$

while $(-i)t$ is defined if and only if t_{i-1} equals $t_i \triangleright s$ for some $s \in W$, and $(-i)(t)$ is then equal to

$$t_0 \ \ldots \ t_{i-2}t_i s t_{i+1} \ \ldots \ t_n.$$

We shall now observe some easy facts. If a is compatible with t, then $|a(t)| = |t|$. The equality $(a \bullet b)(t) = a(b(t))$ holds whenever one of the expressions is defined, and, in the same sense, $(i \bullet j)(t) = (j \bullet i)(t)$ is true for any integers i, j with $||i| - |j|| \geq 2$. Furthermore, observe that $(+i - i)(t)$ equals t whenever $(-i)(t)$ is defined.

Consider the following operations on sequences of integers:

(a) removal of 0;

(b) removal of a factor $+i - i$; and

(c) replacement of a factor ij by a factor ji, when $||i| - |j|| \geq 2$.

If a sequence b is obtained from a sequence a by means of (a), (b) and (c), and a is compatible with a term sequence t, then b is compatible with t as well, and $b(t) = a(t)$ holds.

It is easy to see that every sequence of integers a can be transformed by operations (a), (b) and (c) to a reduced sequence b. Furthermore, using a standard argument, we can also easily verify that if b_k, $k \in \{1, 2\}$, are reduced sequences that have been obtained from an integer sequence a, then b_2 can be obtained from b_1 just by means of (c).

Call integer sequences b_1 and b_2 *equivalent*, if there exists a series of replacements (c) that transform b_1 to b_2. Clearly, b_1 is reduced if and only if b_2 is reduced.

Call an integer sequence a *upward directed*, if it is reduced, and if every two-element factor ij of a satisfies $|j| \geq |i| - 1$. Obviously, each reduced sequence is equivalent to exactly one upward directed sequence, and each factor of an upward directed sequence is upward directed as well.

Say that an integer i *occurs* in an integer sequence $c = i_1 \ldots i_k$ if $i_r = i$ for some r, $1 \leq r \leq k$. Say that $\pm i$ occurs in c if i occurs in c or $-i$ occurs in c.

Finally, call a sequence *separated*, if it is reduced and if $\pm(i-1)$ occurs in b whenever $(-i) \bullet b \bullet (+i)$ is a factor of a, $i > 0$. It is clear that a sequence equivalent to a separated sequence is separated as well, and that each factor of a separated sequence is also separated.

An upward directed sequence a is obtained from a reduced sequence b by shifting a smaller integer to the left whenever possible. We shall now consider shifting to the left integers that are greater than all preceding integers.

Let $a = i_1 \ldots i_k$ be an integer sequence. For every j, $1 \leq j \leq k$, define a sequence a_j in such a way that $a_1 = i_1$ and a_{j+1} is determined by the following rule:

$a_{j+1} = a_j$ if $|i_{j+1}| \geq |i| + 2$ holds for every i that occurs in a_j, and

$a_{j+1} = a_j \bullet i_{j+1}$, if $|i_{j+1}| \leq |i| + 1$ for at least one i that occurs in a_j.

Lemma 1.1. *Assume $a = i_1 \ldots i_k$ and $1 \leq j \leq k$. Then a_j starts with i_1, and it is reduced (or upward directed, or separated) if a has this property.*

Proof. Proceed by induction on j. The case $j = 1$ is clear, and so we can assume $a_{j+1} = a_j \bullet i_{j+1}$. Let $\pm h$ be the rightmost value of a_j, with $h > 0$. Then $h \leq |i_j|$ is clearly true, and thus $|i_j| + 1 \leq |i_{j+1}|$ implies $h + 1 \leq |i_{j+1}|$. We see that a_{j+1} is upward directed if a is upward directed. Suppose now that a_{j+1} is not reduced. Then $i_r = -i_{j+1}$ for some r, $1 \leq r \leq j$, with i_r occuring in a_j and $\big||i| - |i_r|\big| \geq 2$ for all i that occur in a_j right of the occurence of i_r. But then $\big||i_s| - |i_r|\big| \geq 2$ holds for all s with $r < s \leq j$, and we see that a is not reduced. The case of a separated is similar. $\square$

The next lemma is clear.

Lemma 1.2. *Assume $a = i_1 \ldots i_k$, $1 \leq j \leq k$ and $a_j = i_{j_1} \ldots i_{j_r}$. Then $(a_j)_p = i_{j_1} \ldots i_{j_p}$ for every p, $1 \leq p \leq r$.*

The elements i_j that are not included into a_j can be shifted to the left, where they yield a sequence of 'sorted out' elements. Formally, for $a = i_1 \ldots i_k$ put $a'_1 = \varepsilon$ and, furthermore, either $a'_{j+1} = a'_j$, or $a'_{j+1} = a'_j \bullet i_{j+1}$, with the latter taking place if and only if $a_{j+1} = a_j$, $1 \leq j \leq k$.

Lemma 1.3. *Assume $a = i_1 \ldots i_k$ and $1 \leq j \leq k$. Then $i_1 \ldots i_j$ is equivalent to $a'_j \bullet a_j$. If $i_1 \ldots i_j$ is reduced (or upward directed or separated), then a'_j is also reduced (or upward directed or separated). If a'_j is non-empty, then $\min\{|i|; \ i \text{ occurs in } a'_j\} \geq 2 + \min\{|i|; \ i \text{ occurs in } a_j\}$.*

Proof. It is clear that $i_1 \ldots i_j$ is equivalent to $a'_j \bullet a_j$. Thus a'_j is reduced (or separated), if $i_1 \ldots i_j$ is such a sequence. Suppose that $i_1 \ldots i_j$ is upward directed, and use induction on j. The case $j = 1$ is clear. Assume $1 \leq j < k$. If $a'_{j+1} = a'_j$, then there is nothing to prove. Assume $a'_{j+1} = a'_j \bullet i_{j+1}$. If a'_j terminates with i_j, then a'_{j+1} is upward directed by the induction assumption and by $|i_{j+1}| + 1 \geq |i_j|$. If a'_j does not terminate with i_j, then a'_{j+1} can be produced from $i_1 \ldots i_{j-1} i_{j+1}$ and the induction assumption can be thus used.

Finally, note that whenever i_r is sent to a'_r, $1 \leq r \leq j$, then there exists i_s, $1 \leq s < r$, with $|i_s| + 2 \leq |i_r|$. This proves the stated inequality.

By a *subsequence* of a sequence $i_1 \ldots i_k$ we mean any sequence $i_{j_1} \ldots i_{j_r}$ with $1 \leq j_1 < \ldots j_r \leq k$, $r \geq 0$. By an *opposite sequence* to an integer sequence $a = i_1 \ldots i_k$ we mean the sequence $-i_k \cdots - i_1$. It will be denoted by $-a$. Note that a is reduced (or separated) if and only if $-a$ is reduced (or separated).

We have defined sequences a_j and a'_j for any non-empty sequence $a = i_1 \ldots i_k$, $1 \leq j \leq k$. We shall not retain this notation further on, and put $\hat{a} = a_k$. Thus $\hat{a}$ is the subsequence of the sequence a that is obtained by processing a from the left, entry by entry, in such a way that the leftmost entry is always retained and any subsequent entry is removed if and only its absolute value, when diminished by one, is greater than the absolute values of all entries that were retained in the already processed part of the sequence a (the processed part stretches to the left of the given entry). Put also $\hat{\varepsilon} = \varepsilon$. The above lemmas yield:

Proposition 1.4. *Every integer sequence a is equivalent to a sequence $c \bullet \hat{a}$, where both c and $\hat{a}$ are reduced (or separated or upward directed), if a has such a property. Furthermore, $\hat{\hat{a}}$ always equals $\hat{a}$, and both c and $\hat{a}$ are subsequences of a. If c is non-empty, then*

$$\min\{|i|; \ i \text{ occurs in } c\} \geq 2 + \min\{|i|; \ i \text{ occurs in } \hat{a}\}.$$

We shall now characterize sequences that will be later used to define the rewriting rules for the system solving the word problem.

Call an integer sequence a *primitive* if and only if it equals $(-h) \bullet b$ for some $h > 0$, and satisfies:

(a) it is reduced, separated and upward directed;
(b) $\hat{a}$ equals a;
(c) $|i| > h$ holds for every i in b; and
(d) for every occurence of $-i$ in b, where $i > 0$, there exists a positive $j < i$ such that $\pm j$ occurs right of $-i$.

Lemma 1.5. *Let u be a non-empty separated sequence of integers. Then u or $-u$ is equivalent to a sequence $v \bullet a$, with a primitive.*

Proof. Let $k > 0$ be the least integer such that $\pm k$ occurs in u. Choose $u' \in \{u, -u\}$ in such a way that $-k$ occurs in u. We can assume that u' is upward directed. There is no occurence of $-k$ in u' that has an occurence of $+k$ to the right of it. Thus there exist occurences of a negative integer, say $-h$, in u', with $h < |j|$ for every j right of the occurence of $-h$. Consider the rightmost such occurence (with no restriction on the value of $h > 0$), and denote by a the terminating factor of u' that starts with this occurence of $-h$.

Now, a is equivalent to $c \bullet \hat{a}$, where $\hat{a}$ is upward directed, separated, starts with $-h$, and satisfies $\hat{\hat{a}} = a$, by 1.4. The choice of a implies that $\hat{a}$ is really primitive, as $\hat{a}$ is obtained from a by subsequent deletion of integers that are greater by 2 (in absolute values) than all integers to the left of it.

Proposition 1.6. *Every separated sequence of integers u is equivalent to a sequence*

$$-(u_1 \bullet \ldots \bullet u_m) \bullet (v_n \bullet \ldots \bullet v_1),$$

where $m \geq 0$, $n \geq 0$, and all u_i and v_j, $1 \leq i \leq m$, $1 \leq j \leq n$, are primitive.

Proof. Proceed by induction on $|u|$. There is no difference, if one obtains the required form for u or for $-u$, and hence 1.5 and the induction assumption yield the statement in an immediate way.

2. ACTIONS OF SEPARATED SEQUENCES

Lemma 2.1. *Let a be an integer sequence compatible with a term sequence $t = t_0 \ldots t_n$, and let $a(t)$ be equal to $s_0 \ldots s_n$.*

(i) *If ± 1 does not occur in a, then $t_0 = s_0$;*

(ii) *if $+1$ occurs in a and -1 does not occur in a, then t_0 is a proper subterm of s_0; and*

(iii) *if -1 occurs in a and $+1$ does not occur in a, then s_0 is a proper subterm of t_0.*

Proof. The statement can be proved by a straightforward induction on $|a|$.

The above lemma could of course be expressed in a more general way relative to the least value of $|i|$ such that i occurs in a. In fact, future references to the lemma will sometimes pertain to such a more general version (and elsewhere we shall also proceed in a similar spirit).

Upper indices (like a^k) will be used in this section to distinguish between various sequences, and not to indicate a power of a sequence.

Proposition 2.2. *Let $t = t_0 \ldots t_n$ be a sequence of terms in W, $n \geq 0$.*

(i) *If a reduced integer sequence a is compatible with t, then a is separated.*

(ii) *If a^k, $k \in \{1, 2\}$, are two separated upward directed integer sequences, with $a^k(t) = t_0^k \ldots t_n^k$ and $a^k = c^k \bullet b^k$, where b^k is the shortest terminating factor of a^k that contains all occurences of $+1$ and -1, then $t_0^1 = t_0^2$ if and only if $\hat{b}^1 = \hat{b}^2$.*

Proof. We shall prove both (i) and (ii) by a common induction on R, where $R = |a|$ when applied to (i), and $R = |a^1| + |a^2|$ when applied to (ii). The case $R = 0$ is clear, and hence we shall assume $R > 0$.

The induction on R will be, in fact, an inner induction loop, with the outer induction proceeding by n. However, the outer induction loop will be invoked only in the proof of (i). Note that for $n = 1$ one gets (i) immediately.

We shall now prove (i) for $n > 1$. We can assume that a is upward directed. Suppose that a is not separated. By the induction assumption, every proper factor of a has to be separated, and hence a has to be of the form $(-h) \bullet b \bullet (+h)$, with $h > 0$ and neither $\pm h$, nor $\pm(h-1)$ occuring in b. As a is upward directed, we see that $\pm i$ does not occur in b for any i with $0 \le i \le h$. Hence $h = 1$ can be assumed. Put $(b \bullet 1)(t) = t'_0 \ldots t'_n$. Then t'_0 is determined by $1(t)$ and equals $t_0 \triangleright t_1$. As -1 is compatible with $(b \bullet 1)(t)$, we must have $t'_1 = t_0$. Thus $\varepsilon(1(t)) = 1(t)$ starts by $(t_0 \triangleright t_1) \bullet t_0$, and $b(1(t)) = (b \bullet 1)(t)$ starts by $(t_0 \triangleright t_1) \bullet t_0$ as well. Now, ± 1 does not occur in b, but ± 2 does. Define c to be the shortest terminating factor of b which contains all occurences of $+2$ and -2. Then $\hat{c} \ne \varepsilon$, but the induction assumption implies $\hat{c} = \hat{\varepsilon} = \varepsilon$, a contradiction.

Let us now consider (ii). By 1.4, the term t_0^k, $k \in \{1, 2\}$, depends only upon $\hat{b}^k$. Thus $\hat{b}^1 = \hat{b}^2$ implies $t_0^1 = t_0^2$, and to prove the converse implication we can assume $a^k = \hat{b}^k$ by the induction assumption on $R = |a^1| + |a^2|$. Thus a^k, if non-empty, starts by ± 1.

We shall include one more inner induction level, which will run on $\max\{|a^1|, |a^2|\}$, and we shall assume $a^1 \ne a^2$ and $t_0^1 = t_0^2$.

Consider first the case $a^2 = \varepsilon$. Then $a^1 \ne \varepsilon$, and, by Lemma 2.1, both $+1$ and -1 occur in it. As a^1 is separated, all occurences of $+1$ are left to those of -1. Define b and c in such a way that $a^1 = b \bullet (-1) \bullet c$ and -1 does not occur in c. By 2.1, the leftmost elements of $(b \bullet (-1))(c(t)) = a^1$ and $\varepsilon(c(t)) = c(t)$ coincide, and the induction assumption yields $b \bullet (-1) = \varepsilon$, a contradiction, if c is non-empty. If $c = \varepsilon$, then $b = 1$ follows from the coincidence of the leftmost elements of $b(-1(t))$ and $1(-1(t))$, by the induction assumption on $\max\{|a^1|, |a^2|\}$. But b cannot equal 1, and thus we have a contradiction again.

Suppose now that both a^1 and a^2 contain -1. Define a^r, $3 \le r \le 6$, in such a way that $a^k = a^{k+4} \bullet (-1) \bullet a^{k+2}$ for both $k \in \{1, 2\}$, and -1 does not occur in a^{k+2}. Then $+1$ also does not occur in a^{k+2}. Define b^{k+2} and c^{k+2}, $k \in \{1, 2\}$, in such a way that $a^{k+2} = c^{k+2} \bullet b^{k+2}$ and b^{k+2} is the shortest terminating factor of a^{k+2} that contains all the occurrences of ± 2. By Lemma 2.1, t_0 is the leftmost element of $a^{k+2}(t)$. Because -1 is compatible with $a^{k+2}(t)$, we see that the leftmost elements of $a^3(t)$ and $a^4(t)$ coincide. The induction assumption gives $\hat{b}^3 = \hat{b}^4 = d$. Using 1.4, define b^{k+4} so that $b^{k+4} \bullet d$ is equivalent to b^{k+2}, $k \in \{1, 2\}$. Then a^{k+2} is equivalent to $c^{k+2} \bullet b^{k+4} \bullet d$. As a^{k+2} is upward directed, the sequence c^{k+2}, when non-empty, has to terminate with ± 3. It is now easy to see that the sequences $c^{k+2} \bullet b^{k+4}$, $k \in \{1, 2\}$, are upward directed and separated. If d is non-empty, then the induction assumption yields $a^5 \bullet (-1) \bullet c^3 \bullet b^5 = a^6 \bullet (-1) \bullet c^4 \bullet b^6$, and hence also a^1 equals a^2. Suppose that d is empty. Then no ± 2 occurs in a^{k+2}, $k \in \{1, 2\}$, and a^k is equivalent to $a^{k+4} \bullet a^{k+2} \bullet (-1)$. We easily see that a^{k+4} is non-empty, that $a^{k+4} \bullet a^{k+2}$ is upward directed and separated, and that $(a^5 \bullet a^3)(-1(t)) = (a^6 \bullet a^4)(-1(t))$ yields $a^5 \bullet a^3 = a^6 \bullet a^4$ by the induction assumption. As both a^5 and a^6 terminate with ± 1 or ± 2, we obtain $a^1 = a^2$ also in this case.

It remains to consider the case when a^1 does not contain -1. Then t_0 is a proper subterm of t_0^1, by Lemma 2.1. By the same lemma, $+1$ has to occur also in a^2. Therefore both a^k, $k \in \{1, 2\}$, start with $+1$. Define d^k by $a^k = 1 \bullet d^k$. If d^k is non-empty, then its leftmost element equals $+1$ or $+2$ or -2. Since $a^k(t)$ coincide at the leftmost element, $d^k(t)$ have to coincide at the two leftmost elements, $k \in \{1, 2\}$. Note that $\hat{a}^k = a^k$ implies $\hat{d}^k = d^k$, and define e^k and f^k in such a way that $d^k = f^k \bullet e^k$ and e^k is the shortest terminating factor of d^k which contains all the occurrences of ± 1. If both e^1 and e^2 are empty, then $d^1 = d^2$ immediately follows from the induction assumption. Suppose $|e^1| + |e^2| > 0$. The induction assumption then yields $\hat{e}^1 = \hat{e}^2$, and we shall denote this sequence by e. Use 1.4 to define g^k, $k \in \{1, 2\}$, with $g^k \bullet e$ equivalent to e^k. Then $h^k = 1 \bullet f^k \bullet g^k$ is an upward directed separated sequence with $\hat{h}^k = h^k$, for both $k \in \{1, 2\}$. The induction assumption now yields $h^1 = h^2$, and $a^1 = a^2$ follows.

Corollary 2.3. *Let us have primitive sequences* a^k, $k \in \{1, 2\}$, *that start by* $-h_k$. *Put* $h = \min\{h_1, h_2\}$. *Suppose that* a^k *is compatible with a term sequence* $t = t_0 \ldots t_n$ *for both* $k \in \{1, 2\}$ *and that* $a^k(t) = t_0^k \ldots t_n^k$. *Then any of the following equalities implies the other ones:*

$$a^1(t) = a^2(t), \ a^1 = a^2, \ \text{and} \ t_{h-1}^1 = t_{h-1}^2.$$

The following lemma is easy, and we shall omit the proof. However, it seems worth noting that not all separated sequences can act on a term sequence.

Lemma 2.4. *If* a *is an integer sequence* $-2\,1\,-2\,1$ *or* $-2\,1\,1\,2\,1$ *or* $-2\,3\,1\,2\,1$, *then* a *is compatible with no term sequence* t.

3. THE QUASIORDER

In this section there will be defined a linear quasiorder on an absolutely free algebra of terms $W(\Sigma)$ over a set of variables X. We shall impose no restrictions upon the set of operation symbols Σ, and we thus assume here a different level of generality than in other sections. The subset of Σ consisting of operation symbols with arity n will be denoted by Σ_n. If $F \in \Sigma_n$, $t_1, \ldots, t_n \in W(\Sigma)$, then $t = Ft_1 \ldots t_n$ belongs to $W(\Sigma)$, and we define $o_1(t)$ as F and $o_2(t)$ as the sequence $t_1 \ldots t_n$. If t is a variable, then $o_1(t) = t$ and $o_2(t)$ equals the empty sequence ε.

A relation $\leq$ is said to be a *linear quasiorder* on a set S, if it is transitive and if for every $u, v \in S$ there is $u \leq v$ or $v \leq u$.

If $t \in W(\Sigma)$ is a term, then $|t|$ gives its size.

Let $\leq_0$ be a linear quasiorder on $X \cup \Sigma_0$. Define a relation $\leq$ on W so that $s \leq t$, where $o_2(s) = s_1 \ldots s_m$ and $o_2(t) = t_1 \ldots t_n$, takes place if and only if

(1) $s \leq t_j$ for some j, $1 \leq j \leq n$; or

(2) $s > t_j$ for all j, $1 \leq j \leq n$, and $t > s_i$ for all i, $1 \leq i \leq m$, and $o_1(s) <_0 o_1(t)$; or

(3) $s > t_j$ for all j, $1 \leq j \leq n$, and $t > s_i$ for all i, $1 \leq i \leq m$, and $o_1(s) \equiv_0 o_1(t)$ and $o_2(s) \leq o_2(t)$ lexicographically.

To avoid misunderstanding, let us make clear that $u > v$ is, as usual, a shortcut for "$u \geq v$ and not $v \leq u$", and that $s_1 \ldots s_m \leq t_1 \ldots t_m$ holds lexicographically if and only if

(a) there exists i, $1 \leq i \leq \min(m,n)$, with $s_j \equiv t_j$ for every j, $1 \leq j < i$, and $s_i < t_i$; or

(b) $m \leq n$ and $s_i \equiv t_i$ for every i, $1 \leq i \leq m$.

Proposition 3.1. *The relation $\leq$ is a linear quasiorder, and $s < t$ holds whenever s is a proper subterm of t.*

We can easily verify that $u \leq u$ holds for all $u \in W(\Sigma)$. The rest of the proof will be divided into five steps.

Step 1. *$s \leq t$ holds if s is a subterm of t.*

Proof. Proceed by induction on $|s| + |t|$. We clearly have $s \leq t$ if $s = t$ or s is an argument of t. In other cases one can express t as $Ft_1 \ldots t_n$, with s a subterm of t_j, $1 \leq j \leq n$. Then $s \leq t_j$ by the induction assumption, and $s \leq t$ by (1).

Step 2. *If $v \leq w$ and u is a subterm of v, then $u \leq w$.*

Proof. Proceed by induction on $|v| + |w|$. The case $u = v$ is clear. Otherwise there exists j, $1 \leq j \leq n$, with u a subterm of v_j, $v = Fv_1 \ldots v_n$. If $v_j < w$, then $u \leq w$ by $|v_j| + |w| < |v| + |w|$. If $v_j < w$ is not true, then there must be $w = Gw_1 \ldots w_p$ and $v \leq w_k$, $1 \leq k \leq p$, which implies $u \leq w_k$ by $|v| + |w_k| < |v| + |w|$. However, $u \leq w_k$ gives $u \leq w$, by (1).

Step 3. *For any terms s and t, either $s \leq t$ or $t \leq s$.*

Proof. Proceed by induction on $|s| + |t|$. Assume $o_2(s) = s_1 \ldots s_m$ and $o_2(t) = t_1 \ldots t_n$. For every j, $1 \leq j \leq n$, one has $s \leq t_j$ or $s > t_j$, by the induction assumption. Similar relations hold for s_i and t, $1 \leq i \leq m$, and we see that either at least one of the relations $s \leq t$ and $t \leq s$ can be deduced from (1), or both $s > t_j$ and $t > s_i$ hold for all i and j, $1 \leq j \leq n$ and $1 \leq i \leq m$. However, in the latter case one gets $s \leq t$ or $t \leq s$ by (2) or (3), using the induction assumption and linearity of $\leq_0$.

Step 4. *If $u \leq v \leq w$, then $u \leq w$.*

Proof. Proceed by induction on $|u| + |v| + |w|$. Assume $o_2(u) = u_1 \ldots u_m$, $o_2(v) = v_1 \ldots v_n$ and $o_2(w) = w_1 \ldots w_p$. If $u \leq v_j$, $1 \leq j \leq n$, then we get $u \leq w$ by $u \leq v_j \leq w$ and $|u| + |v_j| + |w| < |u| + |v| + |w|$. If $v \leq w_k$ for some

k, $1 \leq k \leq p$, then $u \leq w_k$ holds by the induction assumption, and $u \leq w$ follows from (1). Thus we can now assume:

$u > v_j$ and $w > v_j$ for all j, $1 \leq j \leq n$;

$v > u_i$ for all i, $1 \leq i \leq m$; and

$v > w_k$ for all k, $1 \leq k \leq p$.

From $w \geq v > u_i$ we get $w \geq u_i$ by the induction assumption. If $u_i \geq w$ were true, then $u_i \geq w \geq v$ would give $u_i \geq v$, again by the induction assumption. But $u_i \geq v$ contradicts $v > u_i$, and hence we see that $w > u_i$ holds for all i, $1 \leq i \leq n$.

If $u \leq w_k$ is true for some k, $1 \leq k \leq p$, then $u \leq w$ follows, as required. By Step 3, $u > w_k$ can be assumed for all k, $1 \leq k \leq p$. We have $o_1(u) \leq_0 o_1(v) \leq_0 o_1(w)$ and $o_1(u) \leq_0 o_1(w)$ follows. If $o_1(u) <_0 o_1(v)$ or $o_1(v) <_0 o_1(w)$, then $o_1(u) <_0 o_1(w)$ holds, and we can use (2) to get $u \leq w$. Assume $o_1(u) \leq_0 o_1(w)$. Then $o_2(u) \leq o_2(v)$ and $o_2(v) \leq o_2(w)$ lexicographically, and using the induction assumption, we see that $o_2(u) \leq o_2(w)$ holds lexicographically as well.

Step 5. *If s is a proper subterm of t, then $s < t$.*

Proof. Proceed by induction on $|s| + |t|$. Suppose $s \geq t$, where s is a proper subterm of t. Then $t = Ft_1 \ldots t_n$ for some $n \geq 1$, and s is a subterm of some t_i, $1 \leq i \leq n$. This means $s \leq t_i$, and thus $s \geq t$ cannot be based on (2) or (3). Hence $s' \geq t$ must take place for a proper subterm s' of s. However, s' is also a subterm of t, and $|s'| + |t| < |s| + |t|$ yields a contradiction with the induction assumption.

Proposition 3.2. *Suppose that $\leq_0$ is a well linear quasiorder on $X \cup \Sigma_0$ with the property:*

> *For every equivalence class α of $\equiv_0$ there exists an integer k such that the arity of any operation symbol $F \in \alpha$ is $\leq k$.*

Then $\leq$ is a well linear quasiorder on $W(X)$.

Proof. As $\leq$ is a linear quasiorder by 3.1, all we need is to show that there are no infinite descending chains. For any sequence $\gamma = u_0 u_1 \ldots u_n$ of terms put

$$P(\gamma) = \{t; \ u_0 u_1 \ldots u_n t \text{ is the beginning of an infinite descending chain}\}.$$

It is clear that $P(\gamma)$ is empty if $u_0 u_1 \ldots u_k$ is not a descending sequence. Suppose now that $P(\varepsilon)$ is non-empty, i.e. that there exists at least one infinite descending chain. Define a sequence of terms $u_1 u_2 \ldots$ and a sequence of finite term sequences $v_0 v_1 \ldots$ in such a way that

(1) $v_0 = \varepsilon$ and $v_i = v_{i-1} \bullet u_i$ for all $i \geq 1$;

(2) $u_i \in P(v_{i-1})$ and $|u_i| = \min\{|t|; \ t \in P(v_{i-1})\}$ for all $i \geq 1$.

Because $P(\varepsilon)$ is non-empty, $P(v_i)$ is non-empty for all $i \geq 0$, and $u_1 u_2 \ldots$ is an infinite descending chain.

If $j > 1$ and t is a term with $P(v_{j-1} \bullet t) \neq \emptyset$, then $|t| \geq |u_j|$ must hold. If we are able to find for any infinite descending chain $u_1 u_2 \ldots$ an index $j \geq 1$ and a term t with $P(v_j \bullet t) \neq \emptyset$, and $|t| < |u_j|$, then we have found a contradiction with our assumption $P(\varepsilon) \neq \emptyset$.

Suppose that there exists $j \geq 1$ with $o_2(u_j) = s_1 \ldots s_m$ and $s_i \geq u_{j+1}$ for some i, $1 \leq i \leq m$. Then the sequence $u_1 \ldots u_{j-1} s_i u_{j+2} u_{j+3} \ldots$ is an infinite descending chain, and the above argument shows that no such j may exist.

Hence we have $o_1(u_1) \geq_0 o_1(u_2) \geq_0 o_1(u_3) \geq_0 \ldots$, and because $\geq_0$ is a well quasiorder, we see that there exists an index $k_0 \geq 1$ such that $o_1(u_j) \equiv_0 o_1(u_{k_0})$ for every $j \geq k_0$. Obviously, we can also assume that no u_j, $j > k_0$, belongs to $X \cup \Sigma_0$. Denote, for a while, the arity of $o_1(u_j)$, $j \geq k_0$, by a_j.

We shall now show that for every $r \geq 0$ there exists k_r with

(a) $k_r \geq k_{r-1} \geq \cdots \geq k_0$;
(b) $a_j \geq r + 1$ for all $j \geq k_r$; and
(c) if $u_{k_r} = F s_1 \ldots s_{a_{k_r}}$, $j \geq k_r$, and $u_j = G t_1 \ldots t_{a_j}$, then $s_q \equiv t_q$ for all q with $1 \leq q \leq r$.

The conditions (a)-(c) are satisfied for $r = 0$, and we can continue by induction on r. Thus we shall assume $r \geq 0$ and prove (a)-(c) for $r' = r + 1$.

Put $k = k_r$, $u_k = F s_1 \ldots s_{a_k}$, and consider $j \geq k$ with $u_j = G t_1^j \ldots t_{a_j}^j$. If $1 \leq q \leq r$, then $s_q \equiv t_q^j$, by (c). Suppose that there exists an infinite sequence $k = i_1 < i_2 < \ldots$ with $t_{r+1}^{i_1} > t_{r+1}^{i_2} > \ldots$. Then $u_1 \ldots u_{k-1} t_{r+1}^{i_1} t_{r+1}^{i_2} \ldots$ is an infinite descending chain, and $t_{r+1}^k = s_{r+1}$ is a proper subterm of u_k, a contradiction. Hence there exists $k' \geq k$ such that $t_{r+1}^j \equiv t_{r+1}^{k'}$ for every $j > k'$, and we see that $a_j > r + 1$ must hold for every $j \geq k' + 1$. Therefore we can put $k_{r+1} = k' + 1$.

The operation symbols $o_1(u_{k_r})$, $r \geq 0$, are equivalent with respect to $\equiv_0$. But they have unbounded arities, which contradicts our initial assumption.

Lemma 3.3. *Suppose that $u = F u_1 \ldots u_{j-1} s u_{j+1} \ldots u_n$ and that $v = F u_1 \ldots u_{j-1} t u_{j+1} \ldots u_n$. If $s > t$, then $u > v$.*

Proof. If $s \geq v$, then $u > v$ follows from $u > s$. Assume $s < v$. Then $u > v$ follows from (3).

Corollary 3.4. *Suppose that a term v is obtained from a term u by replacing a subterm s by t. If $s > t$, then $u > v$.*

4. Confluency

Define a directed graph on the set of term sequences W^* in such a way that $t \to s$ holds if and only if $s = a(t)$ for some primitive integer sequence a.

Lemma 4.1. *Suppose that $t_2 = a(t_1)$, where t_1 and t_2 are term sequences, and a is an integer sequence compatible with t_1. Then, for some term s, there exist directed paths from t_1 to s and from t_2 to s.*

Proof. We can assume that a is separated. Hence it is, by Proposition 1.6, equivalent to a sequence $-(u_1 \bullet \ldots \bullet u_m) \bullet (v_n \bullet \ldots \bullet v_1)$, where all u_i, $1 \leq i \leq m$, and v_j, $1 \leq j \leq n$, are primitive. We can thus put $s = v_n \bullet \ldots \bullet v_1(t_1) = u_m \bullet \ldots \bullet u_1(t_2)$.

Corollary 4.2. *The graph* $(W^*, \rightarrow)$ *is locally confluent.*

Define now a directed graph $\rightarrow$ on W by setting $u \rightarrow v$ if and only if v can be obtained from u by replacing a subterm $t_0(t_1(\ldots(t_n t_{n+1})))$ of u, $n \geq 1$, with a subterm $s_0(s_1(\ldots(s_n s_{n+1})))$, where $t_{n+1} = s_{n+1}$ and $t_0 \ldots t_n \rightarrow s_0 \ldots s_n$ in $(W^*, \rightarrow)$.

Denote by $\sim$ the least congruence of W with $u \cdot (v \cdot w) \sim (u \triangleright v) \cdot (u \cdot w)$ for all $u, v, w \in W$.

Lemma 4.3. *Let u and v be terms in W. Then $u \equiv v$ holds if and only if there exists an undirected path from u to v in the graph* $(W, \rightarrow)$.

Proof. If a subterm $(t_1 \triangleright t_2) \cdot (t_1 \cdot t_3)$ of u is replaced by $t_1(t_2 t_3)$, and the resulting term is denoted by v, then $u \rightarrow v$ is true. Hence there exists an undirected path from u to v whenever $u \sim v$. On the other hand, $u \rightarrow v$ clearly implies $u \sim v$.

Lemma 4.4. *The graph* $(W, \rightarrow)$ *is locally confluent.*

Proof. Using standard arguments, we see immediately that it is enough to prove local confluency in the case when a subterm $t_0(t_1(\ldots(t_n t_{n+1})))$ is replaced by $s_0^k(s_1^k(\ldots(s_n^k s_{n+1}^k)))$, $k \in \{1, 2\}$, with $t_{n+1} = s_{n+1}^1 = s_{n+1}^2$ and $t_0 \bullet \ldots \bullet t_n \rightarrow s_0^k \bullet \ldots \bullet s_n^k$ for both $k \in \{1, 2\}$. But then both terms can be sent to a term provided by Lemma 4.2.

Put $\Sigma = \{\alpha_n; \ n \geq 2\} \cup \{\triangleright\}$ and define a linear quasiorder $\leq$ on $W(\Sigma)$ by the method of Section 3, starting from the linear quasiorder $\leq_0$ on $X \cup \Sigma$ that satisfies $x \leq_0 y <_0 \triangleright <_0 \alpha_2 <_0 \alpha_3 < \ldots$ for all $x, y \in X$. By 3.2, $\leq$ is a well quasiorder.

Define now an injective mapping $\nu : W \rightarrow W(\Sigma)$ by $\nu(x) = x$ for all $x \in X$, by $\nu(s \triangleright t) = \nu(s) \triangleright \nu(t)$ for all $s, t \in W$, and by

$$\nu(t_0(t_1(\ldots(t_n t_{n+1})))) = \alpha_{n+2}(\nu(t_0), \nu(t_1), \ldots, \nu(t_{n+1}))$$

whenever $t_0, \ldots, t_{n+1} \in W$, $n \geq 0$ and t_{n+1} is not of the form $u \cdot v$.

Define a well linear quasiorder $\leq$ on W by $u \leq v$ if and only if $\nu(u) \leq \nu(v)$.

Lemma 4.5. *If u is a proper subterm of $v \in W$, then $u < v$.*

Lemma 4.6. *Let t_i, $0 \leq i \leq n+1$, $n \geq 0$, be terms from W, and suppose that an integer sequence a is compatible with the term sequence $t = t_0 \ldots t_n$, and $a(t) = t_0' \ldots t_n'$. Then $t_0(t_1(\ldots(t_n t_{n+1}))) > t_j'$ holds for every j, $0 \leq j \leq n$.*

Proof. Put $w = t_0(t_1(\ldots(t_n t_{n+1})))$ and proceed by induction on $|a|$. The case $a = \varepsilon$ follows from 4.5. Assume $a = i \bullet b$. If $j \neq |i|$, then the induction assumption can be used immediately, and if $i = -j$, then it can be used together with 4.5. Assume $j = i$. Then $t'_j = u \triangleright v$, where $u < w$ and $v < w$. There is $\triangleright <_0 \alpha_{n+2}$, and therefore $\nu(u \triangleright v) \geq \nu(w)$ cannot hold.

Corollary 4.7. *Let t_i, $0 < i \leq n+1$, $n \geq 0$, be terms from W, and suppose that a primitive integer sequence a is compatible with the term sequence $t = t_0 \ldots t_n$. Put $a(t) = t'_0 \ldots t'_n$. Then $t'_0(t'_1(\ldots (t'_n t_{n+1}))) < t_0(t_1(\ldots (t_n t_{n+1})))$.*

Proof. By 4.6, the lexicographical comparison of $t'_0 \ldots t'_n$ and $t_0 \ldots t_n$ is needed to decide the mutual relation of the above terms. The term t'_0 is a proper subterm of t_0, and hence 4.5 can be used.

Theorem 4.8. *The graph $(W, \rightarrow)$ is convergent.*

Proof. Combine 4.4, 4.7, 3.4 and 3.2.

5. CONCLUSIONS AND CONNECTIONS

The infinite rewriting system that has been constructed in previous sections gives us a certain amount of control over the free structures satisfying the identity $x(yz) = (x \triangleright y)(xz)$. While it seems likely that this rewriting system solves the word problem for such structures, we have not completely proved it. The missing step is a proof that for every term sequence $t = t_1 \ldots t_n$ there exist only finitely many primitive integer sequences a that act on t, and that there exists a recursive bound on $|a|$.

Let us now consider again the connections with free conjugacy closed loops. Let $Q(\cdot)$ be such a loop and let X be its free base. Then Q is the closure of X with respect to operations $\cdot$, $/$, $\backslash$ and 1. Denote by A the closure of X just with respect to $\cdot$ and $\triangleright$, where $b \triangleright a = (ba)/b$. As we observe below, the operations $\cdot$ and $\triangleright$ satisfy also some identities that cannot be deduced from $x(yz) = (x \triangleright y)(xz)$. Hence $A(\cdot, \triangleright)$ is not a free structure with respect to this identity. However, it is possible that by adding some other identities we shall be able to construct a free structure with two binary operations ($\triangleright$ and $\cdot$) and a unit element, that can be isomorphically embedded into Q. Starting from A one might then obtain Q through an infinite number of extensions (such procedures are quite usual in the loop and quasigroup theories, see e.g. [11]). Applying the same approach to both-sided conjugacy closed loops could then lead to a solution of the main open problem in conjugacy closed loops (i.e. the existence of the non-trivial nucleus - cf.[9]).

In the loop $Q(\cdot)$ one clearly has $a \triangleright a = a$ for all $a \in Q$. It is not difficult to prove that by adding the rewriting rule $x \triangleright x \rightarrow x$ to the rewriting rules induced by primitive sequences, we still retain the convergency of the rewriting system. However, the next lemma shows that our rewriting rules need a much more dramatic change to suit the needs of left conjugacy closed loops.

Lemma 5.1. *Let $Q(\cdot)$ be a left conjugacy closed loop and define a binary operation $\triangleright$ on Q by $b \triangleright a = (ba)/b$ for all $a, b \in Q$. Then the operation $\triangleright$ is left distributive (i.e. $c \triangleright (b \triangleright a) = (c \triangleright b) \triangleright (c \triangleright a)$ holds for all $a, b, c \in Q$.)*

Proof. Using the equality $x \triangleright y = (x(yz))/(xz)$ verify $u \triangleright (v \triangleright w) = (u((v \triangleright w)v)/(uv) = (u \cdot vw)/(uv)$ and $(u \triangleright v) \triangleright (u \triangleright w) = ((u \triangleright v)((u \triangleright w)u))/((u \triangleright v) \cdot u) = (((uv)/u) \cdot (uw))/(uv) = (u \cdot vw)/(uv)$.

Compare now actions of integer sequences $1\,2\,1$ and $2\,1\,2$ on a term sequence $u\,v\,w = u \bullet v \bullet w$. We obtain

$$(u \triangleright v) \triangleright (v \triangleright w) \bullet u \triangleright v \bullet u \text{ and } u \triangleright (v \triangleright w) \bullet u \triangleright v \bullet u,$$

respectively, and we see that we have, in fact, a partial action of the braid group B_∞ on sequences of elements of a left conjugacy closed loop (only left distributivity of $\triangleright$ is needed for that - the braid group B_∞ acts on sequences of any algebraic structure that obeys the left distributive law - cf.[3][13][16])

The author hopes that by working along these lines it will be possible to utilize the known connections of the braid group and left distributive structures. The ordering by natural numbers (which was important for defining primitive sequences) will be probably replaced by the ordering of braids [5], or, more likely, by the ordering of the group $\tilde{B}_\infty$ [3]. However, some additional preliminary work might be necessary, as the equational theory of free idempotent left distributive structures has not been fully worked out yet [8][14]. It is also possible that an approach using rewriting rules can help to solve some of the remaining problems in free non-idempotent left distributive structures (there has been immense progress in the last ten years - e.g. [1][2][3][4][15][16]).

References

[1] P. Dehornoy: *Free distributive groupoids*, J. Pure Appl. Algebra **61**(1989), 123–146.

[2] P. Dehornoy: *Structural monoids associated to equational varieties*, Proc. Amer. Math. Soc. **117** (1993), 293–304.

[3] P. Dehornoy: *Braid groups and left distributive operations*, Trans. Amer. Math. Soc. **345** (1994), 115–121.

[4] P. Dehornoy: *A normal form for the free left distributive law*, Int. J. Algebra and Computation **4** (1994), 499–528.

[5] P. Dehornoy: *A fast method for comparing braids*, Advances in Mathematics **125** (1997), 200–235.

[6] N. Dershowitz: *Termination of rewriting*, Lecture Notes in Computer Science 355, 1989, Springer-Verlag, 69–116.

[7] N. Dershowitz, J.-P. Jouannaud: *Rewrite systems*, Chapter 6 in J. van Leeuwen, ed., Handbook of Theoretical Computer Science, B: Formal methods and Semantics, North Holland, Amsterdam, 1990.

[8] A. Drápal, T. Kepka and M. Musílek: *Group conjugation has non-trivial LD-identities*, Comment. Math. Univ. Carolinae **35** (1994), 219–222.

[9] E. G. Goodaire and D. A. Robinson: *A class of loops which are isomorphic to all loop isotopes*, Canadian J. Math **34** (1982), 662–672.

[10] E. G. Goodaire and D. A. Robinson: *Some special conjugacy closed loops*, Canadian Math. Bull. **33** (1990), 73–78.

[11] T. Kepka: *Notes on left distributive groupoids*, Acta Univ. Carolinae - Math. et Ph. **22** (1981), 23–37.

[12] K. Kunen: *The structure of conjugacy closed loops*, Trans. Amer. Math. Soc., to appear.

[13] D. Larue: *On braid words and irreflexivity*, Algebra Universalis **31** (1994), 104–112.

[14] D. Larue: *Left-distributive and left-distributive idempotent algebras*, Ph.D.Thesis, University of Colorado, Boulder (1994).

[15] R. Laver: *The left distributive law and the freeness of an algebra of elementary embeddings*, Advances in Mathematics **91** (1992), 209–231.

[16] R. Laver: *Braid group actions on left distributive structures and well-orderings in the braid group*, J. Pure Appl. Algebra **108** (1996), 81–98.

Aleš Drápal,
Department of Mathematics,
Charles University,
Sokolovská 83,
186 75 Praha 8,
Czech Republic.
drapal@karlin.mff.cuni.cz

ON CERTAIN FINITE GENERALIZED TETRAHEDRON GROUPS

MARTIN EDJVET, GERHARD ROSENBERGER, MICHAEL STILLE AND RICHARD M. THOMAS

ABSTRACT. An "ordinary" tetrahedron group is a group with a presentation of the form

$$\langle x, y, z : x^{e_1} = y^{e_2} = z^{e_3} = (xy^{-1})^{f_1} = (yz^{-1})^{f_2} = (zx^{-1})^{f_3} = 1 \rangle$$

where $e_i \geq 2$ and $f_i \geq 2$ for each i. Following Vinberg, we call groups defined by presentations of the form

$$\langle x, y, z : x^{e_1} = y^{e_2} = z^{e_3} = R_1(x,y)^{f_1} = R_2(y,z)^{f_2} = R_3(z,x)^{f_3} = 1 \rangle,$$

where each $R_i(a,b)$ is a cyclically reduced word involving both a and b, *generalized tetrahedron groups*. These groups appear in many contexts, not least as subgroups of generalized triangle groups.

In this paper, we build on work of Fine, Levin, Roehl and Rosenberger, and consider the question of the finiteness of these groups. We survey some of the results already obtained in this area and give some new sufficient conditions for the groups to be infinite.

1. INTRODUCTION

An "ordinary" tetrahedron group is a group G with a presentation of the form

$$\langle x, y, z : x^{e_1} = y^{e_2} = z^{e_3} = (xy^{-1})^{f_1} = (yz^{-1})^{f_2} = (zx^{-1})^{f_3} = 1 \rangle$$

where $e_i \geq 2$ and $f_i \geq 2$ for each i. The group admits an automorphism of order 2 inverting each generator, and so we may form the semidirect product of G with a cyclic group $\langle t \rangle$ of order 2 to get

$$\langle t, x, y, z : t^2 = x^{e_1} = y^{e_2} = z^{e_3} = (xtyt)^{f_1} = (ytzt)^{f_2} = (ztxt)^{f_3} =$$
$$(xt)^2 = (yt)^2 = (zt)^2 = 1 \rangle.$$

Introducing generators $u = xt$, $v = yt$ and $w = zt$, and then deleting x, y and z, yields the presentation

$$\langle t, u, v, w : t^2 = u^2 = v^2 = w^2 = (tu)^{e_1} = (tv)^{e_2} = (tw)^{e_3} =$$
$$(uv)^{f_1} = (vw)^{f_2} = (uw)^{f_3} = 1 \rangle.$$

The authors are grateful to Tim Hsu for the suggestion that led to the argument presented in Section 3. They are also very grateful to the referee for his/her constructive comments and suggestions. The fourth author would like to thank Hilary Craig for all her help and encouragement.

This group is finite if and only if the matrix

$$\begin{pmatrix} 1 & -\cos(\frac{\pi}{e_1}) & -\cos(\frac{\pi}{e_2}) & -\cos(\frac{\pi}{e_3}) \\ -\cos(\frac{\pi}{e_1}) & 1 & -\cos(\frac{\pi}{f_1}) & -\cos(\frac{\pi}{f_3}) \\ -\cos(\frac{\pi}{e_2}) & -\cos(\frac{\pi}{f_1}) & 1 & -\cos(\frac{\pi}{f_2}) \\ -\cos(\frac{\pi}{e_3}) & -\cos(\frac{\pi}{f_3}) & -\cos(\frac{\pi}{f_2}) & 1 \end{pmatrix}$$

has positive determinant (see [2], [3], [4]).

Following Vinberg, we call a group $G(e_1, e_2, e_3, f_1, f_2, f_3)$ defined by a presentation of the form

(1) $\langle x, y, z : x^{e_1} = y^{e_2} = z^{e_3} = R_1(x, y)^{f_1} = R_2(y, z)^{f_2} = R_3(z, x)^{f_3} = 1 \rangle$,

where each $R_i(a, b)$ is a cyclically reduced word involving both a and b, a *generalized tetrahedron group*. These groups appear in many contexts, not least as subgroups of generalized triangle groups.

The case where each $R_i(u, v)$ is a word of the form $u^a v^b$ was considered by Tsaranov in [14], [15], and a complete classification of the finite groups in that case was obtained there. We build on this, and on the work of Fine, Levin, Roehl and Rosenberger in [7], and consider the question of the finiteness of the generalized tetrahedron groups.

A generalized tetrahedron group is an example of a triangle of groups. We use this fact in Section 3 to prove the following result:

THEOREM 1.1. *A generalized tetrahedron group* $G(e_1, e_2, e_3, f_1, f_2, f_3)$ *is infinite if* $\frac{1}{f_1} + \frac{1}{f_2} + \frac{1}{f_3} \leq 1$.

We also have the following:

THEOREM 1.2. *A generalized tetrahedron group* $G(e_1, e_2, e_3, f_1, f_2, f_3)$ *is infinite if* $\frac{1}{e_1} + \frac{1}{e_2} + \frac{1}{f_1} < 1$.

This is proved in Section 4. Of course, by symmetry, the generalized tetrahedron group is infinite if either $\frac{1}{e_2} + \frac{1}{e_3} + \frac{1}{f_2} < 1$ or $\frac{1}{e_3} + \frac{1}{e_1} + \frac{1}{f_3} < 1$. As far as the case $\frac{1}{e_1} + \frac{1}{e_2} + \frac{1}{f_1} = 1$ is concerned, we have the following result:

THEOREM 1.3. *A generalized tetrahedron group* $G(e_1, e_2, e_3, 2, 2, 2)$ *is infinite if* $\frac{1}{e_1} + \frac{1}{e_2} \leq \frac{1}{2}$.

By symmetry we also have that the group $G(e_1, e_2, e_3, 2, 2, 2)$ is infinite if $\frac{1}{e_2} + \frac{1}{e_3} \leq \frac{1}{2}$ or $\frac{1}{e_3} + \frac{1}{e_1} \leq \frac{1}{2}$.

In subsequent papers we will prove further results concerning the finite generalized tetrahedron groups and, as a result, we will be able to remove the hypothesis that $f_1 = f_2 = f_3 = 2$ from Theorem 1.3 and deduce that any generalized tetrahedron group with $\frac{1}{e_1} + \frac{1}{e_2} + \frac{1}{f_1} \leq 1$ is infinite.

2. ESSENTIAL REPRESENTATIONS

A *generalized triangle group* G is a group defined by a presentation of the form

$$\langle x, y : x^p = y^q = R(x,y)^r = 1 \rangle,$$

where $p \geq 2$, $q \geq 2$, $r \geq 2$ and where $R(x,y)$ is a cyclically reduced word in x and y. These groups have been extensively studied; we will only mention some aspects that are relevant to our study of generalized tetrahedron groups. In particular, we should mention that a complete classification as to which generalized triangle groups are finite may be found in [8] and [10]; some further information on these groups may be found in [9].

If L is a linear group and $\rho : G \to L$ is a representation of G, we say that ρ is *essential* if $x\rho$, $y\rho$ and $R(x,y)\rho$ have orders p, q and r respectively in $G\rho$. It is known (see [1] or [6]) that every generalized triangle group has an essential representation in $PSL(2, \mathbb{C})$.

The existence of essential representations still holds when we consider generalized tetrahedron groups. Suppose that G is defined by the presentation

$$\langle x, y, z : x^{e_1} = y^{e_2} = z^{e_3} = R_1(x,y)^{f_1} = R_2(y,z)^{f_2} = R_3(z,x)^{f_3} = 1 \rangle.$$

In a similar vein to the generalized triangle groups, we say that a representation ρ of G is *essential* if $x\rho$, $y\rho$, $z\rho$, $R_1(x,y)\rho$, $R_2(y,z)\rho$ and $R_3(z,x)\rho$ have orders e_1, e_2, e_3, f_1, f_2 and f_3 respectively in $G\rho$. The following result was proved in [7] (see Theorem 1):

THEOREM 2.1. *Every generalized tetrahedron group admits an essential representation in* $PSL(2, \mathbb{C})$.

As an immediate consequence of Theorem 2.1 and [13], one can deduce

COROLLARY 2.2. *The generalized tetrahedron group* $G(e_1, e_2, e_3, f_1, f_2, f_3)$ *is infinite if*

$$\kappa = \frac{1}{e_1} + \frac{1}{e_2} + \frac{1}{e_3} + \frac{1}{f_1} + \frac{1}{f_2} + \frac{1}{f_3} \leq 2.$$

In fact, one can say rather more here. If $\kappa < 2$ then G has a subgroup of finite index which maps onto a free group of rank 2, so that G is SQ-universal, while, if $\kappa = 2$, then G has a subgroup of finite index that maps onto $\mathbb{Z}$; see Theorem 3 of [7].

One can strengthen Theorem 2.1. If we look at its proof in [7], we can deduce the following result:

THEOREM 2.3. *Let G be the generalized tetrahedron group defined by the presentation*

$$\langle x, y, z : x^{e_1} = y^{e_2} = z^{e_3} = R_1(x,y)^{f_1} = R_2(y,z)^{f_2} = R_3(z,x)^{f_3} = 1 \rangle,$$

and let T be the generalized triangle group defined by the presentation

$$\langle x, y : x^{e_1} = y^{e_2} = R_1(x,y)^{f_1} = 1 \rangle.$$

Suppose that ρ is an essential representation of T in $PSL(2, \mathbb{C})$, that $X = x\rho$ and $Y = y\rho$, and that one of the following two possibilities occurs:

1. $\mathrm{tr}[X, Y] \neq 2$;
2. $\langle X, Y \rangle$ *is an infinite metabelian subgroup of $PSL(2, \mathbb{C})$ and $(e_3, f_2, f_3) \neq (2, 2, 2)$.*

Then there is an essential representation $\tilde{\rho} : G \to PSL(2, \mathbb{C})$ such that $x\tilde{\rho} = X$ and $y\tilde{\rho} = Y$. In particular, if $\langle X, Y \rangle$ is infinite, then G is infinite.

Proof. Theorem 2.3 is essentially Corollary 1 of [7]. However, that result had the stronger hypothesis that $(f_2, f_3) \neq (2, 2)$ in part 2. We can amend the proof of Theorem 1 of [7] so that we can obtain Theorem 2.3 as follows. If $(f_2, f_3) = (2, 2)$ and $e_3 = r \geq 3$, then we replace $\mathrm{tr}(A_3) = 2\cos(\frac{\pi}{r})$ in (3.1) by $\mathrm{tr}(A_3) = -2\cos(\frac{\pi}{r})$, and then the construction goes through as for $f_2 \geq 3$ or $f_3 \geq 3$. $\square$

The following fact is very useful:

REMARK 2.4. *If X and Y are as in Theorem 2.3 and $\langle X, Y \rangle$ is an abelian group other than the elementary abelian group of order 4, then there always exists an essential representation $\sigma : T \to PSL(2, \mathbb{C})$ such that $\langle x\sigma, y\sigma \rangle$ is an infinite metabelian group.* $\square$

We will also need the following result:

PROPOSITION 2.5. *Let G be the group defined by the presentation*

$$\langle x, y : x^n = y^m = R(x,y)^2 = 1 \rangle,$$

where $R(x,y)$ is a cyclically reduced word involving both x and y which is not a proper power in the free product of the cyclic groups generated by $\langle x \rangle$ and $\langle y \rangle$. Let $p(t)$ be the trace polynomial for an essential representation of G into $PSL(2, \mathbb{C})$. Assume that $p(t)$ has a root of multiplicity greater than one. Then the group H defined by the presentation

$$\langle x, y : x^n = y^m = R(x,y)^2 = R(x^{-1}, y^{-1})^2 = 1 \rangle$$

is infinite.

Proof. We refer the reader to the construction in [8] and make the following additional observations. We may choose an essential representation $\rho : G \to PSL(2, \Lambda)$, where Λ is a ring of the form $\mathbb{C}[s]/((s - \gamma)^2)$ with $\gamma \in \mathbb{C}$, such that $\mathrm{tr}(R(x,y)\rho) = 0$ and $G\rho$ is infinite as in [8]. Since $\mathrm{tr}(R(x,y)\rho) = \mathrm{tr}(R(x^{-1}, y^{-1})\rho)$, we have $\mathrm{tr}(R(x^{-1}, y^{-1})\rho) = 0$, and hence

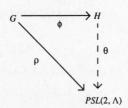

FIGURE 1. Map factorization

that $R(x^{-1}, y^{-1})\rho$ has order 2 in $PSL(2, \Lambda)$. So, if ϕ is the natural homomorphism from G to H, then ρ factors through ϕ, say $\rho = \phi\theta$ where θ is a homomorphism from H to $PSL(2, \Lambda)$ (see Figure 1). Since $G\rho$ is infinite by Lemma 2.4 and Theorem 2.5 of [8], $H\theta$ is infinite, and so H is infinite as required. $\square$

3. PROOF OF THEOREM 1.1

The following is taken from [12]. (A similar approach is given in [11]).

Given two subgroups A and B of a group H, the inclusions $A \hookrightarrow H$ and $B \hookrightarrow H$ determine a homomorphism $\phi : A * B \to H$. If ϕ is injective, then the *angle* $(H; A, B)$ between A and B is defined to be 0; otherwise, $(H; A, B)$ is defined to be $\frac{\pi}{n}$ where $2n$ is the minimal length (in the free product sense) of an element of $\ker(\phi)$.

Now a generalized tetrahedron group

$$\langle x, y, z : x^{e_1} = y^{e_2} = z^{e_3} = R_1(x, y)^{f_1} = R_2(y, z)^{f_2} = R_3(z, x)^{f_3} = 1 \rangle$$

can be realized as a *triangle of groups*, that is as the colimit of the diagram of groups and injective homomorphisms shown in Figure 2 in which

$$X = \langle x_1, y_1 : x_1^{e_1} = y_1^{e_2} = R_1(x_1, y_1)^{f_1} = 1 \rangle,$$
$$Y = \langle y_2, z_1 : y_2^{e_2} = z_1^{e_3} = R_2(y_2, z_1)^{f_2} = 1 \rangle,$$
$$Z = \langle z_2, x_2 : z_2^{e_3} = x_2^{e_1} = R_3(z_2, x_2)^{f_3} = 1 \rangle,$$
$$A = \langle y_1 \rangle \cong \langle y_2 \rangle, \ B = \langle z_1 \rangle \cong \langle z_2 \rangle, \ C = \langle x_1 \rangle \cong \langle x_2 \rangle.$$

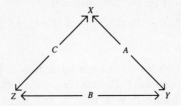

FIGURE 2. Triangle of groups

The triangle of groups in Figure 2 is said to be *spherical* if

$$(X; C, A) + (Y; A, B) + (Z; B, C) > \pi$$

and *non-spherical* otherwise. In [12] it is shown that, if H is the colimit of a non-spherical triangle of groups, then the vertex groups embed in H via the natural maps and that, if S is a finite subgroup in H, then S is conjugate in H to a subgroup of a vertex group.

We have not been able to deduce as an immediate consequence of these results that H must then be infinite. However, following a suggestion by Tim Hsu, we can prove directly that such a group H is infinite using van Kampen diagrams.

PROPOSITION 3.1. *If H is the colimit of a non-spherical triangle of groups then H is infinite.*

Proof. Let H be the colimit of the triangle of groups shown in Figure 2. Since the triangle is assumed to be non-spherical we can find non-trivial elements $a \in A$, $b \in B$ and $c \in C$; we claim that the product abc has infinite order in H.

Assume, by way of contradiction, that K is a van Kampen diagram over H whose boundary label is a power of abc. Let D be an extremal disc of K. Observe that, since the triangle of groups is non-spherical, the boundary label of D must involve a, b and c. If D contains two regions over X that share at least one edge, then amalgamate them into a single region; we continue in this way to get a *maximal X-region*. We do this throughout D to all the X-regions, and, similarly, form maximal Y-regions and maximal Z-regions. Since the vertex groups embed, it can be assumed (without loss of generality) that these maximal regions are simply connected; we call the resulting diagram $\hat{D}$. Note that there may be vertices of degree 2 on the boundary of $\hat{D}$.

We use $\hat{D}$ to obtain a tesselation of the sphere as follows. Place $\hat{D}$ on the southern hemisphere so that the boundary of $\hat{D}$ is the equator and take its dual D^*. The *distinguished vertex* v_0 will be the vertex placed in the northern hemisphere in performing this construction. Partition the regions of D^* into *interior* regions, that is those regions that do not involve v_0, and *non-interior* regions. Observe that each interior region of D^* is a 3-gon.

We denote the degree of a vertex v of D^* by $\deg(v)$ and of a region Δ of D^* by $\deg(\Delta)$. We give each corner at a vertex of D^* of degree d the angle $\frac{2\pi}{d}$. The curvature $c(\Delta)$ of a region Δ of D^* of degree k whose vertices have degrees $d_1, d_2, \ldots, d_k$ is then defined by

$$c(\Delta) = (2 - k)\pi + 2\pi \sum_{i=1}^{k} \frac{1}{d_i}.$$

It follows from Euler's formula that the sum of the curvatures of the regions of D^* is 4π.

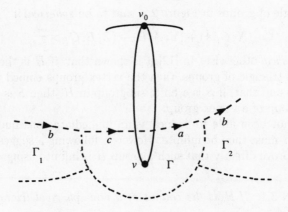

FIGURE 3. Non-interior region of degree 2

Let Δ be an interior region of D^* so that $\deg(\Delta) = 3$. Let $(X; C, A) = \frac{\pi}{k_1}$, $(Y; A, B) = \frac{\pi}{k_2}$ and $(Z; B, C) = \frac{\pi}{k_3}$. It follows that

$$c(\Delta) \leq (2 - 3)\pi + 2\pi(\frac{1}{2k_1} + \frac{1}{2k_2} + \frac{1}{2k_3}) \leq 0$$

since the triangle of groups is non-spherical. It must be the case, therefore, that $\sum c(\Delta) \geq 4\pi$, where the sum is taken over all non-interior regions of D^*.

Let Δ be a non-interior region of D^* and let $\deg(v_0) = k_0$. It can happen that $\deg(\Delta) = 2$ and this is illustrated in Figure 3, where Γ is a maximal X-region. If, in Figure 3, $\deg(v) > 2k_1$, then simply identify the two edges of Δ, noting that any interior region that contains v will still have non-positive curvature. If, however, $\deg(v) = 2k_1$, then Γ_1 must be a Y-region and Γ_2 must be a Z-region. A similar argument occurs if Γ is a Y-region or a Z-region.

A simple argument now shows that the curvature is maximal when the regions are of degree 2 and 3 and are arranged consecutively around v_0 and when k_0 is divisible by 6. An example is given in Figure 4 in which $k_0 = 18$. But the total curvature will be at most

$$\frac{k_0}{6}[(-1 + \frac{2}{k_0} + \frac{1}{k_1} + \frac{1}{k_2}) + (-1 + \frac{2}{k_0} + \frac{1}{k_2} + \frac{1}{k_3}) + (-1 + \frac{2}{k_0} + \frac{1}{k_3} + \frac{1}{k_1})$$
$$+(\frac{2}{k_0} + \frac{1}{k_1}) + (\frac{2}{k_0} + \frac{1}{k_2}) + (\frac{2}{k_0} + \frac{1}{k_3})]\pi \leq 2\pi,$$

a contradiction. $\square$

Now Theorem 1.1 can be deduced from Proposition 3.1 together with the following result:

PROPOSITION 3.2: *Let* $H = \langle a, b : a^p = b^q = R(a, b)^m = 1 \rangle$, $A = \langle a \rangle$ *and* $B = \langle b \rangle$, *where* $m \geq 2$ *and* $R(a, b)$ *is a cyclically reduced word of length at least 2 in the free product* $A * B$. *Then* $(H; A, B) \leq \frac{\pi}{m}$.

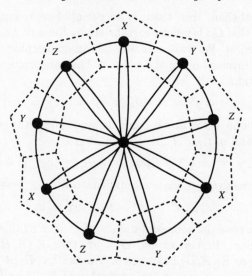

FIGURE 4. Maximal curvature

Proof. If $m \geq 6$ then the result follows from Theorem 3.1 of [5]; if $m < 6$ the result follows from the proof of Theorem 6 of [7]. $\square$

4. PROOF OF THEOREM 1.2

In this section we give a proof of Theorem 1.2. Let G be a generalized tetrahedron group defined by a presentation

$$\langle x, y, z : x^{e_1} = y^{e_2} = z^{e_3} = R_1(x,y)^{f_1} = R_2(y,z)^{f_2} = R_3(z,x)^{f_3} = 1 \rangle,$$

where $\frac{1}{e_1} + \frac{1}{e_2} + \frac{1}{f_1} < 1$ and let T be the generalized triangle group defined by the presentation

$$\langle x, y : x^{e_1} = y^{e_2} = R_1(x,y)^{f_1} = 1 \rangle.$$

Without loss of generality we may assume that $e_1 \leq e_2$. By Theorem 4 (2) of [7], Theorem 1.2 is true if $(f_2, f_3) \neq (2, 2)$. Even if $(f_2, f_3) = (2, 2)$, then Theorem 1.2 is true, with the possible exception of the case $f_1 = 2$ and (e_1, e_2) one of $(3, 8)$, $(3, 10)$, $(4, 5)$, $(4, 6)$, $(4, 8)$ or $(5, 6)$, by Theorem 5 of [7]. We will therefore assume that $(f_1, f_2, f_3) = (2, 2, 2)$ and that (e_1, e_2) is one of these six possibilities. We will assume that G is finite and derive a contradiction.

As in the proof of Theorems 4 and 5 of [7], we can reduce to the case where all the possible representations of T in $PSL(2, \mathbb{C})$ are abelian or infinite metabelian. We let ρ be an essential representation of G in $PSL(2, \mathbb{C})$ and let $X = x\rho$, $Y = y\rho$ and $Z = z\rho$. Then $\langle X, Y, Z \rangle$ is finite and so $\langle X, Y \rangle$ is a finite cyclic group, say $\langle X, Y \rangle = \langle D \rangle$.

If $\langle D, Z \rangle$ is abelian, then there is an essential representation σ of G in $PSL(2, \mathbb{C})$ such that $G\sigma$ is infinite metabelian by Remark 2.4, a contradiction; so $\langle D, Z \rangle$ is dihedral. We deduce that G has an epimorphic image in which z has order 2 and inverts both x and y, i.e. G has epimorphic image $\overline{G}$ defined by the presentation

$$\langle x, y, z : x^{e_1} = y^{e_2} = z^2 = R_1(x, y)^2 = (yz)^2 = (xz)^2 = 1 \rangle,$$

where (e_1, e_2) is one of the above pairs. If we consider the subgroup $\overline{H}$ of $\overline{G}$ generated by x and y, then $\overline{H}$ has presentation

$$\langle x, y : x^{e_1} = y^{e_2} = R_1(x, y)^2 = R_1(x^{-1}, y^{-1})^2 = 1 \rangle.$$

Let U be the generalized triangle group defined by the presentation

$$\langle x, y : x^{e_1} = y^{e_2} = R_1(x, y)^2 = 1 \rangle,$$

and consider an essential representation τ of U in $PSL(2, \mathbb{C})$ with $\langle x\tau, y\tau \rangle$ infinite metabelian. If $A = x\tau$ and $B = y\tau$, then $R_1(A, B)$ has order 2 in $PSL(2, \mathbb{C})$, and so $R_1(A, B)$ has trace 0. We see that $R_1(A^{-1}, B^{-1})$ also has trace 0, so that τ is an essential representation of $\overline{H}$. Thus $\overline{H}$ is infinite, and hence G is infinite, a contradiction.

5. Proof of Theorem 1.3

In this section we give a proof of Theorem 1.3. Let G be a generalized tetrahedron group defined by a presentation

$$\langle x, y, z : x^{e_1} = y^{e_2} = z^{e_3} = R_1(x, y)^2 = R_2(y, z)^2 = R_3(z, x)^2 = 1 \rangle,$$

where $\frac{1}{e_1} + \frac{1}{e_2} \leq \frac{1}{2}$. Without loss of generality, we may assume that $e_1 \leq e_2$. By Theorem 1.2, we may assume that $\frac{1}{e_1} + \frac{1}{e_2} = \frac{1}{2}$ and hence that (e_1, e_2) is either $(3, 6)$ or $(4, 4)$. We assume that G is finite and derive a contradiction.

Let G_1, G_2 and G_3 be the generalized triangle groups defined by the presentations

$$\langle x, y : x^{e_1} = y^{e_2} = R_1(x, y)^2 = 1 \rangle,$$
$$\langle x, z : x^{e_1} = z^{e_3} = R_3(z, x)^2 = 1 \rangle,$$
$$\langle y, z : y^{e_2} = z^{e_3} = R_2(y, z)^2 = 1 \rangle$$

respectively. Since G is finite and $e_1 \geq 3$, $e_2 \geq 3$, Theorem 2.3 and Remark 2.4 show that there is no essential representation $\rho_i : G_i \to PSL(2, \mathbb{C})$ such that $G_i\rho_i$ is abelian or infinite metabelian for $i = 2$ or $i = 3$. We also have no essential representation $\rho_1 : G_1 \to PSL(2, \mathbb{C})$ such that $G_1\rho_1$ is a dihedral group.

Suppose that $(e_1, e_2) = (3, 6)$. Since y has order 6, any essential image of G_1 in $PSL(2, \mathbb{C})$ is abelian or infinite metabelian. If G_1 has image $\langle X, Y \rangle$, then $\mathrm{tr}[X, Y] = 2$. Let $t = \mathrm{tr}(XY)$, so that $1 + 3 + t^2 - \sqrt{3}t - 2 = 2$. We may take $t = 0$ to get a homomorphic image H of G with presentation

$$\langle x, y, z : x^3 = y^6 = z^{e_3} = (xy)^2 = R_2(y, z)^2 = R_3(z, x)^2 = 1 \rangle.$$

By Theorem 2.3 and Remark 2.4, we must have $e_3 = 2$. Since there is no abelian or infinite metabelian essential image of G_2 or G_3, it follows that G_2 must have homomorphic image S_3, A_4, S_4 or A_5, and that G_3 has image the dihedral group of order 12. So H has a homomorphic image K with presentation

$$\langle x, y, z : x^3 = y^6 = z^2 = (xy)^2 = (yz)^2 = (zx)^\beta = 1 \rangle$$

where $2 \leq \beta \leq 5$. So K is an ordinary tetrahedron group and is infinite by the results of Section 1, a contradiction. We may therefore assume that $(e_1, e_2) = (4, 4)$.

First assume that there is an essential representation $\rho_1 : G_1 \to PSL(2, \mathbb{C})$ such that $G_1 \rho_1$ is abelian or infinite metabelian. Then, by Theorem 2.3 and Remark 2.4, we must have that $e_3 = 2$. So we have

$$\langle x, y, z : x^4 = y^4 = z^2 = R_1(x, y)^2 = R_2(y, z)^2 = R_3(z, x)^2 = 1 \rangle.$$

Replacing x by x^{-1} if necessary, we have a factor group with presentation

$$\langle x, y, z : x^4 = y^4 = z^2 = (xy)^2 = R_2(y, z)^2 = R_3(z, x)^2 = 1 \rangle.$$

If we consider the essential representations $\rho_i : G_i \to PSL(2, \mathbb{C})$, $i = 2, 3$, we get the factor group $\overline{G}$ with presentation

$$\langle x, y, z : x^4 = y^4 = z^2 = (xy)^2 = (yz)^n = (zx)^m = 1 \rangle$$

with $2 \leq n, m \leq 4$. In all cases we have an ordinary tetrahedron group which is known to be infinite, a contradiction.

So there is no essential representation $\rho_1 : G_1 \to PSL(2, \mathbb{C})$ such that $G_1 \rho_1$ is abelian or infinite metabelian. We see that G has a presentation of the form

$$\langle x, y, z : x^4 = y^4 = z^r = R_1(x, y)^2 = R_2(y, z)^2 = R_3(z, x)^2 = 1 \rangle,$$

such that, for any essential representation $\rho_1 : G_1 \to PSL(2, \mathbb{C})$, the image $G \rho_1$ is isomorphic to S_4. So, if $\rho : G \to PSL(2, \mathbb{C})$ is an essential representation, we must have that $G\rho$ is isomorphic to S_4, and hence that $r \leq 4$.

We first consider the case $r = 2$ so that we have a presentation of the form

$$\langle x, y, z : x^4 = y^4 = z^2 = R_1(x, y)^2 = R_2(y, z)^2 = R_3(z, x)^2 = 1 \rangle.$$

Then, as above, we have a homomorphic image with presentation

$$\langle x, y, z : x^4 = y^4 = z^2 = R_1(x, y)^2 = (yz)^\alpha = (zx)^\beta = 1 \rangle$$

where $2 \leq \alpha, \beta \leq 3$. Without loss of generality, we may assume that $\alpha \leq \beta$.

If $\alpha = \beta = 2$, then we have the presentation

$$\langle x, y, z : x^4 = y^4 = z^2 = R_1(x, y)^2 = (yz)^2 = (zx)^2 = 1 \rangle.$$

for G. We see that $H = \langle x, y \rangle$ is a normal subgroup of index 2 with presentation

$$\langle x, y : x^4 = y^4 = R_1(x, y)^2 = R_1(x^{-1}, y^{-1})^2 = 1 \rangle.$$

Let $R_1(x, y)$ be the cyclically reduced word $x^{r_1} y^{s_1} x^{r_2} y^{s_2} \ldots x^{r_m} y^{s_m}$. Since the trace polynomial for an essential representation of G has a single root, H is infinite by Proposition 2.5 if $m \geq 2$. If $m = 1$, then, without loss of generality, we may assume that $R_1(x, y) = x^2 y$. The relation $R_1(x^{-1}, y^{-1})^2 = 1$ is then redundant, and we have

$$\langle x, y : x^4 = y^4 = (x^2 y)^2 = 1 \rangle.$$

Adding the relations $x^2 = y^2 = 1$ yields the infinite group $C_2 * C_2$, a contradiction.

If $\beta \geq 3$, we add the relations $x^2 = y^2 = 1$ to get the presentation

$$\langle x, y, z : x^2 = y^2 = z^2 = (xy)^\gamma = (yz)^\alpha = (zx)^\beta = 1 \rangle.$$

If $\gamma = 0$, then the group is infinite and we have finished; so we suppose that $\gamma > 0$. Since G_1 maps onto S_3, we must have that 3 divides γ. In addition, if ϕ is the natural homomorphism from $\langle x : x^4 = 1 \rangle * \langle y : y^4 = 1 \rangle$ to $\langle x : x^2 = 1 \rangle * \langle y : y^2 = 1 \rangle$, then $R_1(x, y)^2 \phi = (xy)^\gamma$. So we have that 2 divides γ, and hence that 6 divides γ. Since $\gamma \geq 6$, the group is infinite by the results of Section 1. This concludes the case $r = 2$.

We next consider the case $r = 3$ so that we have a presentation of the form

$$\langle x, y, z : x^4 = y^4 = z^3 = R_1(x, y)^2 = R_2(y, z)^2 = R_3(z, x)^2 = 1 \rangle.$$

In this case, G_1, G_2 and G_3 all have essential representations with image S_4.

In general, suppose that u and v are elements of order 4 and 3 respectively that generate S_4. Conjugating in S_4, we may assume that $u = (1234)$, and then that $v = (123)$ or that $v = (132)$. In this case, $uv = (1342)$ or (34), $uv^{-1} = (34)$ or (1342), and $u^{-1}v = (14)$ or (1423). So, replacing z by z^{-1} in our presentation if necessary, we may assume that G_3 has an essential representation ρ with image S_4 and $y\rho$, $z\rho$ and $(yz)\rho$ of orders 4, 3 and 2 respectively. Having done this, replacing x by x^{-1} if necessary, we may assume that G_2 has an essential representation σ with image S_4 and $x\sigma$, $z\sigma$ and $(xz)\sigma$ of orders 4, 3 and 2 respectively. Since we have defining relations for S_4 in each case, G has homomorphic image the group with presentation

$$\langle x, y, z : x^4 = y^4 = z^3 = R_1(x, y)^2 = (yz)^2 = (zx)^2 = 1 \rangle.$$

If we add the relations $x^2 = y^2 = 1$, we get a homomorphic image with presentation

$$\langle x, y, z : x^2 = y^2 = z^3 = (xy)^\gamma = (yz)^2 = (zx)^2 = 1 \rangle$$

for some γ. As above, 6 divides γ, so that the group is infinite by the results in Section 1.

Lastly, if $r = 4$, then we have a presentation of the form

$$\langle x, y, z : x^4 = y^4 = z^4 = R_1(x, y)^2 = R_2(y, z)^2 = R_3(z, x)^2 = 1 \rangle.$$

Adding the relation $z^2 = 1$ reduces us to the case $r = 2$, and so we have that G is infinite in this case also.

REFERENCES

[1] G. Baumslag, J. W. Morgan and P. B. Shalen, Generalized triangle groups, *Math. Proc. Cambridge Phil. Soc.* **102** (1987), 25–31.

[2] H. S. M. Coxeter, The polytopes with regular-prismatic vertex figures (part 2), *Proc. London Math. Soc.* **34** (1932), 126–189.

[3] H. S. M. Coxeter, Discrete groups generated by reflections, *Ann. Math.* **35** (1934), 588–621.

[4] H. S. M. Coxeter, The complete enumeration of finite groups of the form $R_i^2 = (R_iR_j)^{k_{ij}} = 1$, *J. London Math. Soc.* **10** (1935), 21–25.

[5] A. J. Duncan and J. Howie, Spelling theorems and Cohen-Lyndon theorems for one-relator products, *J. Pure Applied Algebra* **92** (1994), 123–136.

[6] B. Fine, J. Howie and G. Rosenberger, One-relator quotients and free products of cyclics, *Proc. American Math. Soc.* **102** (1988), 249–254.

[7] B. Fine, F. Levin, F. Roehl and G. Rosenberger, The generalized tetrahedron groups, *in* R. Charney, M. Davis and M. Shapiro (eds.), *Geometric Group Theory*, Ohio State University, Mathematical Research Institute Publications **3**, de Gruyter (1995), 99–119.

[8] J. Howie, V. Metaftsis and R. M. Thomas, Finite generalized triangle groups, *Trans. Amer. Math. Soc.* **347** (1995), 3613–3623.

[9] J. Howie, V. Metaftsis and R. M. Thomas, Triangle groups and their generalisations, *in* A. C. Kim and D. L. Johnson (eds.), *Groups - Korea 1994*, de Gruyter (1995), 135–147.

[10] L. Lévai, G. Rosenberger and B. Souvignier, All finite generalized triangle groups, *Trans. Amer. Math. Soc.* **347** (1995), 3625–3627.

[11] S. J. Pride, Groups with presentations in which each defining relator involves exactly two generators, *J. London Math. Soc.* **36** (1987), 245–256.

[12] J. Stallings, Non-positively curved triangles of groups, *in* E. Ghys, A. Haefliger and A. Verjovsky (eds.), *Group Theory from a Geometrical Viewpoint*, World Scientific (1991), 491–503.

[13] R. M. Thomas, Cayley graphs and group presentations, *Math. Proc. Cambridge Philos. Soc.* **103** (1988), 385–387.

[14] S. V. Tsaranov, On a generalization of Coxeter groups, *Algebras, Groups and Geometries* **6** (1989), 281–318.

[15] S. V. Tsaranov, Finite generalized Coxeter groups, *Algebras, Groups and Geometries* **6** (1989), 421–457.

Martin Edjvet
University of Nottingham
Nottingham NG7 2RD
England

Richard M. Thomas
University of Leicester
Leicester LE1 7RH
England

Gerhard Rosenberger and
Michael Stille
Universität Dortmund
44221 Dortmund
Germany

EFFICIENT COMPUTATION IN WORD-HYPERBOLIC GROUPS

DAVID B. A. EPSTEIN AND DEREK F. HOLT

ABSTRACT. We describe briefly some practical procedures for computing the various constants associated with a word-hyperbolic group, and report on the performance of their implementations in the KBMAG package on a number of examples. More complete technical details will be published elsewhere.

1. INTRODUCTION

During the mid 1980's, there was a growing demand for efficient computational methods for solving some of the classical decision problems, such as the word, conjugacy and isomorphism problems, in infinite finitely presented groups. These are of course theoretically undecidable in general, but some of them have been proved to be solvable in various special classes of groups, such as braid groups, Coxeter groups and knot groups. At that time, the only existing general technique for attempting to solve the word problem on a computer was the Knuth-Bendix algorithm (see, for example [Gilman, 1979]). However, this was not really satisfactory, because most of the time it did not work, and making it work at all for infinite groups usually depended on an appropriate and often non-obvious choice of an ordered generating set for the group.

In 1985, the concept of an *automatic* group was introduced by Thurston, as a formulation in terms of finite state automata of various geometrical properties of discrete cocompact groups of isometries of hyperbolic space that had been proved by Jim Cannon in [Cannon, 84]. Many familiar classes of groups, including word-hyperbolic groups, virtually abelian groups, Coxeter groups, braid groups and most knot groups have been shown to be automatic. The word problem is solvable (by reduction of words to a normal form) in quadratic time in these groups. The conjugacy problem is solvable (but only in exponential time) in the more restricted class of biautomatic groups.

Algorithms for performing these computations were implemented in Warwick during the following few years by the authors of this paper and Sarah Rees. See the multi-author book [Epstein *et al.*, 1992] for the general theory of automatic groups, and [Epstein, Holt, Rees, 1991] and [Holt, 1996] for a detailed description of the algorithms. The latest implementation is part of the second author's package KBMAG [Holt, 1995]. These methods make use of the Knuth-Bendix process, and **KBMAG** includes a standalone implementation of this. It also includes an extensive library of routines for performing operations on finite state automata.

This article is concerned with the more restricted class of *word-hyperbolic* groups, which were first defined by Gromov; see [Gromov, 1987]. They represent an algebraic or geometrical formulation of the idea of a negatively curved group, and they include small-cancellation groups and the discrete cocompact isometries of hyperbolic space studied by Jim Cannon. For a detailed exposition of the definitions and properties of these groups, the reader is referred to [Alonso et al, 1991]. In particular, it is shown in Theorem 2.18 of that article that the word problem is solvable by a Dehn algorithm in linear time.

A number of potentially efficient and useful algorithms have been proposed recently for word-hyperbolic groups. The first author has devised an $n \log(n)$ solution of the conjugacy problem, which represents a substantial improvement over the exponential time method for biautomatic groups. Mike Shapiro has proposed (in outline) a procedure for reducing words to short-lex normal form in linear time, improving on the general quadratic time method for automatic groups. Eric Swenson has proposed a method for testing a quasiconvex subgroup of a word-hyperbolic group for abnormality. (None of these are published yet.)

However, it is not yet clear whether it will be possible to implement these methods efficiently, because they all depend on the knowledge of various constants that are associated with the Cayley graph of a word-hyperbolic group. In this article, we shall describe in outline some algorithms for computing these constants. More complete details have been or will be published elsewhere. Most of them are either already fully available in **KBMAG** (which is freely available by ftp), or have experimental implementations in that system.

A group G is word-hyperbolic if geodesic triangles in the Cayley graph Γ of the group are δ'-slim, in the sense that there is a global constant δ' such that any point on a side of the triangle lies within a distance δ' in Γ of the union of the other two sides. (This definition turns out to be independent of the choice of generating set of G.) There is an equivalent definition, which we shall state precisely in Section 2, in terms of a usually slightly larger constant δ, known as the *thinness* constant of Γ, which specifies that all geodesic triangles are δ-thin. The most difficult algorithm that we shall describe is a method of estimating δ for a given group G. Our current experimental implementation works on easy examples, such as the Von Dyck triangle groups and two-dimensional space groups.

It was proved by Papasoglu in [Papasoglu, 1994], that G is hyperbolic if and only if all geodesic bigons in Γ are γ-thin for some global constant γ. It is much easier to compute γ than δ, so this provides us with a practical method of verifying that a group really is word-hyperbolic. Our method for finding γ

is very similar to an algorithm described in [Wakefield, 1997], but it contains a simplification which appears to improve the performance substantially.

One can also consider geodesic bigons in which one or both of the vertices lies on the boundary of Γ (see Chapter 4 of [Alonso et al, 1991]). These correspond to pairs of infinite geodesic paths in Γ such that all points on the first path lie within a uniformly bounded distance $\hat{\gamma}$ of the other path. If one of the vertices of the bigon is a vertex of Γ, then the paths both start at that vertex, and if both vertices lie on the boundary of Γ then the paths are infinite in both directions. There is a fixed constant $\hat{\gamma}$ such that all such bigons satisfy this criterion, and we shall describe a method for estimating it. (In all of the examples that we have investigated, $\hat{\gamma}$ is no larger than γ.)

Each of these procedures follows a general philosophy of group-theoretical algorithms that construct finite state automata. This approach was originally proposed in the algorithms described in Chapter 5 of [Epstein *et al.*, 1992] for computing automatic structures, and employed in their implementations for short-lex structures described in [Holt, 1996]. The idea is first to find a method of constructing likely candidates for the required automata, which we shall call the *working* automata, The second step is to construct other (usually larger) *test* automata of which the sole purpose is to verify the correctness of the working automata. If this verification fails, then it should be possible to use words in the language of the test automata to construct improved versions of the working automata. One practical difficulty with this approach is that experience shows that incorrect working automata and the resulting test automata are much larger than the correct ones, so it can be extremely important to find good candidates on the first pass.

The remainder of this article is organised as follows, In Section 2 we summarise the required definitions and notation. In Section 3 we describe our method of verifying hyperbolicity by computing γ, and of finding the related constant $\hat{\gamma}$. In Section 4 we describe briefly our proposal for estimating δ. Finally, we discuss the performance of our implementations on some examples in Section 5.

2. NOTATION

G will denote a group with a given finite generating set X. Let $A = X \cup X^{-1}$, and let A^* be the set of all words in A. For $u, v \in A^*$, we denote the image of u in G by $\overline{u}$, and $u =_G v$ will mean the same as $\overline{u} = \overline{v}$. For a word $u \in A^*$, $l(u)$ will denote the length of u and $u(i)$ will denote the prefix of u of length i, with $u(i) = u$ for $i \geq l(u)$.

Let $\Gamma = \Gamma_X(G)$ be the Cayley graph of G with respect to X. We make Γ into a metric space in the standard manner, by letting all edges have unit length, and defining the distance $\partial(x, y)$ between any two points of Γ to be the minimum length of paths connecting them. (The points of Γ include both the vertices, and points on the edges of Γ.) This makes Γ into a *geodesic* space, which means that for any $x, y \in \Gamma$ there exist geodesics (i.e. shortest paths) between x and y. For $g \in G$, $l(g)$ will denote the length of a geodesic path from the base vertex 1_G of Γ to g.

A *geodesic triangle* Δ in Γ consists of three not necessarily distinct points a, b, c together with three directed geodesic paths u, v, w joining bc, ca and ab, respectively. (So $l(u) = \partial(b, c)$, etc.) The vertices a, b, c of Δ are not necessarily vertices of Γ; they might lie in the interior of an edge of Γ.

The most convenient definition of word-hyperbolic groups is in terms of thin triangles. Let Δ be a geodesic triangle as above. Let $\rho(a) = (l(v)+l(w)-l(u))/2$ and define $\rho(b), \rho(c)$ correspondingly. Then $\rho(b) + \rho(c) = l(u)$, so any point d on u satisfies either $\partial(d, b) \le \rho(b)$ or $\partial(d, c) \le \rho(c)$, and similarly for v and w. The points d, e, f on u, v, w with $\partial(d, b) = \rho(b)$ and $\partial(d, c) = \rho(c)$, etc., are known as the *meeting points* of the triangle (see Fig. 1).

In a constant curvature geometry (the euclidean plane, the hyperbolic plane or the sphere), the meeting points of a triangle are the points where the inscribed circle meets the edges. In more general spaces, such as Cayley graphs, the term *inscribed circle* has no meaning, but the meeting points can still be defined.

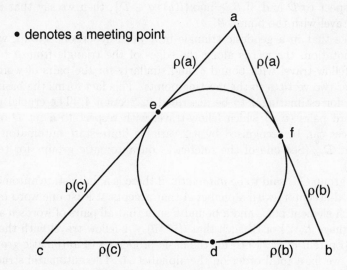

Figure 1. This picture shows the meeting points as the intersection of the inscribed circle with the edges of the triangle in the case of constant curvature geometry.

Suppose a, b and c are vertices in the Cayley graph. Then the meeting points are also vertices if and only if the perimeter $l(u) + l(v) + l(w)$ is even. In our algorithm for estimating δ we will avoid the resulting unpleasantness in the case of triangles of odd perimeter by shifting the meeting points by a distance of $1/2$ to a neighbouring vertex.

Let $\delta \in \mathbb{R}_+$. Then we say that Δ is δ-*thin* if, for any $r \in \mathbb{R}$ with $0 \le r \le \rho(x)$, the points p and q on v and w with $\partial(p, a) = \partial(q, a) = r$ satisfy $\partial(p, q) \le \delta$, and similarly for the points within distance $\rho(b)$ of b and $\rho(c)$ of c. We call such points p and q *companions*. Normally companions are distinct, but there can be many situations where they coincide—for example two geodesics sides of a triangle could have an intersection consisting of a disjoint union of three intervals. Mostly points on the triangle have exactly one companion, but the meeting points normally have two companions— once again, in degenerate situations two or all three of the meeting points may coincide.

The group G is called word-hyperbolic if there exists a δ such that all geodesic triangles in Γ are δ-thin. (It turns out that this definition is independent of the generating set X of G, although the minimal value of δ does depend on X.)

If $u, v \in A^*$, then we call the set $\mathcal{D} = \{ \overline{u(i)}^{-1} \overline{v(i)} \mid i \in \mathbb{Z}, i \ge 0 \}$ the set of word-differences arising from (u, v). We say that u and v *fellow-travel* with respect to $\mathcal{D}$ and, if $\beta \ge \max \{ l(g) \mid g \in \mathcal{D} \}$, then we say that u and v fellow-travel with the bound β.

Notice that in a geodesic triangle in a word-hyperbolic group, with the above notation, the words along the edges of the triangle from a to e and a to f fellow-travel with bound δ, and similarly for the pairs of words from the other two vertices to the meeting points. This fact forms the basis of our method for estimating δ to be described in Section 4. The crucial point is that word pairs (u, v) which fellow-travel with respect to a set $\mathcal{D}$ of word-differences can be recognised by a 2-variable finite state automaton having state set $\mathcal{D}$. (See one of the references on automatic groups for technical details.)

The group G is said to be *automatic*, if there is a finite state automaton W (the word-acceptor) with alphabet A that accepts at least one word mapping onto each element of G, and a bound β such that all pairs of words u and v in the language $L(W)$ of W such that $l(\overline{u}^{-1}\overline{v}) \le 1$ fellow-travel with the bound β. See [Epstein *et al.*, 1992] for the basic properties of automatic groups.

Now we fix a total order on the alphabet A. The automatic structure is called *short-lex* if $L(W)$ consists of the short-lex least representatives of each element $g \in G$; that is the lexicographically least among the shortest words in A^* that map onto g. The existence of such a structure for a given group G depends in general on the generating set X of G, but word-hyperbolic groups are known to be short-lex automatic for any choice of generators. (This is Theorem 3.4.5 of [Epstein *et al.*, 1992].)

The group G is called *strongly geodesically* automatic with respect to X if there is an automatic structure in which $L(W)$ is the set of all geodesic words

$u \in A^*$; that is, those u for which $l(u) = l(\overline{u})$. It is shown in Theorem 3.2.1 of [Epstein *et al.*, 1992] that this property is equivalent to the existence of a constant γ such that all geodesic words $u, v \in A^*$ with $l(\overline{u}^{-1}\overline{v}) \leq 1$ fellow-travel with bound γ. In this case, γ is also a bound on the width of geodesic bigons in Γ, and we shall describe how to compute such a γ, together with the set of associated word-differences of length at most γ, in Section 3. It is proved in Corollary 2.3 of [Papasoglu, 1994] that G is *strongly geodesically automatic* if and only if G is word-hyperbolic, and so this calculation also provides us with an efficient verification procedure for the word-hyperbolicity of G.

We shall assume throughout that our group $G = \langle X \rangle$ is short-lex automatic with respect to X, and that we have already computed the corresponding short-lex word-acceptor W and the set of word-differences $\mathcal{D}$ arising from the pairs of words $u, v \in L(W)$ such that $l(\overline{u}^{-1}\overline{v}) \leq 1$. This data, which is known as the short-lex automatic structure of G with respect to X, can be used to reduce arbitrary words in A^* to their short-lex least equivalent word in G in quadratic time. The above computations can all be carried out using KBMAG.

3. Verifying hyperbolicity

As explained above, to verify hyperbolicity, we need to find a constant γ such that all geodesic word pairs (u, v) with $l(\overline{u}^{-1}\overline{v}) \leq 1$ fellow-travel with bound γ. Since we have already proved short-lex automaticity, it follows easily from the definition of automaticity that it is sufficient to find such a constant γ' for those pairs in which $u =_G v$ and v is short-lex minimal. The calculation of γ and the associated word-difference set from γ' is then a straightforward composite operation on finite state automata (see [Holt, 1996]). We can also construct an automaton GW that accepts all geodesic words in Γ; that is, the word-acceptor in the geodesic automatic structure.

Assuming that such a γ' exists, there is a finite set $\mathcal{WD}$ of word-differences of length at most γ' associated with all such geodesic word pairs for which $u =_G v$ and v is short-lex minimal. The idea is to construct candidates $\mathcal{WD}_n$ for $\mathcal{WD}$, where $\mathcal{WD}_0 = \mathcal{D}$, as defined above. We form $\mathcal{WD}_{n+1}$ from $\mathcal{WD}_n$ by adjoining any extra word-differences that we find. If G is word-hyperbolic and $\mathcal{WD}$ is finite, then this process will halt when $\mathcal{WD}_n$ contains $\mathcal{WD}$, but if G is not word-hyperbolic, then it will not halt, and $\mathcal{WD}_n$ will grow indefinitely large. We also maintain a candidate GW_n for the geodesic word-acceptor GW. This is constructed from $\mathcal{WD}_n$ as the set of all geodesic words u which fellow-travel with their short-lex equivalent v with respect to $\mathcal{WD}_n$.

The principal step in the process is to construct a test automaton T that accepts all geodesic words that fellow travel with some word in GW_n with respect to $\mathcal{WD}_n$. If all such words already lie in $L(GW_n)$ then the procedure halts and we define GW and $\mathcal{WD}$ to be the current candidates. On the other hand, if we find one or more words u in $L(T)\backslash L(GW_n)$, then we reduce them to their short-lex equivalent words v, and adjoin the word-differences associated with (u,v) which are not already in $\mathcal{WD}_n$ to $\mathcal{WD}_n$. (By construction of GW_n, they cannot all lie in $\mathcal{WD}_n$.) The resulting extended set then becomes $\mathcal{WD}_{n+1}$ amd the process is repeated with this in place of $\mathcal{WD}_n$.

It is not difficult to prove by induction on $l(u)$ that any geodesic word $u \in A^*$ must eventually be accepted by some GW_n, and so the process will halt if $\mathcal{WD}$ is finite. From GW and $\mathcal{WD}$, it is straightforward, using the composite operation on automata, to construct an automaton GP such that $L(GP)$ is equal to all pairs of geodesic words (u,v) for which $l(\overline{u}^{-1}\overline{v}) \leq 1$. For full technical details, see [Epstein & Holt 1998] or the similar procedure described in [Wakefield, 1997]

Now consider two infinite geodesic paths u,v in Γ, which either have a vertex of Γ as common starting point and are infinite in one direction only, or are infinite in both directions. We say that u and v are *neighbouring paths* with respect to the constant $\hat{\gamma}$, if every point on one of the paths lies at distance at most $\hat{\gamma}$ from the other path. Such pairs of paths correspond to geodesic bigons in Γ in which one or both vertices lie on the boundary of the group, and it can be shown that there is in fact a single constant $\hat{\gamma}$ such that all pairs of neighbouring paths in Γ are neighbours with respect to $\hat{\gamma}$. As mentioned in the introduction, for several future potential applications, it is important to be able to estimate $\hat{\gamma}$ as well as γ.

In the case of one-way infinite paths, where u and v have a common starting vertex, which might as well be the base point of Γ, the set of word-differences $\mathcal{NGD} = \{\overline{u(t)}^{-1}\overline{v(t)} \,|\, t \in \mathbb{Z},\ t \geq 0\}$ is finite, and we would like to calculate this set. Then we can take $\hat{\gamma}$ to be the maximum length of an element of $\mathcal{NGD}$. For two-way infinite paths u,v, it is can be shown that there are vertices $u(0)$ and $v(0)$ of Γ on u and v such that the elements $\overline{u(t)}^{-1}\overline{v(t)}$ lie in the same set $\mathcal{NGD}$ for all $t \in \mathbb{Z}$. (Here $u(t)$ represents the element of G corresponding to the point on the path u at distance t from $u(0)$, and similarly for $v(t)$.) Hence the same constant $\hat{\gamma}$ will serve also for two-way infinite neighbouring paths.

To find $\mathcal{NGD}$, we construct automata NG_n $(n > 0)$ which accept all pairs of (finite) geodesic words u,v of equal length l with the following two properties:

(i) There exist one-way infinite neighbouring paths $\hat{u},\hat{v}$ such that $u = \hat{u}(l)$ and $v = \hat{v}(l)$ are prefixes of $\hat{u}$ and $\hat{v}$, respectively;

(ii) There exist finite geodesic words u',v' such that $u = u'(l)$, $v = v'(l)$ and $l(\overline{u'}^{-1}\overline{v'}) \leq n$.

It is straightforward to construct NG_1 from GP. We just make all states accepting (which has the effect that all prefixes of accepted words become accepted words), and then (repeatedly) remove any states which are not the source of any transition. This ensures that Condition (i) is satisfied. In general, we construct NG_n by taking a composite of NG_1 and NG_{n-1}, and then removing all states which are not the source of any transition. Since all pairs of words accepted by some NG_n are prefixes of pairs of neighbouring geodesics, all such pairs must fellow travel with respect to $\hat{\gamma}$, and so eventually we will get $NG_{n+1} = NG_n$ and the process will terminate. Then $\mathcal{NGD}$ is simply the set of all word-differences arising from pairs of words accepted by NG_n.

In principle, this procedure might be expected to get more and more difficult as n grows larger, but in fact in all of the examples that we have tried so far, the process has terminated on the first pass with $NG_2 = NG_1$ and so $\hat{\gamma} = \gamma$. We are not aware of any theoretical reason why this should be the case in general, however.

4. FINDING THE THINNESS CONSTANT

Throughout this section, we assume that $G = \langle X \rangle$ is a word-hyperbolic group and that $\delta > 0$ is a constant such that all geodesic triangles in $\Gamma_X(G)$ are δ-thin. The aim is to devise a practical algorithm to find such a δ, preferably as small as possible. As before, we assume that we have already calculated the short-lex automatic structure for G with respect to X; that is the word-acceptor W and word-differences $\mathcal{D}$. We also need to construct the automaton W^R which accepts a word w if and only if the reversed word $w^R \in L(W)$. Full technical details of the material in this section can be found in [Epstein & Holt 1998].

Let us recall the notation for δ-thin geodesic triangles with vertices a, b, c and edges u, v, w in the Cayley graph $\Gamma = \Gamma_X(G)$ that was introduced in Section 2. The meeting points on the sides u, v and w are denoted by d, e and f, respectively. We shall call such a geodesic triangle *short-lex geodesic*, if its vertices are vertices of Γ and if the words A^* corresponding to the edges of the triangle (which we shall also denote by u, v, w) all lie in $L(W)$. We prefer to work with short-lex geodesic triangles, because in general there are far more geodesic triangles and consideration of all of these is likely to make an already difficult computational problem impossible.

Our algorithms are designed to compute the minimal $\delta \in \mathbb{N}$ such that all short-lex geodesic triangles are δ-thin. Let β be the maximum length of an element in $\mathcal{D}$, and let γ and γ' be as defined in Section 3. Then if δ is the thinness constant for short-lex geodesic triangles, it can be shown by an elementary geometrical argument that the thinness constant for arbitrary geodesic triangles in Γ is at most $\delta + 2(\beta + \gamma') + 3$ or, alternatively, at most $\delta + 2\gamma + 3$.

Let $\mathcal{GT}$ be the set of word-differences arising from those pairs of words (u_a, v_a), where u_a and v_a are the words labelling those parts of the two edges of a short-lex geodesic triangle that go from a to the meeting points on the two edges. In other words u_a is the prefix of w labelling the word from a to

f, and v_a is the prefix of v^R labelling the word from a to f. The value of δ is then the maximum length of an element of $\mathcal{GT}$. (As we mentioned before, in the case of triangles with odd perimeter, where the meeting points lie in the middle of edges of Γ, we shift each meeting point by a distance half clockwise around the triangle, to bring it to a neighbouring vertex.)

The general philosophy of our tactics is the same as for the algorithm for finding γ' in Section 3. At any stage, we have a candidate $\mathcal{GT}_n$ for $\mathcal{GT}$, and we then attempt to decide whether $\mathcal{GT}_n$ contains $\mathcal{GT}$ and, if not, to adjoin some missing elements. Unfortunately, this testing process is much more difficult and heavy on computing resources (particularly memory) than it is for the computations described earlier.

As a candidate for $\mathcal{GT}_0$ we could choose $\mathcal{WD}$ as defined in Section 3, which is essentially the corresponding set for geodesic bigons. However, we have found it more effective to produce $\mathcal{GT}_0$ by taking pairs u and v of random elements of $L(W)$ (up to some prescribed length). These can be used to calculate the unique $w \in L(W)$ which completes the shortlex geodesic hyperbolic triangle, and the elements of $\mathcal{GT}$ which arise from this triangle are then computed. We can do this for a large number (perhaps 100000) of random pairs u, v and take $\mathcal{GT}_0$ to be the set of all word-differences in $\mathcal{GT}$ obtained in this fashion.

One might hope to get all of $\mathcal{GT}$ fairly quickly by this method, but unfortunately this does not seem happen in practice, and there are typically some word-differences that are hard to find by random methods. It would make the process more efficient if we could find a more refined method for guessing $\mathcal{GT}$, because the correctness testing process works more effectively on correct data.

To perform the correctness test, we use W and $\mathcal{GT}_n$ to construct an automaton called FRD_n, which stands for 'forward, reverse, difference'. This accepts all pairs of words (u_a, v_a) as described above. Its name comes from the fact that its states have three components, the first a state of $L(W)$ arising from reading w, the second a state of $L(W^R)$ arising from v^R, and the third keeping track of the word-difference of the pair in $\mathcal{GT}_n$.

From this, we form an associated automaton FRD_n^3, which consists of three copies of FRD_n. The three pairs of words read by FRD_n^3 are accepted when they are the three pairs of edges emerging from the three vertices of some short-lex geodesic triangle, ending and meeting at the meeting points of the triangle. The states of FRD_n^3 are just triples of states of FRD_n, but deciding which of them are accept states is quite complicated.

Roughly speaking, for this to be true, the first two components of the three states of FRD_n in the triple (that is, the 'forward' and 'reverse' components) have to match up in such a way that the words u, v and w corresponding to the edges of the triangle all lie in $L(W)$ and, in addition, the product of the three third components, which are the word-differences in $\mathcal{GT}_n$ has to be equal to the identity in G. Again, we refer the reader to [Epstein & Holt 1998] for full technical details. This list of accept states is typically very large.

The idea then is to compute a two-variable finite state automaton GP (geodesic pairs) of which the language is the subset of $A^* \times A^*$ defined by the expression

$$\{(w_1, w_2) \in A^* \times A^* \mid$$
$$\exists (u_a, v_a, u_b, v_b, u_c, v_c) \in L(FRD_n^3) : w_1 = u_a v_b^R, w_2 = v_a u_c^R\}.$$

Then GP accepts the set of pairs of sides (w, v^R) emerging from the vertex a in the triangles that are accepted by FRD_n^3. Thus $\mathcal{GT}_n$ contains $\mathcal{GT}$, and the process terminates if and only if $L(GP) = L(W) \times L(W^R)$. Since checking for equality of the languages of automata is easy, we can perform this check provided that we can construct GP.

If the check fails, then our definition of GP ensures that $L(GP) \subset L(W) \times L(W^R)$. So we can find one or more specific word pairs $(w_1, w_2) \in L(W) \times L(W^R) \backslash L(GP)$ and then compute the word-differences arising from the short-lex geodesic triangle having w_1 and w_2^R as two of its sides. We can then adjoin these to $\mathcal{GT}_n$ to form $\mathcal{GT}_{n+1}$.

The construction of GP can be carried out in theory, but because of the large number of quantified variables involved in the above expression, a naive implementation would be hopelessly expensive in memory usage. Our implementation uses a number of refinements, but it remains very heavy in memory usage. However, it does work for some straightforward examples.

5. EXAMPLES

The procedure for finding the geodesic fellow-traveller constant γ has been run successfully, using the implementation in **KBMAG**, on a variety of examples. These include the following.

(i) Two dimensional surface groups $T_m (m > 1)$, $P_m, K_m (m > 0)$ defined by the presentation

$$\langle a_1, b_1, \ldots, a_m, b_m, x_1, \ldots, x_k \mid \textstyle\prod_{i=1}^m (a_i^{-1} b_i^{-1} ab) \prod_{j=1}^k x_j^2 \rangle,$$

where $k = 0, 1$ or 2 for the groups T_m, P_m and K_m, respectively. Here $\gamma = 2m$ for T_m, P_m and $2m + 2$ for K_m.

(ii) Hyperbolic von Dyck triangle groups, such as

$$\langle a, b \mid a^3 = b^n = (ab)^2 = 1 \rangle$$

for $n \geq 7$. Here γ is about $n/2$.

(iii) The symmetry group of a certain tessellation by dodecahedra of hyperbolic 3-space as featured in the video 'Not Knot' ([Not Knot]), with presentation

$$G_4 = \langle a, b, c, d, e, f \mid a^4 = b^4 = c^4 = d^4 = e^4 = f^4 =$$
$$aba^{-1}e = bcb^{-1}f = cdc^{-1}a = ded^{-1}b = efe^{-1}c = faf^{-1}d = 1\rangle.$$

Here $\gamma = 4$.

(iv) Hyperbolic Fibonacci groups, such as

$$\langle a_1, \ldots a_m \mid a_i a_{i+1} = a_{i+2} \ (1 \le i \le m) \rangle,$$

where m is even, $m \ge 8$, and the indices in the relations are taken modulo m. We have tried $m = 8, 10$ and 12, in which cases $\gamma = 3$.

Of these, the examples in (i) and (ii) were very quick and easy, while those of (iii) and (iv) were rather harder. As we mentioned above, in all of the examples that we have looked at to date we found quickly that $\hat{\gamma} = \gamma$.

We completed the calculation of δ without undue difficulty in each of the examples in (i) and (ii) that we tried. Generally δ was not much larger than γ, with $\delta - \gamma$ at most 3. To date we have not succeeded in computing δ for the examples (iii) and (iv). Of these, we have some hope of eventually succeeding with (iii), but we will either need to refine our implementations, or acquire the use of a computer with more memory.

REFERENCES

[Alonso et al, 1991]   J. Alonso, T. Brady, D. Cooper, V. Ferlini, M. Lustig, M. Mihalik, M. Shapiro and H. Short, *Notes on word-hyperbolic groups*, in E. Ghys, A. Haefliger and A. Verjovsky, eds., Proceedings of the Conference *Group Theory from a Geometric Viewpoint* held in I.C.T.P., Trieste, March 1990, World Scientific, Singapore, 1991.

[Cannon, 84]   J.W. Cannon, *The combinatorial structure of cocompact discrete hyperbolic groups*, Geom. Dedicata 16, (1984),123–148 .

[Epstein *et al.*, 1992]   D.B.A. Epstein, J.W. Cannon, D.F. Holt, S.V.F. Levy, M.S. Paterson, W. Thurston (1992). *Word Processing in Groups*, AKPeters, Natick, Mass.

[Epstein & Holt 1998]   D. B. A. Epstein and D. F. Holt, *Computation in word-hyperbolic groups*, preprint.

[Epstein, Holt, Rees, 1991]   D. B. A. Epstein, D. F. Holt and S. E. Rees, *The use of Knuth-Bendix methods to solve the word problem in automatic groups*, J. Symbolic Computation 12 (1991), 397–414.

[Holt, 1995]   Derek F. Holt, KBMAG—*Knuth-Bendix in Monoids and Automatic Groups*, software package (1995), available by anonymous ftp from ftp.maths.warwick.ac.uk in directory people/dfh/kbmag2.

[Holt, 1996]   D.F. Holt, *The Warwick automatic groups software*, in *Geometrical and Computational Perspectives on Infinite Groups* DIMACS Series in Discrete Mathematics and Theoretical Computer Science, vol. 25, ed. G. Baumslag et. al., 1996, 69–82.

[Holt & Hurt, 1997]   D.F. Holt and D.F. Hurt, *Automatic coset systems and subgroup presentations*, to appear in J. Symbolic Computation.

[Gilman, 1979]   R.H. Gilman, *Presentations of groups and monoids, J. Algebra* 57 (1979), 544-554.

[Gromov, 1987]   M. Gromov, *Hyperbolic groups*, in *Essays in Group Theory*, ed. S.M. Gersten, M.S.R.I. Publ. 8, Springer, 1987, 75–263.

[Kapovich, 1995]   Ilya Kapovich, *Detecting quasiconvexity: algorithmic aspects*, in: *Geometric and Computational Perspectives on Infinite Groups*, DIMACS Series in Discrete Mathematics and Theoretical Computer Science vol. 25, ed. Gilbert Baumslag et. al., 1995, 91–99.

[Not Knot]   Charlie Gunn and Delle Maxwell, *Not Knot*, video ISBN 0-86720-240-8, published by AKPeters, Natick, Mass., with a booklet by D.B.A. Epstein and C. Gunn.

[Papasoglu, 1994]   P. Papasoglu, *Strongly geodesically automatic groups are hyperbolic*, (Warwick preprint) (1994) (to appear).

[Wakefield, 1997]   P. Wakefield, *Procedures for Automatic groups*, Ph.D. Thesis, University of Newcastle upon Tyne, 1997.

David B. A. Epstein and
Derek F. Holt
Mathematics Institute
University of Warwick
Coventry CV4 7AL
dbae@maths.warwick.ac.uk
dfh@maths.warwick.ac.uk

CONSTRUCTING HYPERBOLIC MANIFOLDS

B. EVERITT AND C. MACLACHLAN

ABSTRACT. The Coxeter simplex with symbol ○═○─○─○═○ is a compact hyperbolic 4-simplex and the related Coxeter group Γ is a discrete subgroup of Isom($\mathbb{H}^4$). The Coxeter simplex with symbol ○─○─○═○ is a spherical 3-simplex, and the related Coxeter group G is the group of symmetries of the regular 120-cell. Using the geometry of the regular 120-cell, Davis [3] constructed an epimorphism $\Gamma \to G$ whose kernel K was torsion-free, thus obtaining a small volume compact hyperbolic 4-manifold $\mathbb{H}^4/K$.

In this paper we show how to obtain representations $\Gamma \to G$ of Coxeter groups Γ acting on $\mathbb{H}^n$ to certain classical groups G. We determine when the kernel K of such a homomorphism is torsion-free and thus $\mathbb{H}^n/K$ is a hyperbolic n-manifold. As an example, this is applied to the two groups described above, with G suitably interpreted as a classical group. Using this, further information on the quotient manifold is obtained.

1. INTRODUCTION

Let M^n be an n-dimensional hyperbolic manifold, that is, an n-dimensional Riemannian manifold of constant sectional curvature -1. Thus M^n is isometric to a quotient space $\mathbb{H}^n/K$ of $\mathbb{H}^n$ by the free action of a discrete group $K \cong \pi_1(M^n)$ of hyperbolic isometries.

This paper presents a method of constructing such groups K as the kernels of representations $\Gamma \to G$ of hyperbolic Coxeter groups Γ into finite classical groups. The homomorphisms arise by first representing Γ as a subgroup of the orthogonal group of a quadratic space over a number field k which preserves a lattice. Then reducing modulo a prime ideal in the ring of integers in the number field yields a representation into a finite classical group given as an orthogonal group of a quadratic space over a finite field.

Such kernels K will act freely if and only if they are torsion free. The volume of the resulting manifold $M^n = \mathbb{H}^n/K$ will be $N \times \text{Vol}(P)$, where N is the order of the image in the finite classical group and P is the polyhedron defining the Coxeter group Γ. Starting from a suitable Coxeter group Γ, the method yields infinitely many examples of manifolds. There has been some interest lately in constructing small volume examples when $n \geq 4$ (see [10, 11]). In dimension 4, the compact Davis manifold [3] is constructed by a geometric technique using the existence of a regular compact 120-cell in $\mathbb{H}^4$, which has volume $26 \times 4\pi^2/3$. As an application of our method, we construct a

1991 *Mathematics Subject Classification.* 57M50.

compact 4-manifold M_0 of the same volume which turns out to be isometric to the Davis manifold. With the help of computational techniques, our method gives additional information, producing a presentation for the fundamental group from which we obtain that $H_1(M_0) = \mathbb{Z}^{24}$.

2. Finite representations of hyperbolic Coxeter groups

Consider an $(n + 1)$-dimensional real space V equipped with a quadratic form q of signature $(n, 1)$. Thus with respect to an orthogonal basis,

$$(1) \qquad q(\mathbf{x}) = -x_{n+1}^2 + \sum_{i=1}^{n} x_i^2.$$

The quadratic form q determines a symmetric bilinear form

$$(2) \qquad B(\mathbf{x}, \mathbf{y}) := q(\mathbf{x} + \mathbf{y}) - q(\mathbf{x}) - q(\mathbf{y}),$$

with $B(\mathbf{x}, \mathbf{x}) = 2q(\mathbf{x})$. The Lobachevski (or hyperboloid) model of $\mathbb{H}^n$ is the positive sheet of the sphere of unit imaginary radius in V (see [12]). Equivalently, we take the projection of the open cone $C = \{\mathbf{x} \in V \mid q(\mathbf{x}) < 0 \text{ and } x_{n+1} > 0\}$ with the induced form.

The isometries of the model are the positive $(n + 1) \times (n + 1)$ Lorentz matrices, that is, the orthogonal maps of V, with respect to q, that map C to itself.

A hyperplane in $\mathbb{H}^n$ is the image of the intersection with C of a Euclidean hyperplane in V. Each hyperplane is the projective image of the orthogonal complement $\mathbf{e}^\perp = \{\mathbf{x} \in C \mid B(\mathbf{x}, \mathbf{e}) = 0\}$ of a vector $\mathbf{e}$ with $q(\mathbf{e}) > 0$. Such a vector is said to be space-like, and it is convenient to normalise so that $q(\mathbf{e}) = 1$. The map $r_{\mathbf{e}} : V \to V$ defined by

$$r_{\mathbf{e}}(\mathbf{x}) = \mathbf{x} - B(\mathbf{x}, \mathbf{e})\mathbf{e},$$

when restricted to $\mathbb{H}^n$, is the reflection in the hyperplane corresponding to $\mathbf{e}$.

A polyhedron P in $\mathbb{H}^n$ is the intersection of a finite collection of half-spaces, that is, the image of

$$\Lambda = \{\mathbf{x} \in C \mid B(\mathbf{x}, \mathbf{e}_i) \leq 0, i = 1, 2, \ldots, m\},$$

for some space-like vectors $\mathbf{e}_i$. The intersections of the hyperplanes $\mathbf{e}_i^\perp$ with P are the faces of the polyhedron. The dihedral angle θ_{ij} subtended by two intersecting faces of P is determined by $-2\cos\theta_{ij} = B(\mathbf{e}_i, \mathbf{e}_j)$. On the other hand, non-intersecting faces of P have a common perpendicular geodesic of length η_{ij}, where $-2\cosh\eta_{ij} = B(\mathbf{e}_i, \mathbf{e}_j)$.

All of this information is encoded in the Gram matrix $G(P)$ of P, an $m \times m$ matrix with (i, j)-th entry $a_{ij} = B(\mathbf{e}_i, \mathbf{e}_j)$. Let Γ be the group generated by the reflections $r_i := r_{\mathbf{e}_i}$ in the faces of P so that Γ is a subgroup of the isometry group $\mathrm{Isom}(\mathbb{H}^n)$. Moreover, Γ is discrete exactly when all the dihedral angles θ_{ij} of P are integer submultiples π/n_{ij} of π [12], and in this case, Γ is a hyperbolic Coxeter group. The polyhedron P is depicted by means of its

Coxeter symbol, with a node for each face, two nodes joined by $n - 2$ edges when the corresponding faces subtend a dihedral angle of π/n and other pairs of nodes joined by an edge labelled with the geodesic length between the faces. We use the Coxeter symbol to denote both P and the group Γ arising from it.

Using the Lobachevski model, such a hyperbolic Coxeter group Γ is a subgroup of $O(V, q)$. Using this, Vinberg [13] gave necessary and sufficient conditions for such a group to be arithmetic. We adopt Vinberg's method to conveniently describe the groups Γ, although they need not be arithmetic.

We first give this method and some general notation which we will use throughout. Let M be a finite-dimensional space over a field F. Equipped with a quadratic form f which induces a symmetric bilinear form (as for instance in (1) and (2)), M is a quadratic space over F. The group $O(M, f)$ of orthogonal maps consists of linear transformations $\sigma : M \to M$ such that $f(\sigma(m)) = f(m)$ for all $m \in M$.

Consider the Gram matrix $G(P) = [a_{ij}]$, and for any $\{i_1, i_2, \ldots, i_r\} \subseteq \{1, 2, \ldots, m\}$, define

$$b_{i_1 i_2 \cdots i_r} = a_{i_1 i_2} a_{i_2 i_3} \cdots a_{i_r i_1},$$

and let $k = \mathbb{Q}(\{b_{i_1 i_2 \cdots i_r}\})$. Take the space-like vectors in V defined by

$$\mathbf{v}_{i_1 i_2 \cdots i_r} = a_{1 i_1} a_{i_1 i_2} \cdots a_{i_{r-1} i_r} \mathbf{e}_{i_r}.$$

Let M be the k-subspace of V spanned by the $\mathbf{v}_{i_1 i_2 \ldots i_r}$. A simple calculation gives

$$(3) \qquad r_i(\mathbf{v}_{i_1 i_2 \cdots i_r}) = \mathbf{v}_{i_1 i_2 \cdots i_r} - \mathbf{v}_{i_1 i_2 \cdots i_r i},$$

and

$$(4) \qquad B(\mathbf{v}_{i_1 i_2 \cdots i_r}, \mathbf{v}_{j_1 j_2 \cdots j_s}) = b_{1 i_1 \cdots i_r j_s \cdots j_1}.$$

Thus, M is a quadratic space over k under the restriction of q, and from (3) and (4)

$$B(r_i(\mathbf{v}_{i_1 i_2 \cdots i_r}), r_i(\mathbf{v}_{j_1 j_2 \cdots j_s})) = B(\mathbf{v}_{i_1 i_2 \cdots i_r}, \mathbf{v}_{j_1 j_2 \cdots j_s}).$$

It follows that $\Gamma \to O(M, q)$.

Lemma 1. M is an $(n + 1)$-dimensional space over $k = \mathbb{Q}(\{b_{i_1 i_2 \cdots i_r}\})$.

Proof. If P has finite volume then the vectors $\mathbf{e}_i$ span V and the Gram matrix is indecomposable [13]. So for each i, there is a $j \neq i$ such that $a_{ij} \neq 0$. Successively choose indices $1 = i_0, i_1, \ldots$ such that the i_k-th row contains a non-zero entry in the (i_{k+1})-st column, for $k \geq 1$. We can ensure that the i_k are distinct. For, if the only non-zero entries of the k-th row are those in the columns with indices $1, i_1, \ldots, i_k$, throw away i_k and go back to the (i_{k-1})-st row to rechoose a different column. Eventually, by discarding and moving backwards, we must be able to rechoose, in the i_j-th row, an index different from all the $i_{j+1}, \ldots, i_k$ discarded. Otherwise, $\{\mathbf{e}_1, \ldots, \mathbf{e}_k\}$ are orthogonal to the other basis vectors, contradicting indecomposability. In this way we must arrive at a sequence $1 = i_0, i_1, \ldots, i_{m-1}$ of length m. Hence, for any i, $\mathbf{e}_i = \mathbf{e}_{i_k}$ for some i_k, and $\mathbf{v}_{i_1 \cdots i_k} = a_{1 i_1} a_{i_1 i_2} \cdots a_{i_{k-1} i_k} \mathbf{e}_{i_k}$ with coefficient non-zero. Thus, the vectors $\mathbf{v}_{i_1 i_2 \cdots i_r}$ span V over $\mathbb{R}$ and hence M is $(n+1)$-dimensional over $\mathbb{R}$. Now, if $\{\mathbf{v}_1, \ldots, \mathbf{v}_{n+1}\}$ is an $\mathbb{R}$-basis for M and $\mathbf{v} = \sum x_i \mathbf{v}_i \in M$, then the system of equations $B(\mathbf{v}, \mathbf{v}_j) = \sum x_i B(\mathbf{v}_i, \mathbf{v}_j)$ has a unique solution, since the matrix with (i, j)-th entry $B(\mathbf{v}_i, \mathbf{v}_j)$ is invertible. But the solutions $x_i \in k$, since $B(\mathbf{u}, \mathbf{v}) \in k$ for all $\mathbf{u}, \mathbf{v} \in M$. Thus $\{\mathbf{v}_1, \ldots, \mathbf{v}_{n+1}\}$ is a k-basis for M. $\qquad\square$

We make a number of simplifying assumptions which hold for many examples. Suppose that k is a number field and let $\mathcal{O}$ denote the ring of integers in k. Suppose furthermore that all $b_{i_1 \cdots i_r} \in \mathcal{O}$. Finally let N be the $\mathcal{O}$-lattice in M spanned by the elements $\mathbf{v}_{i_1 \cdots i_r}$ and assume that N is a free $\mathcal{O}$-lattice. This will hold, in particular, when $\mathcal{O}$ is a principal ideal domain.

By (3) above, N is invariant under Γ so that

(5) $$\Gamma \subset O(N, q) := \{\sigma \in O(M, q) \mid \sigma(N) = N\}.$$

With the restriction of q, N is a quadratic module over $\mathcal{O}$. If $\mathfrak{P}$ is any prime ideal in $\mathcal{O}$, let $\bar{k} = \mathcal{O}/\mathfrak{P}$. Reducing modulo $\mathfrak{P}$, we obtain a quadratic space $\bar{N}$ over $\bar{k}$ with respect to $\bar{q}$ and an induced map $\Gamma \to O(\bar{N}, \bar{q})$.

The groups $O(\bar{N}, \bar{q})$ are essentially the finite classical groups referred to earlier. However, the quadratic space $(\bar{N}, \bar{q})$ may not be a regular quadratic space, in which case we must factor out the radical to obtain a regular quadratic space (see Section 4 below). This will occur if the discriminant of $\bar{N}$ is zero. Since the discriminant of $\bar{N}$ is the image in $\bar{k}$ of the discriminant of N this will only occur for finitely many prime ideals $\mathfrak{P}$.

We now attend to the matter of when the kernel of a representation of Γ is torsion-free. In certain circumstances, this can be decided arithmetically using a small variation of a result of Minkowski (see for example [6, page 176]).

Lemma 2. *Let k be a quadratic number field, whose ring of integers $\mathcal{O}$ is a principal ideal domain. Let p be a rational prime. Let $\alpha \in \mathcal{O}$ be such that $\alpha \nmid 2$, and, if 3 is ramified in the extension $k \mid \mathbb{Q}$, then $\alpha \nmid 3$. If $A \in GL(n, \mathcal{O})$ is such that $A^p = I$ and $A \equiv I \pmod{\alpha}$, then $A = I$.*

Proof. Suppose $A \neq I$ so that $A = I + \alpha E$ where $E \in M_n(\mathcal{O})$ and we can take the g.c.d. of the entries of E to be 1. From $(I + \alpha E)^p = I$ we have

$$pE + \frac{p(p-1)}{2}\alpha E^2 \equiv 0 \pmod{\alpha^2}.$$

Thus $pE \equiv 0 \pmod{\alpha}$ and so $p \equiv 0 \pmod{\alpha}$. Since $\alpha \nmid 2$, p is odd. Suppose p is unramified in the extension $k \mid \mathbb{Q}$, then either $p = \alpha$ or $p = \alpha\alpha'$ with $\alpha' \in \mathcal{O}$ and $(\alpha, \alpha') = 1$. So $pE \equiv 0 \pmod{\alpha^2}$ so $E \equiv 0 \pmod{\alpha}$. This is a contradiction.

Now suppose that p is ramified. Then $p = u\alpha^2$ where $u \in \mathcal{O}^*$ and by assumption $p \neq 3$. Expanding as above, but to three terms, gives

$$u\alpha^2 E + u\alpha^2 \frac{p-1}{2}\alpha E^2 + u\alpha^2 \frac{(p-1)(p-2)}{6}\alpha^2 E^3 \equiv 0 \,(\mathrm{mod}\ \alpha^3).$$

This yields the contradiction $E \equiv 0 \,(\mathrm{mod}\ \alpha)$. □

Corollary 1. *If $\alpha \nmid 2$ and if 3 is ramified in $k \mid \mathbb{Q}$, $\alpha \nmid 3$, then the kernel of the mapping on $GL(n, \mathcal{O})$ induced by reduction $(\mathrm{mod}\ \alpha)$ is torsion-free.*

More generally, a geometrical argument allows us to determine when *any* representation has torsion-free kernel, albeit by expending a little more effort. Suppose $v \in P$ is a vertex of the polyhedron P, and Γ_v is the stabiliser in Γ of v. For P of finite volume, v is either in $\mathbb{H}^n$ or on the boundary, and v is called finite or ideal respectively. We have the following "folk-lore" result,

Lemma 3. *Suppose Γ is a discrete group generated by reflections in the faces of some polyhedron P in n-dimensional Euclidean space $\mathbb{E}^n$ or n-dimensional hyperbolic space $\mathbb{H}^n$. If $\gamma \in \Gamma$ is a torsion element, then for some vertex $v \in P$, γ is Γ-conjugate to an element of Γ_v.*

Notice that in the situation described in the lemma, a Coxeter symbol for Γ_v is obtained in the following way: take the sub-symbol of Γ with nodes (and their mutually incident edges) corresponding to faces of P containing v. For brevity's sake, when we say torsion element from now on, we will mean *non-trivial* torsion element.

Corollary 2. *The kernel of a representation $\Gamma \to G$ is torsion-free exactly when every torsion element of every vertex stabiliser Γ_v has the same order as its image in G.*

At this point the situation bifurcates into two cases: if $v \in P$ is a finite vertex, then Γ_v is isomorphic to a discrete group acting on the $(n-1)$-sphere S^{n-1} centered on v, hence is finite. Thus, the conditions of the corollary are satisfied exactly when Γ_v and its image in G have the same order.

If v is ideal, then consider a horosphere Σ based at v, and restrict the action of Γ_v to Σ. Then Σ is isometric to an $(n-1)$-dimensional Euclidean space $\mathbb{E}^{n-1}$, and Γ_v acts on it discretely with fundamental region P', the intersection with Σ of P. Any torsion element of Γ_v is then Γ_v-conjugate by Lemma 3 to the stabiliser (in Γ_v!) of a vertex v' of P'. Write $\Gamma_{v,v'}$ for this stabiliser, and observe that it is isomorphic to a discrete group acting on the $(n-2)$-sphere S^{n-2} in $\mathbb{E}^{n-1}$, centered on v', and hence is also finite. The conditions of the corollary are satisfied exactly when for each $v' \in P'$, the group $\Gamma_{v,v'}$ and its image in G have the same order.

Summarising,

Proposition 1. *Suppose* Γ *is a hyperbolic Coxeter group generated by reflections in the faces of a polyhedron P as above. For each finite vertex v of P, take the stabiliser Γ_v. For each ideal vertex, take the stabilisers $\Gamma_{v,v'}$ for each vertex v' of the Euclidean polyhedron P'. Then* $\ker(\Gamma \to G)$ *is torsion-free if and only if each such Γ_v and $\Gamma_{v,v'}$ has the same order as its image in G.*

It is an elementary process to verify the conditions of the proposition. For, each vertex stabiliser is a finite spherical reflection group of some lower dimension, hence from the well-known list (see [5], Section 2.11 for their orders). To find the orders of their images in G, the computational algebra package MAGMA is enlisted.

3. POLYHEDRA IN $\mathbb{H}^n$

Let P be a polyhedron in $\mathbb{H}^n$, thus the image of

$$\Lambda = \{\mathbf{x} \in C \mid B(\mathbf{x}, \mathbf{e}_i) \le 0, i = 1, 2, \ldots, m\},$$

for some space-like vectors $\mathbf{e}_i$. On occasion, a connected union of several copies of P will yield another polyhedron of interest. In particular, we may want to glue copies of P onto its faces using some of the reflections r_i as glueing maps.

Lemma 4. *If r_i is a reflection in a face of P, then*

$$\Lambda \cup r_i(\Lambda) = \Lambda' := \{\mathbf{x} \in C \mid B(\mathbf{x}, \mathbf{e}_j) \text{ and } B(\mathbf{x}, r_i(\mathbf{e}_j)) \le 0, \text{ for all } j \ne i\}.$$

Proof. If $\mathbf{x} \in \Lambda'$, then either $B(\mathbf{x}, \mathbf{e}_i) \le 0$, in which case $\mathbf{x} \in \Lambda$, or $B(\mathbf{x}, \mathbf{e}_i) > 0$, in which case $B(\mathbf{x}, r_i(\mathbf{e}_i)) \le 0$, hence $\mathbf{x} \in r_i(\Lambda)$. Conversely, if $\mathbf{x} \in \Lambda$, then $B(\mathbf{x}, \mathbf{e}_j) \le 0$ for all j. If $j \ne i$, then

$$B(\mathbf{x}, r_i(\mathbf{e}_j)) = B(\mathbf{x}, \mathbf{e}_j) - B(\mathbf{x}, \mathbf{e}_i)B(\mathbf{e}_i, \mathbf{e}_j) \le 0,$$

since all three terms are ≤ 0. A similar argument deals with the $\mathbf{x} \in r_i(\Lambda)$. $\square$

We illustrate the lemma by considering the situation in four dimensions. In particular, if P is a compact simplex it has Coxeter symbol one of the five depicted in Figure 1 (see [5], Section 6.9). In fact, and this explains the idiosyncratic numbering, $\text{Vol}(\Delta_i) < \text{Vol}(\Delta_j)$ if and only if $i < j$ (see [7]).

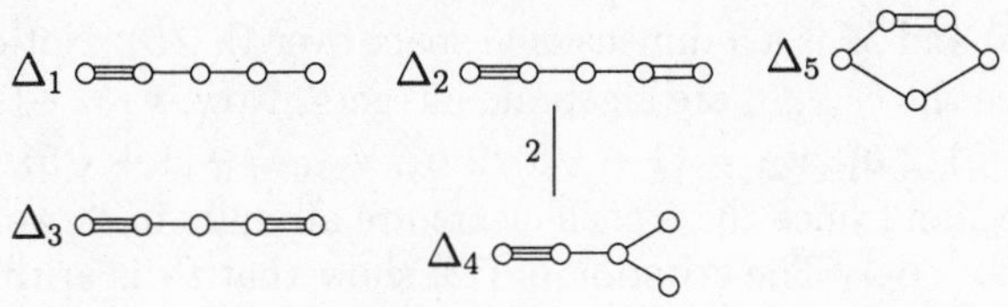

FIGURE 1

Suppose the nodes of Δ_2, read from left to right, correspond to hyperplanes $\mathbf{e}_i^\perp$ for $i = 1, \ldots, 5$. If $r_5 = r_{\mathbf{e}_5}$, we have

$$\Delta_2 \cup r_5(\Delta_2) = \{\mathbf{x} \in C \mid B(\mathbf{x}, \mathbf{e}_i) \leq 0, i = 1, \ldots, 4, \text{ and } B(\mathbf{x}, r_5(\mathbf{e}_4)) \leq 0\},$$

since $r_5(\mathbf{e}_i) = \mathbf{e}_i$ for $i = 1, 2$ and 3. Now, $B(\mathbf{e}_3, r_5(\mathbf{e}_4)) = -2\cos\pi/3$ and $B(\mathbf{e}_4, r_5(\mathbf{e}_4)) = -2\cos\pi/2$, so $\Delta_2 \cup r_5(\Delta_2)$ is a simplex with Coxeter symbol Δ_4. Thus, if Γ_i is the group generated by the reflections in the faces of Δ_i, we have that Γ_4 has index two in Γ_2. By comparing the volumes of the simplices using the results of [7], the only other possible inclusions are Γ_4 and Γ_3 as subgroups of indices 17 and 26 respectively in Γ_1. But a low index subgroups procedure in MAGMA shows that Γ_1 has no subgroups of these indices. Thus Figure 1 is a complete picture of the possible inclusions.

4. AN EXAMPLE

In this section, we apply our method in dimension 4 starting with the Coxeter simplex Δ_3 and related group Γ_3 described above. If P is a finite volume Coxeter polyhedron in $\mathbb{H}^4$, then $\mathrm{vol}(P) = \chi(P)4\pi^2/3$ where $\chi(P)$ is the Euler characteristic of P (see [4]), which coincides with the Euler characteristic of the associated group. This is readily computed from the Coxeter symbol [1, page 250], [2]. For Δ_3, the Euler characteristic is $26/14400$. The vertex stabilisers are $\circ\!-\!\circ\!-\!\circ\!=\!\circ$, $\mathbb{Z}_2 \times \circ\!-\!\circ\!=\!\circ$, $\circ\!=\!\circ \times \circ\!=\!\circ$, $\circ\!=\!\circ\!-\!\circ \times \mathbb{Z}_2$ and $\circ\!=\!\circ\!-\!\circ\!-\!\circ$, having orders $14400, 240, 100, 240$ and 14400 (see [5], Section 2.11). Thus the minimum index any torsion free subgroup of Γ_3 can have is 14400, and we show that there is a normal torsion free subgroup of precisely this index. The corresponding manifold then has Euler characteristic 26, making it the same volume as the Davis manifold [3]. Indeed, it has been shown in [9] that Γ_3 has a unique torsion-free normal subgroup of index 14400. It follows that this manifold is the Davis manifold.

Note that

$$G(\Delta_3) = \begin{pmatrix} 2 & -2\cos\pi/5 & 0 & 0 & 0 \\ -2\cos\pi/5 & 2 & -1 & 0 & 0 \\ 0 & -1 & 2 & -1 & 0 \\ 0 & 0 & -1 & 2 & -2\cos\pi/5 \\ 0 & 0 & 0 & -2\cos\pi/5 & 2 \end{pmatrix},$$

so $k = \mathbb{Q}(\sqrt{5})$ and M is a 5-dimensional space over $\mathbb{Q}(\sqrt{5})$. Notice that all the a_{ij}, and hence the $b_{i_1 i_2 \cdots i_r}$, are algebraic integers. Now, $\mathbf{v}_2 = -(1+\sqrt{5})/2\,\mathbf{e}_2$, $\mathbf{v}_{21} = (3+\sqrt{5})/2\,\mathbf{e}_1$, $\mathbf{v}_{23} = (1+\sqrt{5})/2\,\mathbf{e}_3$, $\mathbf{v}_{234} = -(1+\sqrt{5})/2\,\mathbf{e}_4$, $\mathbf{v}_{2345} = (3+\sqrt{5})/2\,\mathbf{e}_5$, and since these coefficients are all units in $\mathcal{O}$, we have $N = \mathcal{O}$-span of $\{\mathbf{e}_1, \ldots, \mathbf{e}_5\}$. The criterion of [13] show that Γ_3 is arithmetic.

Let $\mathfrak{P} = \sqrt{5}\mathcal{O}$, so that $\bar{k}$ is the field $\mathbb{F}_5$ of five elements. By Lemma 2, the kernel of $\Gamma \to O(\bar{N}, \bar{q})$ is torsion free. Since $(1+\sqrt{5})/2 \equiv -2 \pmod{\mathfrak{P}}$, we

have that
$$\bar{q}(\mathbf{x}) = x_1^2 + 2x_1x_2 + x_2^2 - x_2x_3 + x_3^2 - x_3x_4 + x_4^2 + 2x_4x_5 + x_5^2.$$

We use the same letters $\{\mathbf{e}_1, \mathbf{e}_2, \ldots, \mathbf{e}_5\}$ for the basis of $\bar{N}$. The images of the generating reflections of Γ_3 are then 5×5 matrices with entries in $\mathbb{F}_5$. The computational system MAGMA then shows that the group they generate has order 14400 so that the kernel has index 14400 in Γ_3 as required.

The index 14400 is too large to allow MAGMA to implement the Reidemeister-Schreier process to obtain a presentation for K. However, closer examination of the image group allows this process to be implemented by splitting into two steps. We will deal with this now.

The bilinear form $\bar{B}$ on $\bar{N}$ is degenerate and there is a one-dimensional radical $\bar{N}^\perp$ spanned by $\mathbf{v}_0 = \mathbf{e}_1 - \mathbf{e}_2 + \mathbf{e}_4 - \mathbf{e}_5$. Thus $\bar{N} = W \oplus \bar{N}^\perp$. If $\mathbf{w} \in W$ and $\sigma \in O(\bar{N}, \bar{q})$, then $\sigma(\mathbf{w}) = \mathbf{w}' + t\mathbf{v}_0$ where $\mathbf{w}' \in W$ and $t \in \mathbb{F}_5$. The induced mapping $\bar{\sigma}$ defined by $\bar{\sigma}(\mathbf{w}) = \mathbf{w}'$ is easily seen to be an orthogonal map on W and we obtain a representation $\Gamma_3 \to O(W, \bar{q})$. We now identify $O(W, \bar{q})$ as one of the classical finite groups using the notation in [8]. Let $\mathbf{g}_i, \mathbf{h}_i \in W$, for $i = 1, 2$ be defined by $\mathbf{g}_1 = \mathbf{e}_1 - \mathbf{e}_2$, $\mathbf{h}_1 = -\mathbf{e}_1 + \mathbf{e}_2 + \mathbf{e}_3$, $\mathbf{g}_2 = \mathbf{e}_1 + 2\mathbf{e}_5$ and $\mathbf{h}_2 = -\mathbf{e}_1 + 2\mathbf{e}_5$. Then $\bar{q}(\mathbf{g}_i) = \bar{q}(\mathbf{h}_i) = 0$ and $\bar{B}(\mathbf{g}_i, \mathbf{h}_j) = \delta_{ij}$. Thus $O(W, \bar{q}) \cong O_4^+(5)$. There is a chain of subgroups
$$1 \overset{2}{\subset} Z \overset{3600}{\subset} \Omega_4^+(5) \overset{2}{\subset} SO_4^+(5) \overset{2}{\subset} O_4^+(5),$$
where Z is the largest normal soluble subgroup of $\Omega_4^+(5)$ and

(6)
$$\Omega_4^+(5) \cong \frac{SL(2,5) \times SL(2,5)}{\langle(-I, -I)\rangle}.$$

The image of Γ_3 is isomorphic to a subgroup of index 2 in $O_4^+(5)$, different from $SO_4^+(5)$ and the orientation-preserving subgroup Γ_3^+ maps onto $\Omega_4^+(5)$. The target group has a normal subgroup of index 60 with quotient isomorphic to $PSL(2, 5)$ and hence so does Γ_3^+. Using MAGMA we find a presentation for this subgroup K_1 with three generators and nine relations. The group K is then the kernel of the induced map from K_1 onto $SL(2, 5)$. Again using MAGMA, we obtain a presentation for K on 24 generators and several pages of relations. The abelianisation of K is $\mathbb{Z}^{24}$. This agrees with the homology calculations in [9].

This calculation is readily carried out once the images of the generators of Γ_3^+ are identified with pairs of matrices. We sketch the method of obtaining this description.

Let V be a two dimensional space over $\mathbb{F}_5$ with symplectic form f defined with respect to a basis $\mathbf{n}_1, \mathbf{n}_2$ by
$$f\left(\sum x_i\mathbf{n}_i, \sum y_i\mathbf{n}_i\right) = x_1y_2 - x_2y_1.$$

Let $U = V \otimes V$ and define g on U by
$$g(\mathbf{v}_1 \otimes \mathbf{v}_2, \mathbf{w}_1 \otimes \mathbf{w}_2) = f(\mathbf{v}_1, \mathbf{w}_1)f(\mathbf{v}_2, \mathbf{w}_2).$$

Then g is a symmetric bilinear form on U and $O(U, g) \cong O_4^+(5)$. Note that $SL(2,5) \times SL(2,5)$ acts on U by

$$(\sigma, \tau)(\mathbf{v} \otimes \mathbf{w}) = \sigma(\mathbf{v}) \otimes \tau(\mathbf{w}),$$

and this action preserves g with $(-I, -I)$ acting trivially. This describes the group $\Omega_4^+(5)$. Additionally, the mapping $\rho : U \to U$ given by $\rho(\mathbf{v} \otimes \mathbf{w}) = \mathbf{w} \otimes \mathbf{v}$ also lies in $O(U, g)$ and has determinant -1. Let H be the subgroup generated by $\Omega_4^+(5)$ and ρ.

We identify the image of Γ_3 with H, by first identifying U and W by the linear isometry induced by

$$\mathbf{n}_1 \otimes \mathbf{n}_1 \mapsto \mathbf{g}_1, \mathbf{n}_1 \otimes \mathbf{n}_2 \mapsto \mathbf{g}_2, \mathbf{n}_2 \otimes \mathbf{n}_1 \mapsto -\mathbf{h}_2, \mathbf{n}_2 \otimes \mathbf{n}_2 \mapsto \mathbf{h}_1.$$

It is now easy to check that the image of r_5 is ρ. Recall that Γ_3^+ is generated by $x = r_5 r_4, y = r_5 r_3, z = r_5 r_2, w = r_5 r_1$. Now determine the images of x, y, z, w as pairs of matrices in $SL(2,5) \times SL(2,5)$.

References

[1] K S Brown. *Cohomology of groups.* Graduate Texts in Mathematics no 87, Springer 1982.

[2] I M Chiswell. The Euler characteristic of graph products and of Coxeter groups, in "Discrete groups and geometry", W J Harvey and C Maclachlan editors, LMS Lecture Note Series 173, 36-46, 1992.

[3] M W Davis. A hyperbolic 4-manifold. *Proc. Amer. Math. Soc.*, 93 (1985), 325–328.

[4] M Gromov. Volume and bounded cohomology. *Inst. Hautes Études Sci. Publ. Math.*, 56 (1982), 5-99.

[5] J E Humphreys. *Reflection groups and Coxeter groups.* Cambridge Advanced studies in Mathematics 29, CUP 1990.

[6] M Newman. *Integral Matrices.* Academic Press, New York 1972.

[7] N W Johnson, R Kellerhals, J G Ratcliffe and S T Tschantz. The size of a hyperbolic Coxeter simplex. *Transformation Groups*, to appear.

[8] P Kleidman and M Liebeck. *The subgroup structure of the finite classical groups.* London Mathematical Society Lecture Notes 129, CUP 1990.

[9] J G Ratcliffe and S T Tschantz. On the Davis hyperbolic 4-manifold. preprint.

[10] J G Ratcliffe and S T Tschantz. Gravitational instantons of constant curvature. preprint.

[11] J G Ratcliffe and S T Tschantz. The volume spectrum of hyperbolic 4-manifolds. preprint.

[12] J G Ratcliffe. *Foundations of hyperbolic manifolds.* Graduate Texts in Mathematics 149, Springer 1994.

[13] E B Vinberg. *Discrete groups in Lobachevskii spaces generated by reflections.* Math. USSR-Sb. 1 (1967), 429–444.

B. Everitt   C. Maclachlan
Department of Mathematics   Department of Mathematical Sciences
University of York   University of Aberdeen
Heslington, York YO1 5DD   Aberdeen AB24 3UE
England   Scotland
bje1@york.ac.uk   cmac@maths.abdn.ac.uk

COMPUTING IN GROUPS WITH EXPONENT SIX

GEORGE HAVAS, M. F. NEWMAN, ALICE C. NIEMEYER
AND CHARLES C. SIMS

ABSTRACT. We have investigated the nature of sixth power relations required to provide proofs of finiteness for some two-generator groups with
exponent six. We have solved various questions about such groups using
substantial computations. In this paper we elaborate on some of the calculations and address related problems for some three-generator groups
with exponent six.

1. INTRODUCTION

Motivated by an aim to get estimates for the number and length of sixth
power relations which suffice to define groups with exponent six, we studied
finiteness proofs for presentations of such groups in [5]. We tried to find
relatively small sets of defining relations for various groups, with a view to
improving our understanding of finiteness proofs.

We denote the free group on d generators with exponent n by $B(d, n)$ and
generally use notation as in [5]. One question we would very much like to
be able to answer is whether $B(2, 6)$ can be defined without using too many
sixth powers. Here we focus on the computational components of the process,
giving sample code which solves some associated problems.

We showed that $B(2, 6)$ has a presentation on 2 generators with 81 relations, which is derived from a polycyclic presentation. Here in Section 3 we
give a program to construct a polycyclic presentation for $B(2, 6)$ which shows
the structure of the group. If only sixth power relations are used, we showed
that M. Hall's finiteness proof [2] yields that fewer than 2^{124} sixth powers
can define $B(2, 6)$ [5, Theorem 2]. On the other hand the best lower bound
we have proved is that at least 22 sixth powers are needed [5, Theorem 1].
We expect that 22 is closer to the truth than 2^{124}. We observed that current
proof methods cannot yield fewer than 2^{95} sixth powers in a defining set of
relators.

It seems unlikely that the computer-based methods now available — coset
enumeration and Knuth-Bendix string rewriting — will succeed in finding

We thank Bettina Eick for help with GAP and we thank Greg Gamble for help with both
GAP and MAGMA programs. The first and second authors were partially supported by the
ANU Institute of Advanced Studies/Other Universities Research Collaboration Scheme.
The first and third authors were partially supported by the Australian Research Council.

'small' defining sets of sixth powers for $B(2,6)$ in the foreseeable future. Because of this we considered presentations related to some quotients of $B(2,6)$ for which coset enumeration and rewriting methods enable progress to be made. In a sense this is what Hall did. For him a major stepping stone was a lemma about a three-generator group with exponent six which proved to be finite of order 54 [2, Lemma 4]. He described this as the hardest lemma in his proof [3, p. 338]. It is easy to prove using coset enumeration or string rewriting.

We considered group presentations $\{X \mid \mathcal{R}\}$ with $X, \mathcal{R}$ finite to which the exponent six condition is added; we write $\{X \mid \mathcal{R}, \exp 6\}$. The groups defined in this way are finite and so a finite set of sixth power relators suffices to enforce the exponent six condition. Thus we studied presentations $\{X \mid \mathcal{R}, \mathcal{S}\}$ where $\mathcal{S}$ is a finite set of sixth powers and in some interesting cases found small sets $\mathcal{S}$ for which the groups $\langle X \mid \mathcal{R}, \mathcal{S} \rangle$ and $\langle X \mid \mathcal{R}, \exp 6 \rangle$ defined by the presentations are isomorphic.

Let $C(r, s)$ denote the largest two-generator group with exponent six generated by elements of orders r and s. To understand $B(2,6)$ better we looked at presentations for the groups $C(2,2)$, $C(2,3)$, $C(3,3)$, $C(2,6)$ and $C(3,6)$. Let $\{a, b\}$ be a generating set for $B(2,6)$. The subgroup $H = \langle a^{6/r}, b^{6/s} \rangle$ of $B(2,6)$ is clearly a quotient group of $C(r, s)$. It turns out that H and $C(r, s)$ are isomorphic. The order of H and the index of the normal closure in $B(2,6)$ of $\langle a^r, b^s \rangle$ are easily computed using a polycyclic presentation for $B(2,6)$. In each case these numbers are the same. Programs for some of this are given in Section 3.

The presentation $\{a, b \mid a^2, b^2, (ab)^6\}$ is a minimal presentation for the group $C(2,2)$, the dihedral group of order 12. A minimal presentation for $C(2,3)$, a group of order 216, is $\{a, b \mid a^2, b^3, (ab)^6, [a, b]^6\}$. Therefore the first challenging case to consider is the group $C(3,3)$. It is a group of order $2^{10}3^3$. It turns out to be quite manageable. In this paper we concentrate on computational issues; the main matters can be explained in the context of $C(3,3)$ (Section 4), so we restrict attention to that case. However, we also have some results for $C(2,6)$ and $C(3,6)$ in [5].

In Section 5 we look at the three-generator group which is the key to Hall's finiteness proof. We finish (Section 6) by providing new results about presentations for another three-generator group with exponent six. The group is the relatively free product of a symmetric group on three symbols and a group of order 3; surprisingly it embeds in $B(2,6)$.

2. Computational tools

The major computational tools we use are outlined in [8]. We exhibit current capabilities of various computational methods and illustrate the use of a powerful new tool, a soluble quotient program, see Niemeyer [11]. Coset enumeration can now routinely handle hundreds of millions of cosets; the implementation used is the one by Havas and Ramsay [6], based on the description in [4]. For rewriting (see [13, Chapter 2]) we use the Rutgers Knuth-Bendix Package (RKBP) written by Sims [14]. Such programs are available as standalone packages [8] and in systems such as GAP [12], MAGMA [1] and **quotpic** [9]. We provide some programs, together with input files for and outputs from these systems.

We present real code used at the time of the Conference to solve the given problems. Thus we give a snapshot of capabilities as at July 1998. With continuing improvements in systems such as GAP and MAGMA we would expect these problems could be addressed entirely within one system.

3. A PRESENTATION FOR $B(2, 6)$

In [5] we gave a polycyclic presentation for $B(2, 6)$ based on a composition series with 53 generators and 1431 relations; this presentation shows much of the group's structure. The presentation is consistent and was obtained using the ANU Soluble Quotient Program (ANU SQ) [11] and GAP. Here we detail a GAP program which computes this presentation. The basic approach is the same as that used by Stephen Glasby, the first person to obtain a polycyclic presentation for $B(2, 6)$. We wanted a presentation which exhibits the structure of the group in a readily visible form, and this needs more effort.

Let F be the free group of rank 2 on $\{a, b\}$. Then F has a normal subgroup F^3 which is the subgroup generated by all cubes in F. Since F/F^3 has order 27, it follows from Schreier's Theorem that F^3 is free of rank 28 and hence F^3/M, where $M = (F^3)^2$, is an elementary abelian 2-group of order 2^{28}. Let G_1 denote F/M. Then G_1 has order $2^{28}3^3$ and has exponent six. Let G_2 denote F/L, where $L = (F^2)^3$. Then G_2 has order $2^2 3^{25}$ and exponent six. The group $B(2, 6)$ is isomorphic to $F/(M \cap L)$ (see [5]). Moreover, $F/(M \cap L)$ embeds into the direct product $F/M \times F/L$. Clearly $F/M \times F/L$ is a group with exponent six. The two-generator subgroup of it generated by (aM, aL) and (bM, bL) is isomorphic to $B(2, 6)$.

Using ANU SQ we computed consistent power conjugate presentations for the groups G_1 and G_2 and epimorphisms $\kappa : F \to G_1$ and $\lambda : F \to G_2$. This is straightforward. We use enough sixth powers to ensure that the computed quotients are isomorphic to G_1 and G_2.

```
##   Use the  ANU SQ  package to compute  consistent power conjugate
##   presentations for  the  groups   G1  and   G2   and  epimorphisms
##   lambda: F -> G1   and   kappa: F -> G2   where F is free of rank 2
##
RequirePackage("anusq");
F := FreeGroup("a", "b");; a:=F.1;; b:=F.2;;
G1 := AgGroupFpGroup( Sq(
            F / [ a^6, b^6 ],
            [ [3, 2], # 2 steps in the exponent-3 series
              [2, 1]  # 1 step in the exponent-2 series
              ] ));
```

G. Havas et al

```
lambda := GroupHomomorphismByImages( F, G1, [F.1, F.2], [G1.1, G1.2] );

a:=F.1;;   # Needed because Sq redefines a
G2 := AgGroupFpGroup( Sq(
          F / [ a^6, b^6, (a*b)^6, (a*b^-1)^6, (a^2*b^2)^6, (a^2*b^-2)^6,
               (a*b*a*b^-1)^6, (a*b*a^-1*b)^6, (a*b*a^-1*b^-1)^6,
               (a^3*b*a*b)^6 , (a^-2*b^2*a*b*a*b)^6 ],
          [ [2, 1], # 1 step in the exponent-2 series
            [3, 3]  # 3 steps in the exponent-3 series
            ] ) );
kappa := GroupHomomorphismByImages( F, G2, [F.1, F.2], [G2.1, G2.2] );
```

It is not too difficult to obtain a presentation for $B(2,6)$ from here. However it is more challenging to find one which shows much of the group's
structure. The GAP code in Appendix A computes such a presentation. It
uses two auxiliary functions Action (Appendix B) and BaseChangeAgGroup
(Appendix C). Function Action rewrites the conjugation action of the generators of a section of a group on an elementary abelian p-subgroup as matrices
over $GF(p)$. Function BaseChangeAgGroup computes an automorphism of a
given soluble group mapping the GAP-computed polycyclic presentation to
one based on a user supplied polycyclic generating set.

We start in Appendix A by computing I, a special polycyclic presentation
for $B(2,6)$ inside the direct product, and chi, a homomorphism from the
free group on 2 generators to I. Next we compute matrices for the action of
generators of $B(2,3)$ on the elementary abelian group of order 2^{28}. We then
completely reduce the module into two trivial, four 2-dimensional and three
6-dimensional submodules.

Next we compute the images in the direct product $G_1 \times G_2$ of generating
sets of these submodules. Then, in a similar fashion, we act with the 2 generators of order 2 on an elementary abelian section of order 3^5 and compute
the images in $G_1 \times G_2$ of the generators of the submodules in the decomposition of this module. Finally, this allows us to define a new polycyclic
presentation for $B(2,6)$ exhibiting this structure. With the GAP function
BaseChangeAgGroup we compute a homomorphism chi from the free group
on a and b into $B(2,6)$.

We are able to compute in such presentations for $B(2,6)$ quite readily. For
example, we can compute the order of a subgroup mentioned in Section 1 as
follows.

```
b1 := Image( chi, chi.source.1 );
b2 := Image( chi, chi.source.2 );
C26 := Subgroup( B26, [ b1^3, b2 ] );
# Order of C(2,6) is
Size(C26);
```

4. THE GROUP $C(3,3)$

As pointed out in the introduction, the first challenging quotient of $B(2,6)$ obtained by applying order conditions on the generators is the factor group $C(3,3)$ given by the presentation $\{a, b \mid a^3, b^3, \exp 6\}$. One effective way of studying this group is to use **quotpic**, which provides useful pictures of groups and their quotients.

The order of $C(3,3)$ can easily be computed using finiteness and the fact [3, Theorem 18.4.6] that the 2-length and the 3-length are 1. From this it follows that a Sylow 3-subgroup of $C(3,3)$ is isomorphic to the Sylow 3-subgroup of $(C(3,3)^2)^3$ and the Sylow 2-subgroup to that of $(C(3,3)^3)^2$. Since $C(3,3)^2 = C(3,3)$, a Sylow 3-subgroup is (isomorphic to) $B(2,3)$ and has order 27.

Give **quotpic** the presentation $\{a, b \mid a^3, b^3\}$. Let F be the group this presentation defines. Then F^3 is the kernel of the largest 3-quotient of class 2. The kernel is free of rank 10. Hence $F/(F^3)^2$ has order 27×2^{10}. Therefore the order of $C(3,3)$ is 27×2^{10}. Using **quotpic** to calculate the intermediate quotients between F^3 and $(F^3)^2$ shows (Figure 1) that $C(3,3)^3$ is the direct sum of three irreducible $B(2,3)$-modules with dimensions 2, 2 and 6.

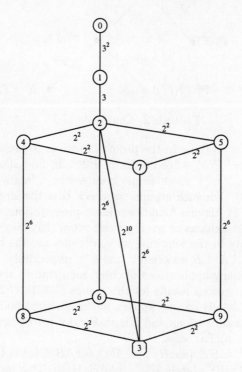

FIGURE 1. F^3 and $(F^3)^2$.

The diagram in Figure 1 is produced by **quotpic** (after judicious movement of vertices), where vertex 2 represents F^3 and vertex 3 represents $(F^3)^2$. The primary interface to **quotpic** is graphical and hard to describe in words. Suffice it to say, the input used to produce this diagram comprised a file with 22 non-whitespace characters plus a small number of mouse clicks (less than a dozen if vertex movement is not counted).

We showed [5, Theorem 9] that every presentation with the two relators a^3 and b^3 and otherwise sixth powers needs at least seven relators. The proof consists of a careful analysis of quotients of the group $F = \langle a, b \mid a^3, b^3 \rangle$ which appear in Figure 2 (which extends the part of Figure 1 without the intermediate quotients).

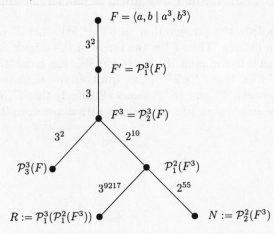

FIGURE 2. Quotients of F

Methods similar to those in the proof allowed us to use a polycyclic presentation for $Q := F/N$, whose order is $3^3 2^{65}$, to find short presentations $\{a, b \mid a^3, b^3, w_1^6, \ldots, w_n^6\}$ which define groups whose largest meta-nilpotent quotient is $C(3,3)$. For such groups, we check that the largest soluble quotient is also $C(3,3)$. Having found candidate presentations with the correct soluble quotient, it remains to see if we can prove that they actually present $C(3,3)$. For brevity in the following we sometimes use the case inverse convention in which A and B denote a^{-1} and b^{-1}, respectively.

In [5] we use many applications of coset enumeration and Knuth-Bendix rewriting to prove various results for the groups $C(3,3)$, $C(2,6)$ and $C(3,6)$. Here we provide details on one application of each method to $C(3,3)$. All successful coset enumerations and Knuth-Bendix rewritings reported in [5] have been done in similar ways.

Let $S_1 = \{(ab)^6, (aB)^6, (abaB)^6, (abAb)^6, (abAB)^6, (ababAB)^6, (abaBAb)^6, (abaBAB)^6, (abAbaB)^6, (abAbAB)^6, (abABaB)^6\}$. This includes one representative of all words of length up to six which are different as potential relators for $C(3,3)$. We proved $\Pi_1 = \{a, b \mid a^3, b^3, \mathcal{S}_1\}$ is a presentation for $C(3,3)$ by coset enumeration and by rewriting. The following MAGMA program does coset enumeration over the subgroup $\langle a, (ab)^3 \rangle$ and prints out the index plus some statistics. The performance of coset enumeration depends very much on the sequence of coset definitions. The values of the parameters Strategy and SubgroupRelations which appear were obtained by guided experimentation. These parameters control the sequence of coset definitions and yield the results quoted here. For further details on coset enumeration parameters see [7].

```
C33<a,b> := Group< a, b | a^3, b^3, (a*b)^6, (a*b^-1)^6, (a*b*a*b^-1)^6,
          (a*b*a^-1*b)^6, (a*b*a^-1*b^-1)^6, (a*b*a*b*a^-1*b^-1)^6,
          (a*b*a*b^-1*a^-1*b)^6, (a*b*a*b^-1*a^-1*b^-1)^6,
          (a*b*a^-1*b*a*b^-1)^6, (a*b*a^-1*b*a^-1*b^-1)^6,
          (a*b*a^-1*b^-1*a*b^-1)^6 >;
H := sub< C33 | a, (a*b)^3 >;
time I,CT,M,T := ToddCoxeter(C33, H : CosetLimit:=5000000,
              Strategy:=<4300,5>, SubgroupRelations:=2 );
print I, M, T;
```

In less than four minutes on a moderate speed workstation we obtain index 1152 with a maximum of 3245801 and total of 3417675 cosets defined. Together with a theoretical argument given in [5], this shows that Π_1 presents $C(3,3)$.

Now consider $\mathcal{S}_2 = \{(ab)^6, (aB)^6, (abaB)^6, (abAb)^6, (abAB)^6, (abaBAb)^6, (aBaBAb)^6, (aBabAB)^6\}$. This yields a ten-relator presentation for $C(3,3)$ which includes 8 sixth powers. By applying RKBP to $\{a, b \mid a^3, b^3, \mathcal{S}_2\}$ we obtain a confluent presentation from which we can compute the group order and much other information. The confluent rewriting system has 5016 rules with longest left and right hand sides of length 22. This calculation is much harder than for $\mathcal{S}_1$, taking about 200 cpu minutes on a moderate speed PC. The following RKBP input together with three auxiliary files suffices to do the computation.

```
echo
input c33
!cat c33.rl1
summary
kb 28 -1 28 28
summary
quit
```

Suitable auxiliary files are shown next in columns headed by file names.

c33.rl1	c33.sb1	c33.sys
3 a^3 0 @	subsystem	system
3 b^3 0 @	rule_type: subword	word_type: group
12 (ab)^6 0 @	index_type: automaton	ngens: 2
12 (aB)^6 0 @	external: len_names	naming_convention: none
24 (abaB)^6 0 @	compressed: no	numbering_convention: none
24 (abAb)^6 0 @	rule_file: c33.rl1	inverse_convention: none
24 (abAB)^6 0 @		generators: a A b B
36 (abaBAb)^6 0 @		inverse_pairs: a A b B
36 (aBaBAb)^6 0 @		weight_convention: constant
36 (aBabAB)^6 0 @		level_convention: constant
		nsub: 1
		subsystems: c33.sb1

5. A GROUP OF ORDER 54

Hall [2, p. 771] wrote:

> If $H = \{x, a, b\}$ is of exponent six, and if $x^2 = 1$, $a^3 = 1$,
> $b^3 = 1$, $xax = a^{-1}$, $xbx = b^{-1}$, then $\{a, b\}$ is of exponent three.

This lemma is critical since we note that $[H : H'] = 2$, and
so if H is finite, then $H' = \{a, b\}$ must be of exponent three
and so of order 27 (or naturally a divisor of 27). Thus if H is
finite, its order divides 54.

Hall takes over three pages to prove the lemma, which is readily proved
by coset enumeration using four sixth powers (see [10]). A MAGMA program
(including an implicit coset enumeration) which shows that this group has
order 54 is trivial:

```
Hall54<x,a,b> := Group< x, a, b | x^2, a^3, b^3, (x*a)^2, (x*b)^2,
            (a*b)^6, (a*b^-1)^6, (x*a*b)^6, (x*(a*b)^3)^6 >;
print Order(Hall54);
```

Following one theme, we ask how few sixth powers are required in this
context. It is straightforward to show that at least three are required using a
module-based argument. Can it be done with three?

Arguments like those in [5, Proof of Theorem 9] tell us that if it can be
done with three sixth powers they will have the forms $(abu)^6$, $(aBv)^6$ and
$(xw)^6$, where both u and v are in the derived subgroup of $\langle a, b \rangle$ and w in
$\langle a, b \rangle$. For simplicity we select u and v trivial and consider possibilities for w.
Let $\Pi_3 = \{\ x, a, b \mid x^2, a^3, b^3, (xa)^2, (xb)^2, (ab)^6, (aB)^6, (xw)^6\ \}$. It is generally easier to experiment with two-generator groups rather than with three-
generator groups. In this case the subgroup $\langle a, b \rangle$ of $\langle \Pi_3 \rangle$ has index 2, so we
study it. Reidemeister-Schreier rewriting shows that this subgroup has the
presentation $\Pi_4 = \{\ a, b \mid a^3, b^3, (ab)^6, (aB)^6, (w\bar{w})^3\ \}$, where $\bar{w}$ is obtained
from w by replacing each symbol in the word w by its inverse. Further Π_3
presents the required group if and only if Π_4 is a presentation for $B(2, 3)$.

We can readily test whether there are candidate presentations of the form
Π_4. We consider words w in the free product F of two groups of order 3

defined by $\{a, b \mid a^3, b^3\}$. We construct words as alternating products of $a^{\pm 1}$ and $b^{\pm 1}$, normalised to start with a.

For all such words w of length less than eight we find that the largest soluble quotient of $\langle \Pi_4 \rangle$ is not $B(2,3)$. At length eight we find 16 words $w_1, \ldots, w_{16}$ which do yield the required soluble quotient. The following simple MAGMA code suffices for this purpose.

```
F<a,b> := FreeGroup(2);
for x2, x4, x6, x8 in [b, b^-1], x3, x5, x7 in [a, a^-1] do
  relns := [ a^3, b^3, (a*b)^6, (a*b^-1)^6,
    (a*x2*x3*x4*x5*x6*x7*x8*(x8*x7*x6*x5*x4*x3*x2*a)^-1)^3 ];
  G := quo< F | relns >;
  Q := pQuotient (G, 3, 3: Print := 0);
  if Order (Q) eq 27 then
    D := ncl < G | (G.2, G.1, G.1), (G.2, G.1, G.1) >;
    x := AbelianQuotientInvariants (D);
    if x eq [] then print "POSSIBLE", G; end if;
  end if;
end for;
```

The 16 words w_i that we obtain are: $abaBAbab$, $abaBABab$, $abAbaBab$, $abAbaBAb$, $abABabaB$, $abABaBAb$, $abABAbab$, $abABAbaB$, $aBabAbaB$, $aBabABaB$, $aBAbabAB$, $aBAbaBab$, $aBAbABab$, $aBAbABaB$, $aBABabaB$, $aBABabAB$. Some of the words $w_i \bar{w}_i$ are conjugate in the free group, meaning that we do not get 16 "essentially different" presentations. However, since coset enumeration can handle presentations with even minor perturbations very differently we chose to consider them all.

We are unable to determine whether any of the groups $\langle \Pi_4 \rangle$ derived from these words is finite. It is interesting that four of the words $w_i \bar{w}_i$ can be written as left-normed commutators of length four. Thus from w_7, w_8, w_{13} and w_{14} we obtain $[a, B, A, B]$, $[a, B, A, b]$, $[a, b, A, B]$ and $[a, b, A, b]$, respectively. In a sense this explains why proving finiteness (should that be the case) is hard. We would need to derive the short relators $(ab)^3$ and $(aB)^3$ which hold in $B(2,3)$ from the initial four relators together with a relator like $[a, B, A, B]^3$.

6. ANOTHER THREE-GENERATOR GROUP

Now that we have looked at one three-generator group with exponent six we consider another which also arises in a natural way and which may contribute to further study of presentations for $B(2,6)$. It is the relatively free product of the symmetric group S_3 on three letters and a group of order 3; we call it $C(S_3, Z_3)$. It has presentation $\Pi_5 = \{a, b, c \mid a^2, b^3, (ab)^2, c^3, \exp 6\}$, and its order is 17496.

It is not immediately obvious that this group embeds in $B(2,6)$ but a few random choices produce the following example. Let x and y be generators for $B(2,6)$. Then $\langle x^3, (x^3 y^3)^2 \rangle$ is an S_3 and with $(xyx)^2$ it generates a subgroup

H which is a quotient of $C(S_3, Z_3)$. We compute that H has order 17496, so $C(S_3, Z_3)$ embeds in $B(2, 6)$.

We now show how to find a small set of sixth powers instead of the exponent six condition in Π_5 to present $C(S_3, Z_3)$. Our process makes use of subtle relations which hold in $C(3, 3)$. A relation is *subtle* if it is shorter than the longest sixth power required for a proof that the relation holds in a suitable context (for more on subtlety see [5]).

Let $G = \langle \Pi_5 \rangle$. Then $\langle b, c \rangle$ is a quotient of $C(3, 3)$. Let $u = (bc)^3$, so $u^2 = 1$. From our knowledge of $C(3, 3)$, we can see that $[b, u]^2 = 1$. An easy enough computation (coset enumeration or Knuth-Bendix) shows that the relations $a^2 = b^3 = (ab)^2 = u^2 = [b, u]^2 = 1$ plus the sixth powers of au, abu, aBu, and $aubu$ imply that $[b, u] = 1$. Thus in G we have $[b, (bc)^3] = 1$. By symmetry, we have $[b, (bC)^3] = 1$ too. Adding these two relations to $C(3, 3)$ gives a group of order 27 in which $(bc)^3 = (bC)^3 = 1$. Once we have these relations, it does not take many sixth powers to get the order of G. Certainly bases of length at most 5 suffice.

Following this proof, but without explicitly requiring the "very subtle" relators $(bc)^3$ and $(bC)^3$ (whose derivation from sixth powers is relatively hard) coset enumeration over the trivial subgroup for the presentation $\{a, b, c, u, v \mid a^2, b^3, c^3, (ab)^2, u = (bc)^3, v = (bC)^3, (bc)^6, (bC)^6, ([c, b][C, b])^2, ([b, c][B, c])^2, [b, u], (au)^6, (abu)^6, (aBu)^6, (aubu)^6, [b, v], (av)^6, (abv)^6, (aBv)^6, (avbv)^6, (ac)^6, [c, a]^6, (abc)^6, (aBc)^6, (abcac)^6, (abCac)^6, (aBcaC)^6\}$ gives index 17496 defining a maximum of 860646 and a total of 1559920 cosets. Here $([c, b][C, b])^2$ and $([b, c][B, c])^2$ are subtle relators from $C(3, 3)$.

Deleting any one of the three base length 5 sixth powers gives a group which is 3 times bigger.

Having proved finiteness this way we can prune the proof by using the result of a different enumeration. The index of $\langle a, c \rangle$ in the group with presentation $\{a, b, c, u, v \mid a^2, u^2, v^2, b^3, c^3, (ab)^2, u = (bc)^3, v = (bC)^3, (bc)^6, (bC)^6, [b, u], [b, v], (ac)^6, [c, a]^6, (abc)^6, (aBc)^6, (abcac)^6, (abCac)^6, (aBcaC)^6\}$ is readily shown to be 81, from which finiteness again follows. Suffice it to say that this leads to a presentation for $C(S_3, Z_3)$ of the form Π_5 with a reasonable number of sixth powers for which we can prove finiteness.

Perhaps the next sensible three-generator problem to consider is short presentations for $C(V_4, Z_3) = \{a, b, c \mid a^2, b^2, (ab)^2, c^3, \exp 6\}$. Investigations so far show this to be more difficult.

REFERENCES

[1] Wieb Bosma, John Cannon and Catherine Playoust. The Magma algebra system I: the user language. *J. Symbolic Comput.* **24** (1997) 235–265.

[2] Marshall Hall Jr. Solution of the Burnside problem for exponent six. *Illinois J. Math.* **2** (1958) 764–786.

[3] Marshall Hall Jr. *The theory of groups.* The Macmillan Co., New York (1959).

[4] George Havas. Coset enumeration strategies. In *ISSAC'91, Proceedings of the 1991 International Symposium on Symbolic and Algebraic Computation.* ACM Press, New York (1991) 191–199.

[5] George Havas, M. F. Newman, Alice C. Niemeyer and Charles C. Sims. Groups with exponent six. *Commun. Algebra* **27** (1999) 3619–3638.

[6] George Havas and Colin Ramsay, *Coset enumeration: ACE version 3* (1999). Available as http://www.csee.uq.edu.au/~havas/ace3.tar.gz.

[7] George Havas and Colin Ramsay. *Experiments in coset enumeration.* Technical Report 13, Centre for Discrete Mathematics and Computing, The University of Queensland, 1999.

[8] George Havas and Edmund F. Robertson. Application of computational tools for finitely presented groups. In *Computational support for discrete mathematics*, DIMACS Ser. Discrete Math. Theoret. Comput. Sci. **15** (1994) 29–39.

[9] Derek F. Holt and Sarah Rees. A graphics system for displaying finite quotients of finitely presented groups. In *Groups and computation*, DIMACS Ser. Discrete Math. Theoret. Comput. Sci. **11** (1993) 113-126.

[10] M. F. Newman. Groups of exponent six. In *Computational group theory*. Academic Press, London (1984) 39–41.

[11] Alice C. Niemeyer. A finite soluble quotient algorithm. *J. Symbolic Comput.* **18** (1994) 541–561.

[12] Martin Schönert et al. GAP – *Groups, Algorithms and Programming*. Lehrstuhl D für Mathematik, RWTH Aachen (ftp://www.gap.dcs.st-and.ac.uk/gap), fifth edition (1995).

[13] Charles C. Sims. *Computation with finitely presented groups*, Encyclopedia of Mathematics and its Applications **48**. Cambridge University Press, Cambridge (1994).

[14] Charles C. Sims. RKBP, Version 1.3 (1995). Available via anonymous ftp from dimacs.rutgers.edu, in directory /pub/rkbp.

APPENDIX A. COMPUTE $B(2,6)$

```
##    Compute the direct product G1xG2 of G1 and G2 (already computed)
##    and identify the image of B(2,6) in it.
F.relators := [];
G1xG2 := DirectProduct( G1, G2 );
em1 := G1xG2.embeddings[1]; em2 := G1xG2.embeddings[2];

delta1 := lambda*em1; delta2 := kappa*em2;
mu := GroupHomomorphismByImages( F, G1xG2, F.generators, [
  Image(delta1,F.1) * Image(delta2,F.1),
  Image(delta1,F.2) * Image(delta2,F.2)] );
immu := Image(mu, F); I := SpecialAgGroup(immu);
chi :=  mu.operations.CompositionMapping( I.bijection^-1, mu );
##    Now  I  is a special ag presentation for  B(2,6)  inside the
##    direct product  G1xG2  and  chi: F -> I  is a homomorphism.
##    Compute matrices for the action of the generators of B(2,3)
##    on the elementary abelian group of order 2^28.
##    Then determine the complete reduction of the module into:
##    T T 2 2 2 6 6 6.
acton := [I.1, I.2];
Append( acton, Sublist( I.generators, [6..31] ) );
mats1 := Action( I, acton, [I.3, I.4, I.5], GF(2) );
```

```
RequirePackage("matrix");
gmod1  := GModule( mats1 ); cp1 := CompositionFactors( gmod1 );

submods := [];
for i in [ 1 .. Length(cp1) ] do
    minsubs := MinimalSubGModules ( cp1[i][1], gmod1, 21 );
    span := []; d := 0;
    for j in [1 .. Length(minsubs)] do
        vs := VectorSpace( Union( span, minsubs[j]), GF(2) );
        if Dimension(vs) > d then
            d := Dimension(vs);
            span := Union( span, minsubs[j] );
            Add( submods, minsubs[j] );
        fi;
    od;
od;
newbI := []; newbI[1] := 0; newbI[2] := 0; newbI[3] := 0; s := 3;
for i in [ 1 .. Length(submods) ] do
    for j in [ 1.. Length(submods[i]) ] do
        x := I.1^Int(submods[i][j][1]) * I.2^Int(submods[i][j][2]);
        x := x * Product(List([3..Length(submods[i][j])],
                x->I.generators[3+x]^Int(submods[i][j][x])) )^3;
        Add(newbI, x );
    od;
od;
two1 := newbI[4]; two2 := newbI[5];
Print("#I mapped the module generators into the direct product \n");

## Now determine the structure of the module where B(2,6)/B(2,6)^2
## acts on the largest elementary abelian 3-quotient of B(2,6)^2;
## it has order 3^5. This is then a module for a group of order 4
## and decomposes into: T T ma mb mab, where a fixes the
## generator of ma, b  fixes the generator of  mb and ab fixes the
## generator of mab (a, b the generators of the group of order 4).

mats2 := Action(I, [I.3, I.4, I.32, I.33, I.34 ], [I.1, I.2], GF(3));
gmod2 := GModule(mats2); cp2 := CompositionFactors( gmod2 ); i := 1;
while cp2[i][2] <> 2 do
    i := i + 1;
od;
Swap( cp2, 1, i);
Print("#I chopped the 5-dimensional module over GF(3) \n");

submods2 := [];
for i in [ 1 .. Length(cp2) ] do
    Add( submods2, MinimalSubGModules( cp2[i][1], gmod2, cp2[i][2] ));
od;

nbs := [];
for i in [ 1 .. Length(submods2) ] do
    for j in [ 1.. Length(submods2[i]) ] do
```

```
         for k in [ 1 .. Length(submods2[i][j]) ] do
             x := I.3^Int(submods2[i][j][k][1])*
                  I.4^Int(submods2[i][j][k][2])*
                  I.32^Int(submods2[i][j][k][3])*
                  I.33^Int(submods2[i][j][k][4])*
                  I.34^Int(submods2[i][j][k][5]);
           Add(nbs, x );
        od;
     od;
od;
Print("#I mapped the module generators into the direct product \n");

# Compute the exponents of the basis elements in the quotient group
# and then lift the elements into the big group.
u1 := nbs[1]; u2 := nbs[2]; u3 := Comm( u2, u1 );
# the 9-dimensional part
u4 := nbs[3];       u5 := nbs[4];       u6 := nbs[5];
u7 := Comm(u4,u1); u8 := Comm(u4,u2); u9 := Comm(u5,u1);
u10:= Comm(u5,u2); u11:= Comm(u6,u1); u12:= Comm(u6,u2);
# the 12-dimensional part
u13 := Comm(u5,u4); u14 := Comm(u6,u4); u15 := Comm(u6,u5);
u16 := Comm(u4,u3); u17 := Comm(u5,u3); u18 := Comm(u6,u3);
u19 := Comm(u7,u5); u20 := Comm(u7,u6); u21 := Comm(u8,u5);
u22 := Comm(u8,u6); u23 := Comm(u9,u6); u24 := Comm(u10,u6);
# the 1-dimensional part
u25 :=  Comm( Comm( u6, u5), u4 );

newbI[1] := u1; newbI[2] := u2; newbI[3] := u3;
Add(newbI,u4^2);  Add(newbI,u5^2);  Add(newbI,u6^2);  Add(newbI,u7^2);
Add(newbI,u8^2);  Add(newbI,u9^2);  Add(newbI,u10^2); Add(newbI,u11^2);
Add(newbI,u12^2); Add(newbI,u13^2); Add(newbI,u14^2); Add(newbI,u15^2);
Add(newbI,u16^2); Add(newbI,u17^2); Add(newbI,u18^2); Add(newbI,u19^2);
Add(newbI,u20^2); Add(newbI,u21^2); Add(newbI,u22^2);
Add(newbI,u23^2); Add(newbI,u24^2); Add(newbI,u25^2);
Print("#I computed a new basis for B(2,6) inside the direct product \n");

alpha := BaseChangeAgGroup( I, newbI );
B26 := alpha.range; chi := chi*alpha;

newb:=[];
newb[1] := B26.1*B26.41^2*B26.42^2*B26.47^2*B26.48*B26.50*B26.51^2;
newb[2] := B26.2*B26.41*B26.42^2*B26.43^2*B26.47*B26.49*B26.51^2*B26.52;
newb[3] := B26.3*B26.47*B26.48^2*B26.49*B26.50*B26.51^2;
for i in [ 4 .. Length(B26.generators)] do
    newb[i] := B26.generators[i];
od;

## Finally, determine the desired presentation for B26,
## and chi a homomorphism from F to B26.
beta := BaseChangeAgGroup( B26, newb);
B26 := beta.range; chi := chi * beta;
```

APPENDIX B. GAP FUNCTION Action

```
Action := function(grp, gens, actgens, F)
    local  x, y, pos, e, mt, act;

    pos := List( gens, i->Position(grp.generators,i));
    act := [];
    for x in actgens do
        mt := [];
        for y in gens do
            e := Exponents(grp, y^x, F);
            Add(mt, List(pos, i->e[i]));
        od;
        Add( act, mt );
    od;

    return act;
end;
```

APPENDIX C. GAP FUNCTION BaseChangeAgGroup

```
BaseChangeAgGroup := function( g, newb )
    local  ser, i, k, igs, u, alpha;

    ser := [];
    for i in [1..Length(newb)] do
        k := Subgroup( g, Sublist(newb,[i..Length(newb)]));
        Cgs( k );  Add( ser, k );
    od;
    Add( ser, TrivialSubgroup(g) );
    alpha := IsomorphismAgGroup( ser );  k := alpha.range;

    igs := List( newb, x -> Image(alpha,x) );
    u := Subgroup( k, igs );  u.igs := igs;
    u.operations.AddShiftInfo(u);
    g := AgGroupFpGroup( FpGroup(u) );
    alpha := GroupHomomorphismByImages( ser[1], g, newb, g.generators );

    return alpha;
end;
```

George Havas,
The University of Queensland,
Queensland 4072,
Australia.

Alice C. Niemeyer,
University of Western Australia,
Nedlands, WA 6907,
Australia.

M. F. Newman,
Australian National University,
Canberra 0200,
Australia.

Charles C. Sims,
Rutgers University,
New Brunswick, NJ 08903,
USA.

REWRITING AS A SPECIAL CASE OF NON-COMMUTATIVE GRÖBNER BASIS THEORY

ANNE HEYWORTH

1. INTRODUCTION

Rewriting for semigroups is a special case of Gröbner basis theory for non-commutative polynomial algebras. The fact is a kind of folklore but is not fully recognised. So our aim in this paper is to elucidate this relationship.

A good introduction to string rewriting is [2], and a recent introduction to non-commutative Gröbner basis theory is [12]. Similarities between the two critical pair completion methods (Knuth-Bendix and Buchberger's algorithm) have often been pointed out in the commutative case. The connection was first observed in [7, 5] and more closely analysed in [3, 4] and more recently in [11] and [10]. In particular it is well known that the commutative Buchberger algorithm may be applied to presentations of abelian groups to obtain complete rewrite systems.

Rewriting involves a presentation $sgp\langle X|R\rangle$ of a semigroup S and presents S as a factor semigroup $X^\dagger/=_R$ where $X^\dagger$ is the free semigroup on X and $=_R$ is the congruence generated by the subset R of $X^\dagger \times X^\dagger$. Non-commutative Gröbner basis theory involves a presentation $alg\langle X|F\rangle$ of a non-commutative algebra A over a field K and presents A as a factor algebra $K[X^\dagger]/\langle F\rangle$ where $K[X^\dagger]$ is the free K-algebra on the semigroup $X^\dagger$ and $\langle F\rangle$ is the ideal generated by F, a subset of $K[X^\dagger]$. Given a semigroup presentation $sgp\langle X|R\rangle$ we consider the algebra presentation $alg\langle X|F\rangle$ where $F := \{l - r : (l,r) \in R\}$. It is well known that the word problem for $sgp\langle X|R\rangle$ is solvable if and only if the (monomial) equality problem for $alg\langle X|F\rangle$ is solvable. Mora [8] recorded that a complete rewrite system for a semigroup S presented by $sgp\langle X|Rel\rangle$ is equivalent to a non-commutative Gröbner basis for the ideal specified by the congruence $=_R$ on $X^\dagger$ in the algebra $\mathbb{F}_3[X^\dagger]$ where $\mathbb{F}_3$ is the field with elements $\{-1, 0, 1\}$.

In this paper we show that the non-commutative Buchberger algorithm applied to F corresponds step-by-step to the Knuth-Bendix completion procedure for R. This is the meaning intended for the first sentence of this paper.

Acknowledgements. This research work was supported by an earmarked EPSRC Research Studentship 'Identities among Relations for Monoids and Categories' in 1995-98. I would like to thank Larry Lambe for pointing out this likely connection, Ronnie Brown for discussion on the paper, and Neil Dennis for encouraging my work.

2. RESULTS

First we note that the relation between the two kinds of presentation is given by the following variation of a result of [8].

Proposition 2.1. *Let K be a field and let S be a semigroup with presentation $sgp\langle X|R\rangle$. Then the algebra $K[S]$ is isomorphic to the factor algebra $K[X^\dagger]/\langle F\rangle$ where F is the basis $\{l - r | (l, r) \in R\}$.*

Proof. Define $\phi : K[X^\dagger] \to K[S]$ by $\phi(k_1 w_1 + \cdots + k_t w_t) := k_1[w_1]_R + \cdots + k_t[w_t]_R$ for $k_1, \ldots, k_t \in K$, $w_1, \ldots, w_t \in X^\dagger$. Define a homomorphism $\phi' : K[X^\dagger]/\langle F\rangle \to K[S]$ by $\phi'([p]_F) := \phi(p)$. It is injective since $\phi'[p]_F = \phi[q]_F$ if and only if $[p]_F = [q]_F$ (using the definitions $\phi(p) = \phi(q) \Leftrightarrow p - q \in \langle F\rangle$). It is also surjective. Let $f \in K[S]$. Then $f = k_1 m_1 + \cdots + k_t m_t$ for some $k_1, \ldots, k_t \in K$, $m_1, \ldots, m_t \in S$. Since S is presented by $sgp\langle X|R\rangle$ there exist $w_1, \ldots, w_t \in X^\dagger$ such that $[w_i]_R = m_i$ for $i = 1, \ldots, t$. Therefore let $p = k_1 w_1 + \cdots + k_t w_t$. Clearly $p \in K[X^\dagger]$ and also $\phi'[p]_F = f$. Hence ϕ' is an isomorphism. □

Now we give our main result.

Theorem 2.2. *Let $sgp\langle X|R\rangle$ be a semigroup presentation, let K be a field and let $alg\langle X|F\rangle$ be the K-algebra presentation with $F := \{l - r : (l, r) \in R\}$. Then the Knuth-Bendix completion algorithm for the rewrite system R corresponds step-by-step to the non-commutative Buchberger algorithm for finding a Gröbner basis for the ideal generated by F.*

Proof. Both the Knuth-Bendix algorithm for R and the Buchberger algorithm for F begin by specifying a monomial ordering on $X^\dagger$ which we denote $>$. The correspondence between terminology in the two cases is

(i)	rewrite system	basis
(ii)	rule	two-term polynomial
(iii)	word	monomial
(iv)	reduction	reduction
(v)	left hand side	leading monomial
(vi)	subword	submonomial
(vii)	right hand side	remainder
(viii)	overlap	match
(ix)	critical pair	S-polynomial
(x)	resolve	reduce to zero
(xi)	reduced critical pair	reduced S-polynomial
(xii)	complete rewrite system	Gröbner basis

The key parts of the correspondence (viii) and (ix) are illustrated diagrammatically in the next section.

In terms of rewriting we consider the rewrite system R which consists of a set of rules of the form (l, r) orientated so that $l > r$. A word $w \in X^\dagger$ may be reduced with respect to R if it contains the left hand side l of a rule (l, r) as a subword i.e. if $w = ulv$ for some $u, v \in X^*$. To reduce $w = ulv$ using the rule (l, r) we replace l by the right hand side r of the rule, and write $ulv \rightarrow_R urv$. The Knuth-Bendix algorithm looks for overlaps between rules. Given a pair of rules (l_1, r_1), (l_2, r_2) there are four possible ways in which an overlap can occur: $l_1 = u_2 l_2 v_2$, $u_1 l_1 v_1 = l_2$, $l_1 v_1 = u_2 l_2$ and $u_1 l_1 = l_2 v_2$. The critical pair resulting from an overlap is the pair of words resulting from applying each rule to the smallest word on which the overlap occurs. The critical pairs resulting from each of the four overlaps are: $(r_1, u_2 r_2 v_2)$, $(u_1 r_1 v_1, r_2)$, $(r_1 v_1, u_2 r_2)$ and $(u_1 r_1, r_2 v_2)$ respectively (see diagram). In one pass the completion procedure finds all the critical pairs resulting from overlaps of rules of R. Both sides of each of the critical pairs are reduced as far as possible with respect to R to obtain a reduced critical pair (c_1, c_2). The original pair is said to resolve if $c_1 = c_2$. The reduced pairs that have not resolved are orientated, so that $c_1 > c_2$, and added to R forming R_1. The procedure is then repeated for the rewrite system R_1, to obtain R_2 and so on. When all the critical pairs of a system R_n resolve (i.e. $R_{n+1} = R_n$) then R_n is a complete rewrite system.

In terms of Gröbner basis theory applied to this special case we consider the basis F which consists of a set of two-term polynomials of the form $l - r$ multiplied by ± 1 so that $l > r$. A monomial $m \in X^\dagger$ may be reduced with respect to F if it contains the leading monomial l of a polynomial $l - r$ as a submonomial, i.e. if $m = ulv$ for some $u, v \in X^*$. To reduce $m = ulv$ using the polynomial $l - r$ we replace l by the remainder r of the polynomial, and write $ulv \rightarrow_F urv$. The Buchberger algorithm looks for matches between polynomials. Given a pair of polynomials $l_1 - r_1$, $l_2 - r_2$ there are four possible ways in which an match can occur: $l_1 = u_2 l_2 v_2$, $u_1 l_1 v_1 = l_2$, $l_1 v_1 = u_2 l_2$ and $u_1 l_1 = l_2 v_2$. The S-polynomial resulting from a match is the difference between the pair of monomials resulting from applying each two-term polynomial to the smallest monomial on which the match occurs. The S-polynomials resulting from each of the four matches are: $r_1 - u_2 r_2 v_2$, $u_1 r_1 - v_1, r_2$, $r_1 v_1 - u_2 r_2$ and $u_1 r_1 - r_2 v_2$ respectively (see diagram). In one pass the completion procedure finds all the S-polynomials resulting from matches of polynomials of F. The S-polynomials are reduced as far as possible with respect to F to obtain a reduced S-polynomial $c_1 - c_2$. Note that reduction can only replace one term with another so the reduced S-ploynomial will have two terms unless the two terms reduce to the same thing $c_1 = c_2$ in which case the original S-polynomial is said to reduce to zero. The reduced S-polynomials that have not been reduced to zero are multiplied by ± 1, so that $c_1 > c_2$, and added to F forming F_1. The procedure is then repeated for the basis F_1, to obtain F_2 and so on. When all the S-polynomials of a basis F_n reduce to zero (i.e. $F_{n+1} = F_n$) then F_n is a Gröbner basis.

A critical pair in R will occur if and only if a corresponding S-polynomial occurs in F. Reduction of the pair by R is equivalent to reduction of the S-polynomial by F. Therefore at any stage any new rules correspond to the new two-term polynomials and $F_i := \{l - r : (l, r) \in R_i\}$. Therefore the completion procedures as applied to R and F correspond to each other at every step. □

3. ILLUSTRATION

This is a picture of the correspondence (viii) and (ix) between critical pairs and S-polynomials and the four ways in which they can occur, as described in the above proof.

possible overlaps	possible matches
of rules	of polynomials
$l_1 \rightarrow r_1$ and $l_2 \rightarrow r_2$	$l_1 - r_1$ and $l_2 - r_2$

$l_1 = u_2 l_2 v_2$ $\qquad$ $l_1 = u_2 l_2 v_2$

$(r_1, u_2 r_2 v_2)$ $\qquad$ $u_2 r_2 v_2 - r_1$

$u_1 l_1 v_1 = l_2$ $\qquad$ $u_1 l_1 v_1 = l_2$

$(u_1 r_1 v_1, r_2)$ $\qquad$ $r_2 - u_1 r_1 v_1$

$l_1 v_1 = u_2 l_2$ $\qquad$ $l_1 v_1 = u_2 l_2$

$(r_1 v_1, u_2 r_2)$ $\qquad$ $u_2 r_2 - r_1 v_1$

$u_1 l_1 = l_2 v_2$ $\qquad$ $u_1 l_1 = l_2 v_2$

$(u_1 r_1, r_2 v_2)$ $\qquad$ $r_2 v_2 - u_1 v_1$

4. REMARKS

The result that the Knuth-Bendix algorithm is a special case of the non-commutative Buchberger algorithm is something that requires further investigation. Rewriting techniques and the Knuth-Bendix algorithm have recently been applied to presentations of Kan extensions over sets [6] and it is not immediately obvious what this will imply for non-commutative Gröbner bases. Another interesting line of investigation would be to attempt to adapt rewriting procedures for constructing crossed resolutions of group presentations [6] to the more general Gröbner basis situation.

REFERENCES

[1] F.Baader and T.Nipkow : Term Rewriting and All That, *Cambridge University Press* (1998).

[2] R.Book and F.Otto : String-Rewriting Systems, *Springer-Verlag, New York*, (1993).

[3] B.Buchberger : Basic Features and Development of the Critical Pair Completion Procedure, in Rewriting Techniques and Applications, J.Jouannaud (ed), *Springer-Verlag, LCNS vol.202 p1-45* (1986).

[4] B.Buchberger : History and Basic Features of the Critical Pair / Completion Procedure, *J.Symbolic Computation, vol.3, p3-38* (1987).

[5] B.Buchberger and R.Loos : Algebraic Simplification, *Computing Supplement, vol.4 p11-43* (1982).

[6] A.Heyworth : Applications of Rewriting Systems and Gröbner Bases to Computing Kan Extensions and Identities Among Relations, *PhD Thesis, University of Wales, Bangor* (1998)

[7] R.Loos : Term Reduction Systems and Algebraic Algorithms, in Proc 5th German Workshop on Artificial Intelligence, J.Siekmann (ed), *Springer-Verlag, Informatik Fachberichte, vol.47 p214-234* (1981).

[8] T.Mora : Gröbner Bases and the Word Problem, *Preprint, University of Genova* (1987).

[9] T.Mora : An Introduction to Commutative and Non-commutative Gröbner Bases, *Theoretical Computer Science vol.134 p131-173*(1994).

[10] B.Reinert : On Gröbner Bases in Monoid and Group Rings *PhD Thesis, Universität Kaiserslautern* (1995).

[11] K.Stokkermans : A Categorical Framework and Calculus for Critical Pair Completion, *Phd Thesis, Royal Institute for Symbolic Computation, Johannes Kepler University, Linz* (1995).

[12] V.Ufnarovski : Introduction to Non-commutative Gröbner Bases Theory, in Gröbner Bases and Applications, B.Buchberger and F.Winkler (eds), *Proc. London Math. Soc. vol.251 p305-322* (1998).

Anne Heyworth
School of Mathematics
University of Wales
Bangor
Gwynedd LL57 1UT
Wales.
map130@bangor.ac.uk

DETECTING 3-MANIFOLD PRESENTATIONS

CYNTHIA HOG-ANGELONI

ABSTRACT. Let $\langle a_1, \ldots, a_g \mid R_1, \ldots, R_h \rangle$ denote a group presentation, where $R_j \in S(a_i)$, the semigroup on the symbols $a_1, a_1^{-1} \ldots a_g, a_g^{-1}$. Let $K_{\mathcal{P}}^2$ denote the corresponding standard 2–complex. It is well known that every compact connected 3–manifold has a spine which is a standard complex, but that not every $K_{\mathcal{P}}^2$ is a spine of some 3–manifold. Neuwirth's algorithm [Neu68] decides which $K_{\mathcal{P}}^2$ embed.

The embeddability of $K_{\mathcal{P}}^2$ depends on the choice of cancelling pairs in the R_j. However, work of Zieschang and Kaneto shows that in order to decide whether insertion or deletion of cancelling pairs in the relators turns $\mathcal{P}$ into a 3-manifold presentation, it suffices to check the embeddability of the reduced presentation in the result of an application of the Whitehead algorithm to $\mathcal{P}$.

For reduced 2–generator presentations, Osborne and Stevens [OS77] exhibit algebraic conditions on the relators that are necessary for being a 3–manifold presentation, and an algorithm involving fewer steps than Neuwirth's. This paper generalizes their results to the case of g generators.

1. NOTATION AND CONVENTIONS

Throughout this paper, 3–*manifold* means *compact, connected, orientable PL–3–manifold*.

There is a well known correspondence between standard 2–complexes and presentations (see [H-AM93], p. 9). To make this work, the relators R_j of a presentation $\langle a_1, \ldots, a_g \mid R_1, \ldots, R_h \rangle$ are considered as h (not necessarily distinct) elements of the semigroup on the symbols $a_1, a_1^{-1} \ldots a_g, a_g^{-1}$. The cyclically reduced word obtained from R_j is denoted by $\hat{R}_j$ throughout this paper. Two group presentations are considered to be the same if their corresponding 2–complexes are isomorphic, i.e. if one can be obtained from the other by relabeling the generators, replacing a generator by its inverse, cyclically conjugating and taking inverses of the relators.

Insertion or deletion of cancelling pairs in the relators of a presentation may change the embedding behaviour of the associated 2-complex. It is discussed in §4 how to reduce the general problem of embeddability to that of embeddability of 2-complexes associated to reduced presentations. Then from §5 onwards, all relators are assumed to be reduced.

2. INTRODUCTION

Given any 3-manifold M^3, we can obtain a presentation of its fundamental group by (removing an open 3–ball if M^3 is closed and) collapsing it to a 2–complex with a single vertex, called a spine [H-AM93, page 17]. Two spines K^2, L^2 of M^3 differ by a 3–deformation $K^2 \nearrow M^3 \searrow L^2$, so M^3 not only determines a group $\pi_1(M^3)$ but a *presentation class* [H-AM93, page 21]. Many homotopy invariants can be derived from any member of this presentation class. For example, Dunwoody's exotic presentation of the trefoil group [Dun76] defines a $K_{\mathcal{P}}$ which is not homotopy equivalent to any 3-manifold spine [HLM85] whereas the underlying group is the fundamental group of the trefoil knot space.

Conversely, given a presentation $\mathcal{P}$ it is often helpful to know whether the corresponding 2–complex embeds into some 3-manifold. E.g. if $K_{\mathcal{P}}^2$ embeds into some 3-manifold and if $\mathcal{P}$ presents a nontrivial free product $G_1 * G_2$, its homotopy type splits into a one point union $K_1^2 \vee K_2^2$ where $\pi_1(K_i) = G_i$. For $\mathcal{P}$ not a 3-manifold presentation, counterexamples are given in [HLM85].

Also, if $K_{\mathcal{P}}^2$ embeds, its second homotopy group is subject to the restrictions given by the existence of prime factorizations for 3-manifolds, see [Hem76].

On the other hand, if $K_{\mathcal{P}}^2$ embeds and $\mathcal{P}$ is a balanced presentation of the trivial group (i.e. $K_{\mathcal{P}}$ is contractible), either $\mathcal{P}$ is in the presentation class of the trivial presentation $\langle a_1, \dots, a_g \mid a_1, \dots, a_g \rangle$ (and is thus ruled out among the candidates for counterexamples to the Andrews-Curtis Conjecture) or M^3 is a counterexample to the Poincare Conjecture in dimension 3.

Neuwirth's algorithm (see [Neu68] or [H-AM93]) decides which $\mathcal{P}$ are 3–manifold presentations. For the case of 2–generator presentations, Osborne and Stevens [OS74] exhibited a condition on the exponents of the relators of a 3–manifold presentation that can be immediately read off. Furthermore in case the condition is fulfilled, a simple drawing (called "railroad system") checks whether the presentation lives in fact on a 3–manifold. Their work is generalized to the case of g generators in this paper.

3. PRELIMINARIES

Let us collect some background material from [H-AM93] for the convenience of the reader.

We wish to visualize a 3-manifold by means of some spine $K_{\mathcal{P}}^2$ with corresponding presentation $\mathcal{P} = \langle a_1, \dots, a_g \mid R_1, \dots, R_h \rangle$. A regular neighbourhood of the 1–skeleton of $K_{\mathcal{P}}^2$ in M^3 is just a solid handlebody V of genus g, each 1–handle corresponding to an a_i. The boundary ∂V intersects $K_{\mathcal{P}}^2$ in a collection of pairwise disjoint simple closed curves k_j, namely the boundaries of the discs $\overline{K^2 - N_{K^2}(K^1)}$ where $N_{K^2}(K^1)$ denotes a regular neighbourhood of the 1–skeleton of K^2. The intersections $k_j \cap m$ of the loops k_j with a complete system $m = \{m_1 \dots m_g\}$ of oriented meridian discs read R_j.

This picture leads to the following criterion (see (51), p. 32 of [H-AM93]):

Criterion 3.1. *$K_{\mathcal{P}}^2$ with presentation $\langle a_1, \ldots, a_g \mid R_1, \ldots, R_h \rangle$ has a 3–dimensional thickening if and only if it is possible to draw a disjoint system of loops k_j on the boundary of a handlebody V with g handles, (one handle for each a_i,) such that k_j reads R_j on V.*

Example: The standard complex associated to the presentation

$$\mathcal{P}_1 = \langle a, b \mid a^{-2}bab, a^{-3}b^5 \rangle$$

of the binary icosahedral group with reduced R_j is a spine of the spherical dodecahedral space. The above criterion gives

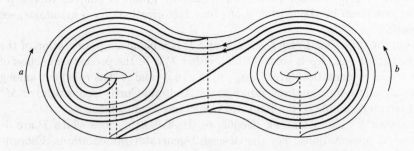

FIGURE 1. $(V; k_1, \ldots, k_h)$ for $\mathcal{P}_1 = \langle a, b \mid a^{-2}bab, a^{-3}b^5 \rangle$

Now cut the handlebody V along the system $\boldsymbol{m}$ of meridian discs specified above. The curves $k_j \subset \partial V$ split into arcs running on a sphere with $2g$ holes, yielding:

Criterion 3.2. *$K_{\mathcal{P}}^2$ with presentation $\langle a_1, \ldots, a_g \mid R_1, \ldots, R_h \rangle$ has a 3–dimensional thickening if and only if it is possible to draw collections k_j of simple arcs λ_{j_l}, all pairwise disjoint, on a 2–sphere with $2g$ holes such that k_j reads R_j. (See Figure 3.)*

There is a further simplification of the picture [Zie88, OS74, H-AM93]: We determine a collection of g tori, called *handles* H_i of V by choosing a system of g disjoint discs B_i properly embedded in the handlebody V which

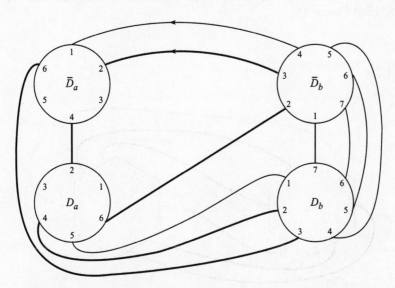

FIGURE 2. (S^2 with $2g$ discs ; $k_1, \ldots, k_h$) for $\mathcal{P}_1$ as in Figure 1. Points on D_i and $\bar{D}_i$ with same number label coincide on V.

cut V into solid tori $H_1, \ldots, H_g$, $B_i \subset \partial H_i$ and a ball $\overline{V - \cup H_i}$. The loops ∂B_i are called *belt curves*; they encompass both attaching discs D_i and $\bar{D}_i$ in Figure 2. Each handle H_i determines uniquely, up to isotopy and reversal of orientation, a meridian m_i of V such that $m_i \cap \partial B_i = \emptyset$. We can assume that the relator curves k_j are transverse to the belt curves. They are cut into *connections*, i.e. pairwise disjoint arcs[1] on the handles H_i, as well as arcs on $S^2 = \partial(\overline{V - \cup H_i})$ with endpoints on the belt curves. We can further assume that the connections cannot be homotoped to ∂B_i^2 on ∂H_i.

It is shown in the above cited papers that only 3 isotopy classes of connections can run on each handle. A detailed analysis shows that each B_i^2 on the 2–sphere $S^2 = \partial(\overline{V - \cup H_i})$, along which a handle has been cut off, should be imagined as a hexagon labelled by the generator a_i. The edges of this hexagon are labelled clockwise (according to an orientation of the 2–sphere) by $n_i, n_i + p_i, p_i, -n_i, -(n_i + p_i), -p_i$; where n_i and p_i are coprime integers[2]. The arcs of intersection of the relator curves k_j with S^2 connect the hexagons. Their endpoints avoid the corners of the hexagon and we get from one arc of intersection of k_j with S^2 to the next one by proceeding along a line segment

---

[1]If some k_j runs entirely on an H_i, isotope a little arc onto $\overline{V - \cup H_i}$

[2]Warning: Zero is not excluded here, e.g. $n_i = 0, p_i = 1$. This will lead to some trouble, see Figure 5

in the hexagon orthogonal to two of its edges and reading the label of the edge. (See Figure 3):

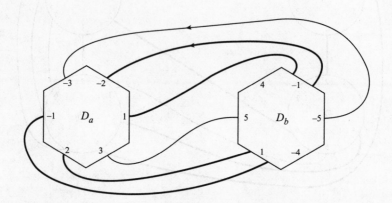

FIGURE 3. (S^2 with g hexagons ; $k_1, \ldots, k_h$) for $\mathcal{P}_1$ as in Figure 1

Criterion 3.3. $K_\mathcal{P}^2$ *with presentation* $\mathcal{P} = \langle a_1, \ldots, a_g \mid R_1, \ldots, R_h \rangle$ *has a 3–dimensional thickening if and only if it is possible to draw collections k_j of pairwise disjoint simple subarcs on a 2–sphere with g labelled hexagons such that k_j reads R_j.*

The picture devised by this criterion is called a *railroad system* by Osborne and Stevens [OS74].

4. ELIMINATION OF CANCELLING PAIRS

Among other things, a choice of meridians was involved in the above criteria. Changes in the system of meridian discs correspond to certain changes of basis of $F(a_i)$, the *Whitehead transformations*, (see [Whi36], [LS77]) defined by:

For some fixed 'multiplier' $x \in \{a_1^{\pm 1}, \ldots a_g^{\pm 1}\}$ replace a_i in the relators throughout by one of $a_i, a_i x, x^{-1} a_i, x^{-1} a_i x$.

Let us draw a picture of this transformation starting from the handlebody that has been cut along meridians as in Figure 2. Denote by i_0 the index of the multiplier and take the union of the following discs:

$$\begin{cases} D_i \text{ if } a_i^{-1} \text{ is the multiplier or if } a_i \text{ is replaced by } x^{-1}a_i \\ \bar{D}_i \text{ if } a_i \text{ is the multiplier or if } a_i \text{ is replaced by } a_i x \\ \text{both } D_i \text{ and } \bar{D}_i \text{ if } a_i \text{ is replaced by } x^{-1} a_i x \end{cases}$$

Connect these discs by ($\#discs - 1$) arcs transversal to the relator curves k_j such that a regular neighbourhood N of this union of discs and arcs is itself a disc D'_{i_0}. Push the interior of D'_{i_0} slightly into the ball to become properly embedded.

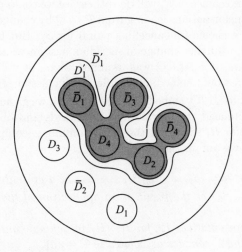

FIGURE 4. $(x = a_1; a_1 \mapsto a_1, a_2 \mapsto a_1^{-1}a_2, a_3 \mapsto a_3a_1, a_4 \mapsto a_1^{-1}a_4a_1)$

Now cut along D'_{i_0} so that $\bar{D}'_{i_0}$ and D_{i_0} lie on the same ball, then identify D_{i_0} with $\bar{D}_{i_0}$. The system $D_1, \ldots, D_g$ minus D_{i_0} together with D'_{i_0} gives a new system of meridian discs. When we add the relator curves k_j in Figure 4 (that have been omitted there for the sake of clarity) it can be seen that entering N adds a letter x^{-1} and leaving N adds a letter x to the relator R_j. The reader should verify that the above change of meridians subjects the R_j to the Whitehead transformation we started with. The price we pay is a bunch of cancelling pairs; e.g. where the relator curves k_j cross the connecting arcs.

Kaneto [Kan82], generalizing a result of Zieschang [Zie70], showed that among all choices for meridian discs there is a system m_0 such that the intersections $k_j \cap m_0$ along k_j read a subword W_j of $\hat{R}_j$, the cyclically reduced representative of R_j. In other words, there is a spine $K_{\mathcal{Q}}^2$ for M^3 with $\mathcal{Q} = \langle a_1, \ldots, a_g \mid W_1, \ldots, W_h \rangle$ such that W_j is obtained from R_j by performing all cancellations and potentially suppressing more letters $a_i^{\pm 1}$. In particular, if $\mathcal{P}$ has the property that the sum over the lengths of the cyclically reduced $\hat{R}_j$ takes on the minimal value among all choices of meridian discs (i.e. $\mathcal{P}$ has minimal Whitehead length, see below), it follows that $W_j = \hat{R}_j$.

Whenever we want to detect for a given $\mathcal{P} = \langle a_1, \dots, a_g | \hat{R}_1, \dots, \hat{R}_h \rangle$ whether there is a choice of cancelling pairs making it the spine of a 3-manifold, we apply the *Whitehead algorithm* [Whi36], see also [LS77, Proposition 4.20], to obtain a presentation $\mathcal{Q} = \langle b_1, \dots, b_g | \hat{S}_1, \dots, \hat{S}_h \rangle$ of minimal Whitehead length. Next we check whether $K_\mathcal{Q}^2$ with reduced $\hat{S}_j$ embeds into some 3-manifold.

If it does not, no choice of R_j will do: otherwise we could perform the chain of Whitehead transformations leading from $\mathcal{P}$ to $\mathcal{Q}$ by changes of meridians as above, obtaining a choice (of cancelling pairs) for S_j. But by Kaneto's work mentioned above, a further change of meridian system would then eliminate all cancelling pairs in S_j, as $\mathcal{Q}$ was assumed to be of minimal Whitehead length. Hence $K_\mathcal{Q}^2$ would embed, contradicting the assumption.

If on the other hand $K_\mathcal{Q}^2$ embeds with reduced $\hat{S}_j$, we geometrically perform the chain of Whitehead transformations backwards and obtain a choice (of cancelling pairs) for R_j such that the corresponding 2–complex embeds.

Summing up, we obtain the

Lemma 4.1. *Let* $\mathcal{P} = \langle a_1, \dots, a_g | \hat{R}_1, \dots, \hat{R}_h \rangle$ *be a presentation and let* $\mathcal{Q} = \langle b_1, \dots, b_g | \hat{S}_1, \dots, \hat{S}_h \rangle$ *be the result of an application of the Whitehead algorithm to* $\mathcal{P}$.

There are representatives R_j *for* $\hat{R}_j$ *(R_j in general containing cancelling pairs) such that the 2–complex associated to* $\mathcal{P} = \langle a_1, \dots, a_g | R_1, \dots, R_h \rangle$ *is a spine, if and only if the 2–complex associated to* $\mathcal{Q} = \langle b_1, \dots, b_g | \hat{S}_1, \dots, \hat{S}_h \rangle$ *(with reduced* $\hat{S}_j$*) is a spine.*

In other words, the embeddability of the result $\mathcal{Q} = \langle b_1, \dots, b_g | \hat{S}_1, \dots, \hat{S}_h \rangle$ ($\hat{S}_j$ reduced) of an application of the Whitehead algorithm to $\mathcal{P}$ completely determines whether there is a choice of (cancelling pairs for) R_j such that the corresponding 2–complex embeds. By this reasoning we can focus our attention on the embeddability of $K_\mathcal{P}^2$ with reduced $\hat{R}_j$ and presentations of minimal Whitehead length; the case of an arbitrary $\mathcal{P}$ is solved by juxtaposition of the Whitehead algorithm with the algorithm in question.

5. ZERO–CROSSES AND RECURRENT ARCS

Recall that from now on all presentations are assumed to be reduced. No adjacent pair xx^{-1} or $x^{-1}x$ shall occur in the relators. Criterion 3.3 suggests that a necessary restriction for the form of the relators of a 3-manifold presentation should be that every generator a_i only occurs with 3 pairwise coprime exponents n_i, p_i and $n_i + p_i$. In this section we will discuss why this is *not* true in general and find assumptions which then guarantee the desired necessary condition.

The point is that a railroad system not only determines a presentation, but also a syllable–decomposition of the relators, see below, and only the latter fulfills the exponent condition.

For example this condition would rule the presentation $\langle a, b | a^2 b, a^4 b \rangle$ out as a 3-manifold presentation, as 2 and 4 are not coprime (see [OS74], p. 464). But $\mathcal{P}$ *is* associated to the spine of a 3-manifold as Figure 5 shows:

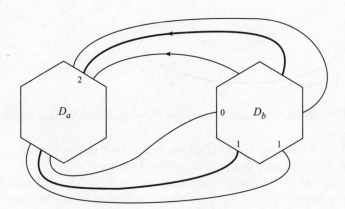

FIGURE 5. Railroad System for $\mathcal{P}_2 = \langle a, b | a^2 b, a^4 b \rangle$

Note that the relator curves k_1 and k_2 literally read $a^2 b$ and $a^2 b^0 a^2 b$: The labels n_b and p_b of the b–hexagon are 0 and 1 and by the formal insertion of the syllable b^0, a no longer occurs with exponent 4.

To overcome the problem of the zero crosses as in the example of Figure 5, Osborne and Stevens threw in the assumption that for each generator a_i there is an a_i–syllable in some R_j whose exponent is not ± 1.

Another way to get more exponents into a 3-manifold presentation works with *recurrent arcs* in the railroad system. If a relator curve leaves some handle H_i crossing its belt curve, runs on the 2–sphere and returns to the *same* handle H_i, the railroad system has an arc with endpoints on the same hexagon (see Figure 5).

Note that the relator curves literally read $a^{-2} c a^2 a^2 b$ and $a^3 c a^{-1} b$: By the formal splitting of the syllable a^4 into the product $a^2 a^2$, a no longer occurs with exponent 4.

If we apply the Whitehead transformation that conjugates c with a^2, the presentation $\mathcal{P}_3$ from Figure 6 turns into $\mathcal{P}_4 = \langle a, b, c | c a^2 b, a^5 c a^{-3} b \rangle$. The sum over the lengths of the relators is equal for $\mathcal{P}_3$ and $\mathcal{P}_4$ (in fact they are Whitehead–reduced) but a syllable a^e has disappeared (other a-syllables have

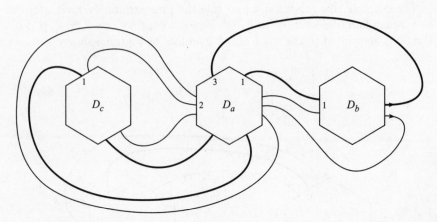

FIGURE 6. Railroad System for $\mathcal{P}_3 = \langle a, b, c | a^{-2}ca^4b, a^3ca^{-1}b \rangle$

changed their exponent). We shall see below that this is the only situation
in which recurrent arcs can help for embeddability.

Each R_j can be uniquely written as a word $R_j = \prod_{l=1}^{k} a_{i_l}^{e_l}$ with nonzero
integers e_l and with $a_{i_l} \neq a_{i_{l+1}}$, called the *natural* syllable–decomposition of
R_j.

Definition 5.1. A presentation $\mathcal{P}$ is not *syllable–reduced* if there is a
(Whitehead–)transformation carrying each a_i into one of a_i or $a_{i_0}^{-m} a_i a_{i_0}^{m}$
($m \in \mathbb{Z}$) such that the natural syllable–decompositions of the $R_j = \prod_{l=1}^{k} a_{i_l}^{e_l}$
turn into $R_j = \prod_{l=1}^{k} a_{i_l}^{e_l'}$ where some e_l' has become zero.

Note that the sequence $(a_{i_1} \ldots a_{i_k})$ is assumed to be the same. No "birth"
of a new syllable is allowed. If $\Gamma_\mathcal{P}$ is the graph having one vertex for each
a_i and an edge between two such vertices whenever there are adjacent sylla-
bles $a_{i_l}^{\pm 1} a_{i_{l+1}}^{\pm 1}$ in some cyclical R_j, then a necessary condition for $\mathcal{P}$ not being
syllable–reduced is that $\Gamma_\mathcal{P}$ has some vertex as articulation point. It is easy
to see that if $\mathcal{P}$ is Whitehead–reduced but not syllable–reduced, there is a
syllable–reduction leading to another Whitehead–reduced presentation. (The
elementary Whitehead transformation τ where τ^m is the transformation from
Definition 5.1., has the property: if τ increases the Whitehead length, then
τ^{-1} decreases it.) We can hence focus our attention on the embeddability
of $K_\mathcal{P}^2$ with reduced and syllable–reduced R_j, the case of an arbitrary $\mathcal{P}$ is
solved by juxtaposition of the "stronger" Whitehead algorithm (that gives a
Whitehead–reduced *and* syllable–reduced presentation) with the algorithm in
question.

6. THE CRITERION

Theorem 6.1. *Let $\mathcal{P} = \langle a_1, \ldots, a_g \mid R_1, \ldots, R_h \rangle$ be reduced and syllable-reduced and assume that for each a_i there is some R_j in which a_i occurs with exponent other than ± 1. If $\mathcal{P}$ is a 3-manifold presentation, then there is a railroad system for the natural syllable-decomposition of the R_j.*

Corollary 6.2. *Let $\mathcal{P} = \langle a_1, \ldots, a_g \mid R_1, \ldots, R_h \rangle$ be reduced and syllable-reduced and assume that for each a_i there is some R_j in which a_i occurs with exponent other than ± 1. If $\mathcal{P}$ is a 3-manifold presentation, then the absolute values of the a_i-exponents take on at most three distinct integers $n_i, n_i + p_i, p_i$ where $\gcd(n_i, p_i) = 1$.* □

Proof of the Theorem. $\mathcal{P}$ being a 3-manifold presentation, it satisfies Criteria 3.1 and 3.2. By the assumption that each generator occurs with exponent $\neq \pm 1$ in some R_j, in Figure 2 for each i there is an arc β_i that is disjoint to the relator curves k_j, joining D_i to $\bar{D}_i$. Cutting along these β_i gives belt curves and thus a railroad system where each hexagon is labelled by $1, 1, 0$. (Recall that a penetration through the hexagon at an edge labelled k contributes a syllable a_i^k to the word we are reading on the handlebody.)

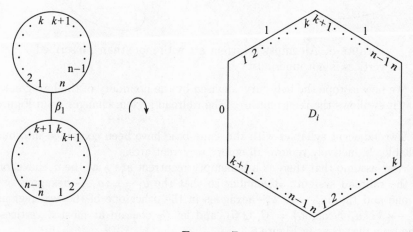

FIGURE 7

Reading the labels while following a relation curve gives an *admissible* syllable-decomposition, i.e. a syllable-decomposition $\prod_{l=1}^{m} a_{i_l}^{e_l} = R_j$ where $(i_{k-1} \neq i_k = \ldots = i_s \neq i_{s+1} \Rightarrow i_k + \ldots + i_s \neq 0)$; in particular $e_l \neq 0$. Note that an admissible syllable-decomposition of R_j, where no two adjacent syllables have the same base, is the natural one. Inductively we want to eliminate arcs which do not constitute an entire relator curve but have both endpoints on the same hexagon; call those *recurrent* arcs. This operation will reduce the number of syllables and maintain admissibility of the syllable–decomposition. A recurrent arc λ is called *empty* if one of the two regions $S^2 - \{\lambda \cup \text{hexagon containing } \partial\lambda\}$ does not contain hexagons. Consider an innermost empty recurrent arc. Note that it has at most 2 inscribed vertices.

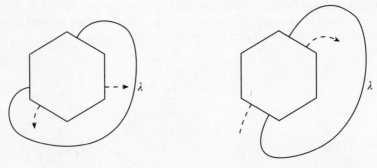

FIGURE 8. An empty recurrent arc with more than 2 inscribed vertices is not innermost

We now isotope the belt curve devised by the boundary of the hexagon so that it swallows the recurrent arc. The railroad system changes as in Figure 6.

Two adjacent syllables with the same base have been combined into one syllable. Inductively remove all empty recurrent arcs.

Now assume that there is a nonempty recurrent arc λ at the a_s–hexagon of the railroad system. Renumber so that the $a_1 - \ldots a_{s-1}$–hexagons are in one and the $a_{s+1} - \ldots a_g$–hexagons in the other one of the two regions $S^2 - \{\lambda \cup a_s\text{–hexagon}\} = \mathcal{G}_1 \cup \mathcal{G}_2$, and let $\mathcal{G}_1$ contain at most 3 vertices. Choose a disc D as in Figure 6

Isotope the hexagons in $\mathcal{G}_1$ through $(D \cap a_s\text{–hexagon})$. (See Figure 6.)

At this point, all arcs in $\mathcal{G}_1$ are empty and can thus be successively removed. Eventually, λ is empty and can be removed.

During this process, every penetration from $\mathcal{G}_2$ into $\mathcal{G}_1$ through the a_s–hexagon has been substituted at most by another a_s–penetration. Hence,

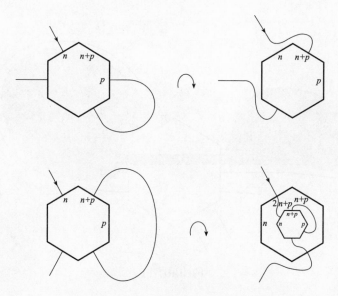

FIGURE 9. $n, p \in \mathbb{Z}$, $gcd(n, p) = 1$.

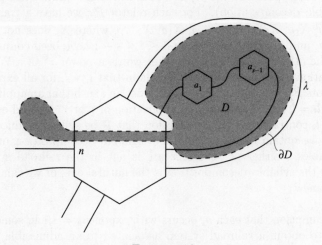

FIGURE 10

some syllables with base a_s may have disappeared (those parallel to the dragging track), some changed their exponent; combining two syllables at the

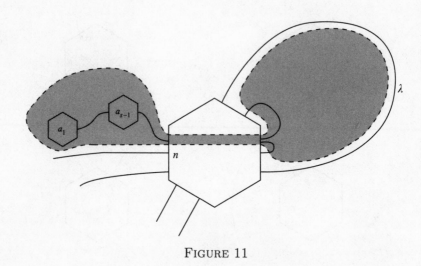

FIGURE 11

beginning and end of λ eventually has decreased the total number of syllables.

What has happened to the R_j as words in the free group (as opposed to their syllable–decomposition)? For each relator R_j, we have a transition of the form $a_s^{l_1} X_1 a_s^{l_2} X_2 \ldots \rightsquigarrow a_s^{\bar{l}_1} X_1 a_s^{\bar{l}_2} X_2 \ldots$, where X_i does not begin or end with a_s and $l_i \neq 0 \; \forall i$. The a_i for $1 \leq i \leq s-1$ have been conjugated by a_s^k while the others stayed fixed, in other words a power τ^k of a Whitehead transformation τ has been applied. We claim that $l_i = \bar{l}_i$ for all exponents l_i in all relators R_j. Otherwise there would exist an i such that an application of τ leads to $\bar{l}_i = l_i \pm 1$. Then iterated application of τ or τ^{-1} would eventually annihilate l_i contradicting the assumption that $\mathcal{P}$ be syllable–reduced.

Hence, we obtain another admissible syllable–decomposition of $\mathcal{P}$ with fewer syllables. Either no recurrent arc is left in the railroad system, in which case the syllable–decomposition is the natural one, or we can start the induction anew. $\square$

The assumption that each a_i occurs with exponent $\neq \pm 1$ in some R_j was necessary to obtain a railroad system associated to an admissible syllable–decomposition. Instead, one can work with "generalized" railroad systems where hexagons are only introduced for those a_i that occur with exponent $\neq \pm 1$ in some R_j, the others keep the two discs as in Criterion 3.2. The notion of "syllable–reduced" (Definition 5.1) has to be extended allowing a power of *any* elementary Whitehead–transformation. Then the proof of Theorem 6.1 carries through to yield:

Theorem 6.3. *Let* $\mathcal{P} = \langle a_1, \ldots, a_g \mid R_1, \ldots, R_h \rangle$ *be reduced and syllable-reduced. If* $\mathcal{P}$ *is a 3-manifold presentation, then there is a generalized railroad system for the natural syllable–decomposition of the* R_j. $\qquad\Box$

Corollary 6.4. *Let* $\mathcal{P} = \langle a_1, \ldots, a_g \mid R_1, \ldots, R_h \rangle$ *be reduced and syllable-reduced. If* $\mathcal{P}$ *is a 3-manifold presentation, then the absolute values of the* a_i*-exponents take on at most three distinct integers* $n_i, n_i + p_i, p_i$ *where* $gcd(n_i, p_i) = 1$. $\qquad\Box$

REFERENCES

[Dun76] M. J. Dunwoody. The homotopy type of a two-dimensional complex. *Bull. London Math. Soc.*, 8:282–285, 1976.

[H-AM93] C. Hog-Angeloni and W. Metzler, Geometric aspects of 2-dimensional complexes. In: *Two-dimensional Homotopy and Combinatorial Group Theory* (C. Hog-Angeloni, W. Metzler, and A. Sieradski, editors), Lond. Math. Soc. Lecture Note Series 197. Cambridge University Press, pages 1-50, 1993.

[Hem76] J. Hempel. *3–Manifolds*, volume 86. Annals of Math. Studies, Princeton University Press, Princeton, New Jersey, 1976.

[HLM85] C. Hog, M. Lustig, and W. Metzler. Presentation classes, 3-manifolds and free products. In *Geometry and Topology Proceedings, 1983-84* (J Alexander and J. Harer, editors), Springer Lecture Notes 1167, pages 154–167, 1985.

[Kan82] T. Kaneto. On simple loops on a solid torus of general genus. *Proc. Amer. Math. Soc. 86*, pages 551–552, 1982.

[LS77] R. Lyndon and P. Schupp. *Combinatorial group theory*. Springer Verlag, Berlin, 1977.

[Neu68] L. Neuwirth. An algorithm for the construction of 3-manifolds from 2-complexes. *Proc. Camb. Phil. Soc. 64*, pages 603–613, 1968.

[OS74] R. P. Osborne and R. S. Stevens. Group presentations corresponding to 3-manifolds, I. *American J. Math. 96 No. 3*, pages 454–471, 1974.

[OS77] R. P. Osborne and R. S. Stevens. Group presentations corresponding to 3-manifolds, II. *Trans. Amer. Math. Soc. 234*, pages 213-243, 1977

[Whi36] J. H. C. Whitehead. On certain sets of elements in a free group. *Proc. London Math. Soc (2) 41*, pages 48–56, 1936.

[Zie70] H. Zieschang. On simple systems of paths on complete pretzels. *Amer. Math. Soc. Transl. (2) 92*, pages 127–137, 1970.

[Zie88] H. Zieschang. On Heegaard diagrams of 3-manifolds. *Asterisque No. 163/164*, pages 247–280, 1988.

Cynthia Hog-Angeloni
Mathematisches Seminar der
Johann Wolfgang Goethe Universität
Robert-Mayer-Straße 6-10
D-60054 Frankfurt am Main
Germany

IN SEARCH OF A WORD WITH SPECIAL COMBINATORIAL PROPERTIES

ŠTĚPÁN HOLUB

ABSTRACT. In Section 2, the notions of canonical and principal solution of an equation in a free monoid are discussed. In Section 3 it is proved that if a non-cyclic solution α of the system of equations $(x_1 \dots x_n)^s = x_1^s \dots x_n^s$, $s = 2, 3$, exists, no $\alpha(x_i)$ is a power of a letter. It is also shown that in such a case the shortest non-cyclic solution is principal and of rank two. In Section 4 some similar results are presented without proof.

1. INTRODUCTION

The equality

$$(1.1) \qquad (u_1 \dots u_n)^s = u_1^s \dots u_n^s$$

trivially holds for all integers s if every pair u_i, u_j commutes. In a free semigroup it means that all words u_i, $1 \le i \le n$, are powers of a common word v. However, for any k there exists an integer n and words $u_1, \dots, u_n$ such that (1.1) holds just for $s = k$ (and $s = 1$). Take for example $n = 2k - 1$ and

$$u_i = \begin{cases} A & i = 2j, \quad 1 \le j \le k - 1 \\ \\ A^{k-j} B A^{j-1} & i = 2j - 1, \quad 1 \le j \le k \end{cases}$$

with some letters A, B.

The question is whether there exists an n-tuple $u_1, \dots, u_n$ such that (1.1) holds for more than one integer $s > 1$ but does not hold for each $s \in \mathbb{N}$. (It is not difficult to see that if (1.1) holds for each $s \in \mathbb{N}$, the $u_1, \dots, u_n$ are powers of a common word.) It was shown (see [3]) that (1.1) holds for all $s \in \mathbb{N}$ as soon as it holds for three integers greater than 1. The question whether there exists an n-tuple of words $u_1, \dots, u_n$ such that (1.1) holds for exactly two integers greater than 1, remains open. We shall give a condition that any such n-tuple must satisfy.

2. EQUATIONS AND THEIR SOLUTIONS

In this section we introduce some concepts regarding equations in free semigroups and their solutions.

2.1. Basic notions. Let A be a finite *alphabet*. Elements of A are called *letters* and sequences of letters are called *words*. The sequence of length zero is called the *empty word*, denoted ε. The set of all words (all non-empty words resp.) is denoted by A^* (A^+, resp.). It is a monoid (semigroup, resp.) under the operation of concatenation. The length of a word u will be denoted by $|u|$. We say that a word u is a factor of a word v if and only if there exist words z, $z' \in A^*$ such that $v = zuz'$.

By a *cyclic factor* of v we shall understand every factor u of vv with $|u| \le |v|$. A factor u of a word v can occur in v in different *instances* (each of those determined by the length of the word preceding u in v). An instance of a cyclic factor will be called a *cyclic instance* of u, and it corresponds to an instance of u in vv that starts within the first copy of v. The set of all cyclic factors of v will be denoted by $C(v)$.

Let X be a finite set of unknowns. Every

$$(e, e') \in X^+ \times X^+$$

we shall call an *equation* in unknowns from X. We shall always assume that X contains only unknowns occuring in (e, e'). For a particular equation (e, e') we shall often use the suggestive notation $e = e'$.

We say that a morphism $\alpha : X^+ \to A^+$ is a *solution* of the system of equations $S \subseteq X^+ \times X^+$ in the semigroup A^+, if and only if for every $(e, e') \in S$ the equality $\alpha(e) = \alpha(e')$ holds. By $alph(\alpha)$ we denote the set of letters occuring in $\alpha[X] = \{\alpha(x); x \in X\}$. We shall always assume that $alph(\alpha) = A$. Two systems of equations S, S' are called *equivalent* if and only if they have the same set of solutions.

We say that a solution $\alpha : X^+ \to A^+$ is *cyclic* if and only if there exists a word $v \in A^+$ such that $\alpha(x)$ is a power of v for every $x \in X$.

We say that two solutions $\alpha : X^+ \to A^+$ and $\beta : X^+ \to B^+$ are isomorphic if and only if there exists a bijection $\theta : A \to B$ such that $\beta = \theta \circ \alpha$.

2.2. Ranks. A subset S of a free semigroup A^+ closed under the operation of concatenation is called a subsemigroup of A^+. A subsemigroup S is not necessarily free. For example the subsemigroup of $\{a, b\}^+$ generated by elements $\{a, ab, ba\}$ is not free as $a(ba) = (ab)a$. However, as the set of free subsemigroups is closed under intersection (see [2], p.6), there exists the smallest free subsemigroup containing a subset $S \subset A^+$, called the *free hull* of S. The cardinality of the basis of the free hull is called the *rank* of S. The rank of a morphism $\alpha : X^+ \to A^+$ is the rank of the set $\alpha[X]$. The basis of the free hull of the set $\alpha[X]$ is called the *basis* of the morphism α. Finally we say that the rank of an equation is the maximal rank of its solutions. The rank of an equation is 1 if and only if the equation admits only cyclic solutions.

2.3. Canonical solution.

Let $\alpha : X^+ \to A^+$ be a solution of an equation (e, e'). Suppose $X = \{x_1, \ldots, x_n\}$. The n-tuple $(d_1, \ldots, d_n)$, with $d_i = |\alpha(x_i)|$, shall be called the *type* of α. The solution α is called a *canonical solution of type* $(d_1, \ldots, d_n)$ (or simply *canonical*) if and only if for every solution $\beta : X^+ \to B^+$ of (e, e') and every mapping $\theta : B \to A$, such that $\alpha = \theta \circ \beta$, the mapping θ is a bijection. In other words, α is canonical if and only if the cardinality of $alph(\alpha)$ is maximal among all solutions of (e, e') of the same type. All non-canonical solutions of a given type result from a canonical solution by identification of some letters. It is easy to see that all canonical solutions are isomorphic.

The notion of canonical solution was introduced by Appel and Djorup in [1], for the particular equation $z_1^n \ldots z_k^n = y^n$. The definition they use is different from that presented above and is more intuitive. In fact it describes the construction of the canonical solution of a given type and can be generalized as follows. Let $(d_1, \ldots, d_n)$ be the type for which we wish to construct the canonical solution of an equation (e, e') in unknowns $\{x_1, \ldots, x_n\}$. We introduce an alphabet Y consisting of new letters $\eta_{i,j}$, $1 \le i \le n$, $1 \le j \le d_i$ and define a morphism $\psi : X^+ \to Y^+$ by equalities

$$\psi(x_i) = \eta_{i,1} \ldots \eta_{i,d_i}, \ 1 \le i \le n.$$

Obviously we assume that $|\psi(e)| = |\psi(e')|$, otherwise the equation has no solution of the type $(d_1, \ldots, d_n)$. We shall call the equation $(\psi(e), \psi(e'))$ *the type equation associated with* (e, e') *and* $(d_1, \ldots, d_n)$.

Let $\sim$ be the smallest equivalence relation on Y such that $\eta_{i,j} \sim \eta_{k,l}$, as soon as $\psi(e) = v\eta_{i,j}w$ and $\psi(e') = v'\eta_{l,k}w'$ for some $v, v', w, w' \in Y^*$ such that $|v| = |v'|$, $|w| = |w'|$. Let A be the set of equivalence classes of $\sim$ and $\pi : Y^+ \to A^+$ the natural projection $\pi(\eta_{i,j}) = [\eta_{i,j}]_\sim$. Then $\pi \circ \psi : X^+ \to A^+$ is a canonical solution of the type $(d_1, \ldots, d_n)$.

Let $\alpha : X^+ \to A^+$ be a canonical solution. Denote by L the set of all left letters, i.e. all letters $a \in A$ such that $\alpha(x) = av$, for some $x \in X$ and $v \in A^+$. Similarly denote by R the set of all right letters. Denote by H the set of all words $v \in A^+$, such that v begins with a left letter, ends with a right letter and does not contain any other left or right letter. (If $a \in R \cap L$, then $a \in H$). It can be proved (see [1]) that H is the basis of α.

2.4. Principal solution.

On the set of all solutions of an equation (e, e') we define a partial ordering $\le$. Given two solutions $\alpha : X^+ \to A^+$ and $\beta : X^+ \to B^+$ we say that $\alpha \le \beta$ if and only if there exists a morphism $\theta : A^+ \to B^+$ such that $\beta = \theta \circ \alpha$. If θ is an isomorphism (i.e. it is generated by a bijection $A \to B$) then both $\alpha \le \beta$ and $\beta \le \alpha$, as $\alpha = \theta^{-1} \circ \beta$. Any minimal element of this ordering is called a *principal solution* of (e, e'). In other words a solution α is principal if and only if θ is an isomorphism as soon as $\theta \circ \alpha$ is a solution of (e, e'). It follows that every principal solution is canonical, but not vice versa. All canonical solutions $\beta : X^+ \to B^+$ can be expressed as $\theta \circ \alpha$, where $\alpha : X^+ \to A^+$ is a principal solution and $\theta : A^+ \to B^+$ is a morphism such that for $a_1, a_2 \in A$, $a_1 \neq a_2$, the words $\theta(a_1)$, $\theta(a_2)$ have no common letter.

The rank of a principal solution $\alpha : X^+ \to A^+$ is equal to the cardinality of A. Indeed, if $\{v_1, \ldots, v_m\}$ is the basis of α, we can introduce an alphabet B, consisting of new letters $b_1 \ldots b_n$, and define a morphism $\theta : B^+ \to A^+$ by equalities $\theta(b_i) = v_i$, $1 \leq i \leq n$. Then $\alpha = \theta \circ \beta$ is a solution of (e, e') and θ is an isomorphism.

One can see that a solution $\alpha : X^+ \to A^+$ is principal if and only if it is canonical and A is the basis of α. In other words α is principal if and only if the following condition is satisfied: Let $\beta : X^+ \to B^+$ be a canonical solution of the same type as α. Then every $b \in B$ is both a right and left letter.

3. SHORTEST COUNTEREXAMPLE

Consider now following system of equations

(3.1)
$$(x_1 \ldots x_n)^2 = x_1^2 \ldots x_n^2,$$
$$(x_1 \ldots x_n)^3 = x_1^3 \ldots x_n^3,$$

in unknowns $X = \{x_1, \ldots, x_n\}$. Henceforward we shall suppose that there exists a positive integer n such that the system (3.1) has a non-cyclic solution (i.e. its rank is greater than one) and let n be the smallest one. Surely $n > 2$, because all equations (e, e') in two unknowns, such that $e \neq e'$, have only cyclic solutions in free semigroups (see e.g. [2], p.164). We say that α is a *shortest counterexample* if and only if it is a non-cyclic solution of (3.1) and for every non-cyclic solution α' the inequality

$$|\alpha(x_1 \ldots x_n)| \leq |\alpha'(x_1 \ldots x_n)|$$

holds.

Lemma 3.1. *Let α be a shortest counterexample. Then it is a principal solution of rank 2.*

Proof. Note that every solution of the same type as α is a shortest counterexample. First, suppose that $\alpha : X^+ \to A^+$ is a canonical solution and let C be a basis of α. We can understand α as a morphism $X^+ \to C^+$ and, thanks to the minimality of $|\alpha(x_1 \ldots x_n)|$, we have $C = A = \{a_1, \ldots, a_m\}$. Suppose that $rank(\alpha) = card(A) = m$ is greater than 2. Denote by A' the set $\{a_1, a_2\}$ and define a morphism $\theta : A^+ \to (A')^*$ by

$$\theta(a_i) = \begin{cases} a_i & 1 \leq i \leq 2 \\ \varepsilon & 3 \leq i \leq m. \end{cases}$$

If we omit all x_i such that $\theta \circ \alpha(x_i) = \varepsilon$ we get a solution $\alpha' = \theta \circ \alpha : X^+ \to (A')^+$ of (3.1) with some $n' \leq n$. If α' is not cyclic, we have a contradiction to the fact that α is a shortest counterexample. Suppose that α' is cyclic, and let $v = b_1 \ldots b_l$, $l \geq 2$, $b_i \in A'$, $1 \leq i \leq l$, be the shortest word such that all $\alpha'(x_i)$, $1 \leq i \leq n'$, are powers of v. It is not difficult to see that α' is canonical (as α is) and therefore $b_i \neq b_j$, $1 \leq i < j \leq l$. It follows that b_1 is not the final letter of any $\alpha(x_i)$, a contradiction to the fact that A is the basis of α. We have proved that if a shortest counterexample is canonical, it is of rank 2.

Suppose now, for a contradiction, that $\alpha : X^+ \to A^+$ is a general shortest counterexample that is not a principal solution. Let $\beta : X^+ \to B^+$ be a solution of (3.1) and $\theta : B^+ \to A^+$ a morphism, such that $\alpha = \theta \circ \beta$ and θ is not an isomorphism. Clearly

$$|\alpha(x_1 \ldots x_n)| \geq |\beta(x_1 \ldots x_n)|$$

and the equality must hold, because α is a shortest counterexample and β is not cyclic. We deduce that θ maps B onto A. Suppose that θ is not injective. Then the cardinality of B must be at least 3. As the cardinality of the alphabet of a canonical solution is maximal among all solutions of given type, the canonical solution of the type $(d_1, \ldots, d_n)$, with $d_i = |\alpha(x_i)| = |\beta(x_i)|$, is of rank greater than 2 and at the same time it is a shortest counterexample, a contradiction to what we proved above. Therefore any shortest counterexample is principal, which implies it is canonical and therefore of rank 2. □

Lemma 3.2. *Let* $\alpha : X^+ \to A^+$, $A = \{a, b\}$, *be a shortest counterexample. Then for all* $1 \leq i \leq n$, *the word* $\alpha(x_i)$ *contains both letters* a *and* b.

Proof. First put

$$u_1 = (x_1 \ldots x_n)^6, \quad u_2 = (x_1^2 \ldots x_n^2)^3, \quad u_1 = (x_1^3 \ldots x_n^3)^2,$$

and note that the system (3.1) is equivalent to the system

$$(u_1, u_2),$$
(3.2)
$$(u_1, u_3).$$

Put $d_i = |\alpha(x_i)|$, $1 \leq i \leq n$. Define Y and $\psi : X^+ \to Y^+$ in such a way that

$$(\psi(u_1), \psi(u_2)),$$
(3.3)
$$(\psi(u_1), \psi(u_3))$$

are the type equations associated with (3.2) and $(d_1 \ldots d_n)$. Let

$$y_i = \eta_{i,1} \ldots \eta_{i,d_i}, \quad 1 \leq i \leq n,$$

and

$$w_i = \psi(u_i), \quad 1 \leq i \leq 3.$$

Denote by $\pi : Y^+ \to A^+$ the morphism satisfying $\pi(y_i) = \alpha(x_i)$. Such a morphism is determined uniquely and $\pi(\eta) \in A$ for all $\eta \in Y$. It implies that

$|\pi(w)| = |w|$. Let $F = \bigcup_{1 \leq i \leq 3} C(w_i)$, the set of all factors that can be found in equations (3.3). We will proceed by contradiction. Suppose (without loss of generality) that for some i, $1 \leq i \leq n$, $\alpha(x_i)$ is a power of a. Let m be the biggest integer for which there exists a word $w \in F$ of length m such that $\pi(w) = ba^m b$, and y_i^3 is a factor of w for some $1 \leq i \leq n$. Let Z be the set of all words $w \in F$ such that $\pi(w) = ba^m b$. If $w \in Z$ then

$$|w| = m + 2 \leq |y_1 \ldots y_n| = \sum_{i=1}^{n} d_i.$$

Now we shall define a disjoint factorization of Z, according to the complexity of its elements. Let $w \in Z$. Denote by $i(w)$ the number of different first indices of letters occuring in w. It is the number of different words y_i affected by w. Denote by $\sigma(w)$ the minimal exponent k such that w is a cyclic factor of $y^k_1 \ldots y_n^k$. Obviously $\sigma(w) \leq 3$ for $w \in Z$. Denote by $W(i,j)$, $1 \leq i \leq 2$, $1 \leq j \leq 3$, the set of all words $w \in Z$, such that $i(w) = i$ and $\sigma(w) = j$. Also denote by $W(3,j)$ the set of all words $w \in Z$, such that $i(w) \geq 3$ and $\sigma(w) = j$. It follows from the definitions that $W(i,j)$, $1 \leq i \leq 3$, $1 \leq j \leq 3$, is really a disjoint factorization of Z.

The definition of the sets $W(i,j)$ is motivated by the fact that w, w' belong to the same set if and only if, in all three words w_1, w_2, w_3, the number of cyclic instances of w is the same as the number of cyclic instances of w'. Indeed if we let (i,j,k) denote the number of cyclic instances of a word $w \in W(i,j)$ in the word w_k, we can easily verify the following values:

$$
\begin{aligned}
(1{,}2{,}1) = (1{,}3{,}1) = (1{,}3{,}2) = (2{,}2{,}1) = (2{,}3{,}1) = (2{,}3{,}2) = & \\
(3{,}1{,}2) = (3{,}1{,}3) = (3{,}2{,}1) = (3{,}2{,}3) = (3{,}3{,}1) = (3{,}3{,}2) \ &= 0 \\
(1{,}3{,}3) = (2{,}1{,}3) = (2{,}2{,}3) = (2{,}3{,}3) = (3{,}3{,}3) \ &= 2 \\
(1{,}2{,}2) = (2{,}1{,}2) = (2{,}2{,}2) = (3{,}2{,}2) \ &= 3 \\
(1{,}2{,}3) \ &= 4 \\
(1{,}1{,}1) = (1{,}1{,}2) = (1{,}1{,}3) = (2{,}1{,}1) = (3{,}1{,}1) \ &= 6.
\end{aligned}
$$

Suppose that $w \in W(3,1)$. Then w has at least one factor

$$\eta_{i-1,d_{i-1}} y_i \eta_{i+1,d_{i+1}},$$

with $1 \leq i \leq n$, and $i-1$, $i+1$ considered modulo n. For such an i we have $\pi(y_i) = a^l$, $l \geq 1$. If we substitute in the word w all such words y_i by y_i^3, we get a word w' such that $\pi(w') = ba^p b$, $p > m$, contradicting the maximality of m. It follows that $W(3,1)$ is empty. Similarly we can see that $W(3,2)$, $W(2,2)$, $W(2,3)$, $W(1,3)$ are also empty.

Denote by P the number of cyclic occurences of the word $ba^m b$ in the word $u = \alpha(u_1) = \alpha(u_2) = \alpha(u_3)$. Looking at a word w_k, $1 \leq k \leq 3$, we can see that P is equal to the total number of cyclic instances of elements from Z in w_k. Thanks to the disjoint factorization of Z, we can express P in three ways (for $1 \leq k \leq 3$) as a sum

$$\sum_{1 \leq i,j \leq 3} |W(i,j)|(i,j,k).$$

Using all the above knowledge we get equalities

$$
\begin{aligned}
P = 6|W(1,1)| + 6|W(2,1)| = & \\
(3.4) \qquad = 6|W(1,1)| + 3|W(2,1)| + 3|W(1,2)| = & \\
= 6|W(1,1)| + 2|W(2,1)| + 4|W(1,2)| + 2|W(3,3)|. &
\end{aligned}
$$

From these equalities it easily follows that $|W(3,3)| = 0$, a contradiction to the definition of the set Z. $\square$

4. FURTHER RESULTS AND REMARKS

The method used to prove Lemma 3.2 is described in the most general form in [3]. Thanks to that method, some other results were achieved. First, the statement of Lemma 3.2 is valid for all systems of equations

$$
(4.1) \qquad
\begin{aligned}
(x_1 \ldots x_n)^r &= x_1^r \ldots x_n^r, \\
(x_1 \ldots x_n)^s &= x_1^s \ldots x_n^s,
\end{aligned}
$$

with $r > s > 1$.

In the proof of Lemma 3.2 the fact that α is a shortest counterexample was not used. For a general solution the lemma can be reformulated as follows.

Theorem 4.1. *Let $\alpha : X^+ \to A^+$ be a solution of (4.1) and v be an element of the basis of α. Then v is a factor of each $\alpha(x_i)$, $1 \leq i \leq n$.*

The proof of this theorem is mutatis mutandis the same as that of Lemma 3.2.

The proof of the following lemma is a bit technical (see [4]).

Lemma 4.2. *Let $\alpha : X^+ \to A^+$ be a solution of (4.1) with $A = \{a, b\}$. Let $v = ba^m b$, $m \geq 1$, be a cyclic factor of $\alpha(x_1 \ldots x_n)$. Then v is a cyclic factor of every $\alpha(x_i)$, $1 \leq i \leq n$.*

Finally, the most important result (see [3]), mentioned in the Introduction, is that the rank of the equational system

$$(x_1^{k_1} \ldots x_n^{k_1})^{k_2 k_3} = (x_1^{k_2} \ldots x_n^{k_2})^{k_1 k_3} = (x_1^{k_3} \ldots x_n^{k_3})^{k_1 k_2},$$

$k_1 > k_2 > k_3 > 1$, is 1.

REFERENCES

1. K. I. Appel, F. M. Djorup, *On the equation $z_1^n z_2^n \ldots z_k^n = y^n$ in a free semigroup*, Trans. Am. Math. Soc. **134** (1968), pp. 461–470.
2. M. Lothaire, *Combinatorics on words*, Cambridge University Press, 1983.
3. Š. Holub, *Local and global cyclicity in free monoids*, submitted.
4. Š. Holub, *O rovnicích $(x_1^s \ldots x_n^s)^r = (x_1^r \ldots x_n^r)^s$ ve volných pologrupách*, M.D. thesis, Charles University, Prague, 1998.

Štěpán Holub,
Department of Mathematics,
Charles University,
Sokolovská 83,
186 75 Praha,
Czech Republic.
holub@karlin.mff.cuni.cz

CANCELLATION DIAGRAMS WITH NON-POSITIVE CURVATURE

GÜNTHER HUCK AND STEPHAN ROSEBROCK

ABSTRACT. We define small cancellation conditions W^* and W representing non-positive curvature in cancellation diagrams. Each of these two conditions, which are dual to each other, generalizes the classical small cancellation conditions C(6), C(4) T(4), C(3) T(6). W also contains the small cancellation condition $W(6)$ of Juhasz. Our main result is the solution of the conjugacy problem for groups with a presentation satisfying W^* or W. Its proof uses the geometry of non-positively curved piecewise Euclidean complexes developed by Bridson. Following a sketch of Gersten, we also give a detailed proof of the solvability of the word problem for such groups by a quadratic isoperimetric inequality.

1. THE CONDITIONS W^* AND W

Let $P = \langle x_1, \ldots, x_n \mid R_1, \ldots, R_m \rangle$ be a finite presentation of the group G. We always assume that each relator R_i is cyclically reduced and no relator is the trivial word or a cyclic permutation of another relator or of the inverse of another relator. (By "relator" we always mean a defining relator of the presentation.) Let F be the free group on the generators. K_P denotes the standard 2-complex modeled on P.

The *Whitehead graph* W_P of K_P is the boundary of a regular neighborhood of the only vertex of K_P. For each generator x_i of P it has two vertices $+x_i$ and $-x_i$ which correspond to the beginning and the end of the oriented loop labeled x_i in K_P. The edges of W_P are the *corners* of the 2-cells of the 2-complex. The *star graph* S_P is the same as the Whitehead graph if no relator of P is a proper power. If P contains relators that are proper powers, the star graph is obtained from the Whitehead graph by identifying edges of W_P that correspond under the periodicity of a relator R_i. More precisely, if the relators R_i of P are of the form $s_i^{k_i}$ with s_i not a proper power, then the star graph S_P is the Whitehead graph of the presentation $\langle x_1, \ldots, x_n \mid s_1, \ldots, s_m \rangle$ and s_i is called the *root* of the relator R_i.

Let $\pi : W_P \to S_P$ be the natural map from the Whitehead graph to the star graph. We use the star graph when dealing with algebraic questions about presentations, such as the word problem and the conjugacy problem, because, algebraically, a rotation of a periodic relator by any multiple of its period is irrelevant. For topological questions, e.g. when dealing with asphericity of 2-complexes, such rotations are quite significant and therefore the Whitehead graph is used instead.

A *combinatorial map* is a cellular map that maps each open cell homeomorphically onto an open cell. A cell complex is said to be *combinatorial* if its attaching maps are combinatorial. Clearly, a 2-complex K_P modeled on a presentation P is combinatorial. A *diagram over* P is a combinatorial map $f : M \to K_P$, where $M = S^2 - \cup_{i \leq \tau} \sigma_i^2$ is a combinatorial cell decomposition of a 2-sphere minus τ disjoint open 2-cells. If $\tau = 0$, M is a sphere and we speak of a *spherical diagram*. If $\tau = 1$, we call the diagram a *disk diagram*, which is the same as a van Kampen diagram. If $\tau = 2$ we speak of an *annular diagram*.

For $\tau > 0$ let δM denote the boundary of M, where we think of the boundary as a collection of boundary paths, namely the boundary paths of the 2-cells σ_i that are deleted from S^2 to obtain M. The *length of the boundary* $l(\delta M)$ is the number of edges in the boundary paths, counting with multiplicity edges that are traversed twice in opposite direction. An oriented edge e in M is labeled by the name x_i^ϵ of its image $f(e)$ in K_P ($\epsilon \in \{-1, +1\}$), and hence every edge path in M is labeled by a word w in F.

A word w in F is trivial in G if and only if there is a disk diagram over P reading w along its boundary. Two words u and v in F represent conjugate elements in G if and only if there is an annular diagram over P such that its boundary paths, oriented parallely, read for suitable choices of start points the words u and v. (A parallel orientation is an orientation of the two boundary paths such that they are freely homotopic in M.)

The Whitehead graph $W(K)$ of an arbitrary combinatorial 2-complex K is the boundary of the regular neighborhood of the 0-skeleton. Its vertices, edges are the components of the intersections of the 1-cells, 2-cells of K with $W(K)$, respectively. A combinatorial map between 2-complexes induces a combinatorial map between the corresponding Whitehead graphs; this applies in particular to maps $f : M \to K_P$ which define diagrams over P. For a diagram M over P the components of the Whitehead graph $W(M)$ are either circles, arcs, or isolated vertices. We say the diagram M is *vertex reduced* if each component $w \subset W(M)$ maps under $\pi \circ f$ onto a path in the star graph S_P that does not pass an edge twice in different directions.

If a diagram M is not vertex reduced, an elementary reduction move can be performed on M, as indicated in Figure 1. In this figure, D_1 and D_2 are distinct 2-cells of M that intersect at least in one vertex Q and map to the same 2-cell in K_P such that the oppositely oriented corners α and β at Q map to the same edge γ in S_P. (This is equivalent to the oppositely oriented boundary paths ν_1 and ν_2 mapping to the same path labeled w in K_P.) The

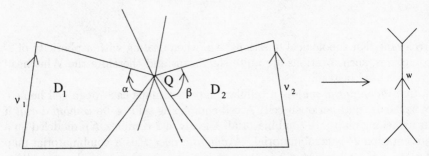

FIGURE 1. Reduction along a vertex

elementary reduction deletes the 2-cells D_1 and D_2 in M and identifies the paths ν_1 and ν_2 to the path ν as sketched in Figure 1.

In general, the situation may be a little more complicated than Figure 1 suggests. An elementary reduction where the boundaries of D_1 and D_2 intersect in more than one component or where the boundary of each D_i has self intersections may result in parts of the diagram being squeezed off as spheres. This splitting off of spheres under an elementary reduction is discussed in detail in [2]. The squeezed off spheres can than be deleted and the following holds: by iterating the above procedure of elementary reduction followed by deleting the squeezed off spherical components, any disk diagram for a word w and any annular diagram for conjugate words u and v (that are not equal to 1 in G) can be transformed into a corresponding vertex reduced diagram that has fewer 2-cells.

We also need the more general notion of a "reduced diagram" which is standard in small cancellation theory and in other applications of diagrams over groups. A diagram is said to be *reduced* if it does not contain a pair of (distinct) 2-cells D_1 and D_2 that intersect in at least one edge x (see Figure 2) such that their oppositely oriented boundary paths $x\nu_1^{-1}$ and $x\nu_2^{-1}$ read the same word in F. Equivalently, we can say M is reduced if no component of the Whitehead graph of M maps under $\pi \circ f$ onto a path in S_P with backtracking, i.e. a path that travels an edge γ immediately followed by its inverse. If a diagram M is not reduced, an elementary reduction can be performed to M that deletes the pair of 2-cells D_1 and D_2 and identifies the paths ν_1 and ν_2. Regarding the details, the same discussion as above applies.

Using the stricter notion of "vertex reduced diagram" instead of the standard notion of "reduced diagram" will allow us to define the small cancellation classes W^* and V^* more generally.

Let the *valence* $d(v)$ of a vertex in a diagram M be the number of edges incident to v, counting edges twice that have both boundary vertices at v. In

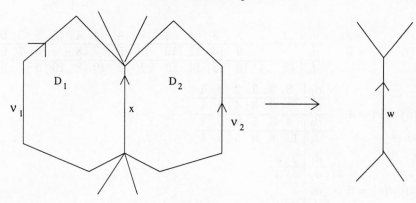

FIGURE 2. Reduction along an edge

all of the following, when we speak of an *inner or interior vertex* of a diagram M we mean an interior vertex of valence $d(v) \geq 3$, i.e. we delete all interior vertices of valence 2, combining the adjacent edges of such a vertex to a single edge. (There are no interior vertices of valence 1 in a diagram, since relators are cyclically reduced.) Under this assumption we define the degree $d(D)$ of a 2-cell D of the diagram to be the number of edges in the boundary path of D, counting with multiplicity edges that are traversed twice (in opposite directions).

A diagram M is said to be of *type W^** if it satisfies the following properties:

(a) Every 2-cell has degree greater than or equal to 3. (This corresponds essentially to the condition $C(3)$ of small cancellation theory)

(b) When every 2-cell D of M of degree d is given the angles of a regular Euclidean d-gon, i.e. every corner of D obtains angle $(1 - 2/d)\pi$, then at every inner vertex the sum of angles is greater than or equal to 2π.

A presentation P is said to be of *type W^** if every vertex reduced diagram over P is of type W^*.

We can express the condition W^* combinatorially: P is of *type W^** if for every vertex reduced diagram M over P and every inner vertex v of M the valence $d(v)$ and the degrees d_i ($i = 1, ..., d(v)$) of the 2-cells incident with v satisfy one of the following four conditions, where the cases $(i), (ii), (iii)$ are to be understood as follows: for each vertex of valence $d(v) = 3, 4, 5$ the degrees d_i of adjacent 2-cells in a suitable order must be greater than or equal to the corresponding numbers in a single column of the table for d_i. (The degree of a 2-cell that has several corners at v occurs with multiplicity):

(i) $d(v) = 3$,

d_1	3	3	3	3	3	3	4	4	4	4	5	5	5	6
d_2	7	8	9	10	11	12	5	6	7	8	5	6	7	6
d_3	42	24	18	15	14	12	20	12	10	8	10	8	7	6

(ii) $d(v) = 4$,

d_1	3	3	3	3	3	4
d_2	3	3	3	4	4	4
d_3	4	5	6	4	5	4
d_4	12	8	6	6	5	4

(iii) $d(v) = 5$,

d_1	3	3
d_2	3	3
d_3	3	3
d_4	3	4
d_5	6	4

(iv) $d(v) \geq 6$, $d_i \geq 3$ for $i = 1, ..., d(v)$

Note that, contrary to the standard non-metric small cancellation conditions (which are included as special cases), arbitrary combinations of different cases or subcases are permitted for the different vertices of the same diagram.

Theorem 1. *Let P be a finite presentation of type W^*. If M is any vertex reduced disc-diagram over P with f 2-cells, then*

$$(1) \qquad f \leq \frac{l^2(\delta M)}{\sqrt{3}\pi}$$

This theorem gives a quadratric isoperimetric inequality and, hence, a solution to the word problem.

In the statement of the following theorem we use the term "piece" as it is understood in small cancellation theory, i.e., in the context of diagrams over a presentation, a *piece* is a word that occurs as the label of an interior edge of some reduced diagram.

Theorem 2. *Let $P = \langle x_1, \ldots, x_n \mid R_1, \ldots, R_m \rangle$ be a finite presentation of type W^* of the group G. Let F be the free group on the generators and p the maximal word length of pieces for the presentation P. If $u, v \in F$ are conjugate in G, then there exists a word $w \in F$ such that $u = wvw^{-1}$ and*

$$(2) \qquad |w| \leq N + 2 \cdot \max\{|u|, |v|\},$$

where N is the number of words in the alphabet $\{x_i^{\pm 1}\}$ of length $\leq 3 \cdot p \cdot \max\{|u|, |v|\}$.

This theorem gives a solution of the conjugacy problem by reducing it to the word problem.

One can dualize the definition of W^* as follows:

A diagram is said to be of *type W* if it satisfies the following:

If at every inner vertex of M of valence k (which is always ≥ 3) every corner is given the angle $(2\pi)/k$, then the sum of the angles of every inner 2-cell D is that of an Euclidean or hyperbolic polygon, i.e. less than or equal to $(d(D) - 2)\pi$. (An inner 2-cell is a closed 2-cell that is contained in the interior of M)

A finite presentation P is said to be of *type W* if every <u>reduced</u> diagram over P is of type W.

Combinatorially this is characterized by the same tables as above (in the case W^*) with valences of vertices and degrees of 2-cells interchanged, i.e. for every inner 2-cell D of degree $d(D) = 3, 4, 5$, the valences d_i of the vertices at the corners of D, in a suitable order, must be greater or equal to the numbers in one column of the corresponding table. If a vertex is incident to several corners of the same 2-cell D its degree occurs with multiplicity.

As for W^*-presentations, the word problem and conjugacy problem for groups which have presentations of type W are solvable.

Theorem 3. *Let P be a finite presentation of type W, such that each generator occurs at least twice in the set $\{s_1, ..., s_m\}$ of the roots of relators (meaning precisely, that the number of occurences of letters from the set $\{x_1, x_1^{-1}, ..., x_n, x_n^{-1}\}$ in the collection of words is ≥ 2). If M is a reduced disk diagram over P with f 2-cells, then*

$$(3) \qquad f \leq \frac{2 \cdot 20^2 \cdot l^2(\delta M)}{\sqrt{3}\pi}$$

Theorem 4. *Let $P = \langle x_1, \ldots, x_n \mid R_1, \ldots, R_m \rangle$ be a finite presentation of type W of the group G, where each generator occurs at least twice in the set $\{s_1, ... s_m\}$ of the roots of relators. Let F be the free group on the generators and r the maximal length of the relators of P. If $u, v \in F$ are conjugate in G, then there exists a word $w \in F$ such that $u = wvw^{-1}$ and*

$$(4) \qquad |w| \leq N + 2\max\{|u|, |v|\},$$

where N is the number of words in the alphabet $\{x_i^{\pm 1}\}$ of length less than or equal to $31 \cdot r \cdot (|u| + |v|)$.

Remark: The technical hypothesis in theorems 3 and 4 turns out to be not restrictive. Let P be a presentation of type W and suppose a generator of P, say x_k, occurs only once in the set of roots of the relators. Then, by a basis transformation, we see that $G = H * \mathbb{Z}_r$ ($\mathbb{Z}_r$ is trivial if $r = 1$), where H has the presentation $P' = \langle x_i \ (i = 1, \ldots, m, i \neq k) \mid R_j \ (j = 1, \ldots, m, j \neq l) \rangle$, which, as a subpresentation of P, is also of type W. But then the solvability of the word and conjugacy problems for G follows from that for H.

There is a similar restriction implicit in the definition of W^* where we use the condition $C(3)$ "positively" (i.e. every relator must be a product of at least 3 pieces). In classical small cancellation theory it is defined negatively (no relator is a product of fewer than 3 pieces) including the possibility that it may not be a product of pieces at all, which happens exactly when a generator occurs only once in the set of roots of the relators. This is the only case where W^* is not more general than the classical non-metric small cancellation conditions. The above discussion shows that this exception is irrelevant.

In [6] it is shown that presentations of type W satisfy the cycle test and are therefore "combinatorially aspherical". This is also true by a quite similar proof for presentations of type W^*.

The following examples show that, in many cases, it may be quite easy to test if a presentation is of type W or W^*.

Example 1. *Let* $P_n = \langle x, y, z \mid z^n = y, yx = xy \rangle$ *for* $n \geq 2$. *We will show that* P_n *satisfies* W^* *but not* W.

To prove that P_n is of type W^* we must analyze the local situations around inner vertices of valence ≤ 6 that can occur in a reduced diagram $f : M \to K_P$. The link of an inner vertex maps under $\pi \circ f$ onto a circuit γ in the star graph S_P. The length of γ gives the valence of v, and γ is reduced, i.e. has no backtracking, since M is reduced.

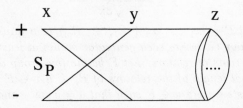

FIGURE 3. star graph of P

The drawing of S_P in Figure 3 shows that there are two reduced circuits of length 5, one of length 4, and none of length 3. Figure 4 (b) shows the local situation around a vertex of valence 4. It is easy to see that the commutator relator can not break into pieces of length > 1. Therefore, the corresponding 2-cells have degree 4 and case (ii) of the definition of W^* holds. Figure 4 (a) shows the local situation around a vertex corresponding to the circuit of length 5 that passes the vertices $+x, +y, +z, -z, -y$ in S_P in that order. The edges of the 2-cells are already combined to pieces of maximal length. Therefore, the degrees of the adjacent 2-cells are 4, 4, and ≥ 3 for the remaining three 2-cells, which is the second subcase of (iii) in the combinatorial definition of

W^*. The other reduced circuit of length 5 yields essentially the same situation as in Figure 4 (a), except that the edges labeled x are oriented downwards.

It is clear that P_n does not satisfy any of the standard small cancellation conditions. It is also not of type W. This can be seen by extending Figure 4 (a) such that Q and R become inner vertices of a vertex reduced diagram, each having as link a circuit of length 5. Then the 2-cell c of degree 3 has, at all its three corners, vertices of valence 5, contradicting the combinatorial definition of W.

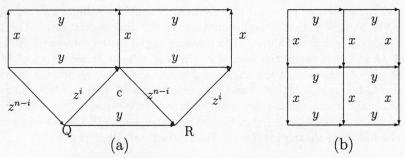

FIGURE 4. neighborhood of an inner vertex

Example 2. *Let $P = \langle x, y \mid y^2 x = xy^2 \rangle$. Juhász shows in [8] that this presentation satisfies W. P is not of type W^**

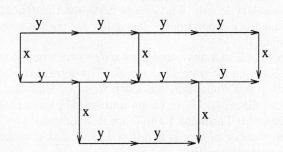

FIGURE 5. An inner vertex of valence 3

Figure 5 shows a neighborhood of an inner vertex of valence 3 in a reduced diagram over P. Since the only possible pieces for this presentation are y^2 and y^{-2}, the degrees of the adjacent 2-cells are 5 or 6. If the combinatorial definition of W^* were true, then the degrees of all three 2-cells would have to be greater than or equal to 6. It is easy to extend Figure 5 to a larger diagram, where the three 2-cells have degree 5.

The group defined by P is the same as the group defined by P_2 of the last example. This shows that W and W^* are attributes of the specific presentation and not of the group, which was to be expected.

2. PROOF OF THEOREMS 1 AND 3

A combinatorial 2-complex is called *piecewise Euclidean* (PE) if each of its closed 2-cells has the metric of a convex polygon in the Euclidian plane and each of its closed 1-cells has the metric of a straight Euclidean line segment and these metrics agree on the overlaps. Let M be a finite PE 2-disk, i.e. a PE 2-complex that is homeomorphic to a 2-dimensional disk. The given metric on each 2-cell allows us to measure angles in the corners of the 2-cells of M and to measure the area of 2-cells. Let v be an inner vertex of M and let $g(v)$ be the sum of the angles that occur around v. Define the curvature at v to be $\kappa(v) := 2\pi - g(v)$ and the *curvature of M* to be $\kappa(M) := \max \kappa(v)$, where the maximum is taken over all inner vertices $v \in M$ (set $\kappa(M) = 0$ if M has no inner vertices). Let l_M be the metric length of the boundary of M. The following theorem is due to Aleksandrov and Reshetnyak [11]:

Theorem 5. *Let M be a PE disk with $\kappa(M) \leq 0$, then*

$$(5) \qquad Area(M) \leq \frac{l_M^2}{4\pi}$$

Here $Area(M)$ is the sum of the areas of the 2-cells of M. This result is certainly plausible. It gives as an upper estimate the area of a euclidean disk of the given boundary length. The idea of using the theorem of Reshetnyak for proving a quadratic isoperimetric inequality for presentations of type W is due to Gersten [4].

For the proofs of Theorems 1 and 3 we may restrict ourselves to the case when the van Kampen diagram M is a disk. The general case (for arbitrary van Kampen diagrams) follows by induction from the following argument: If M consists of two disks M_1, M_2 which are connected by some path or just by a common vertex and if Theorems 1 and 3 are true for disks, i.e. $f_i \leq cl^2(\delta M_i)$ where f_i is the number of 2-cells of M_i ($i = 1, 2$) and c is the constant of formula 1 or 3 respectively, then

$$f = f_1 + f_2 \leq c(l^2(\delta M_1) + l^2(\delta M_2)) \leq c \cdot l^2(\delta M).$$

Proof of Theorem 1: Assume M is a vertex reduced disk-diagram over P of type W^* that is homeomorphic to a disk. We provide M with the structure of a PE-2-complex (PE-structure) by realizing each closed 2-cell of M of degree d as a regular Euclidean d-gon with edges of length 1. If the 2-cell D has identifications on its boundary we must use a suitable stellar subdivision of a regular Euclidean d-gon for realizing D, so that any two edges or two vertices that get identified belong to different 2-cells of the subdivision. This is necessary since a closed 2-cell of a PE-structure cannot have identifications on its boundary. By abuse of notation we call this PE-2-complex also M. The assumption that the diagram M is of type W^* guarantees that the sum of angles around each original interior vertex of M is $\geq 2\pi$. Since the sum of angles at those interior vertices created by the subdivision is equal to 2π, the PE-disk M has non-positive curvature and Theorem 5 applies, yielding that $Area(M) \leq l^2(\delta M)/(4\pi)$. The smallest area of all regular n-gons of edge-length 1 is that of an equilateral triangle which has area $\sqrt{3}/4$. Therefore, if f is the number of original 2-cells in M, $Area(M) \geq \sqrt{3}/4f$ and the quadratic isoperimetric inequality (1) in theorem 1 follows from the fact that the length $l(\delta M)$ of the boundary word is equal to the metric length l_M of the boundary of the PE-disk M. □

In the proofs of Theorems 3 and 4 we will make use of the Curvature Lemma 6 below.

Let M be a surface (with or without boundary) with a combinatorial cell decomposition, and let g be an arbitrary *weight function* for M, i.e. a function from the set of corners of M to the reals (One may think of the weight $g(\gamma)$ of a corner γ as an angle which may assume arbitrary real values). We write $\gamma \prec D, \gamma \prec v$ to indicate that the corner γ belongs to the 2-cell D, is incident with the vertex v, respectively. Let $g(D) := \Sigma_{\gamma \prec D} g(\gamma)$, $g(v) := \Sigma_{\gamma \prec v} g(\gamma)$, and let $\chi(M)$ denote the Euler characteristic of M. For $v \in M^\circ$ (the interior of M) let $\kappa(v) := 2\pi - g(v)$, for $v \in \delta M$ let $\kappa(v) := \pi - g(v)$, and for a 2-cell $D \in M$ let $\kappa(D) := g(D) - (d(D) - 2)\pi$. Think of $\kappa(v)$, $\kappa(D)$ as the local curvature of M at v, in D, respectively. Note that the local curvature at an interior vertex or in a 2-cell is zero exactly when the sum of angles is as in the euclidean case.

Lemma 6 (Curvature Lemma). *If g is an arbitrary weight function for a combinatorial cell decomposition of a surface M, then*

(6) $$\sum_{v \in M} \kappa(v) + \sum_{D \in M} \kappa(D) = 2\pi\chi(M)$$

The proof of 6 is straightforward and omitted.

Proof of Theorem 3: Assume P satisfies the hypothesis of Theorem 3, and assume M is a reduced diagram over P (and hence of type W), and M is homeomorphic to a disk. Let M^* be the dual of M. Then M^* consists, in general, of a collection of disks and arcs, where the disks are connected by arcs or by common points such that the result is simply connected; in other words, M^* looks like a general disk diagram. Also, by the assumption that M is of type W, M^* is of type W^*. Hence, the proof of Theorem 1 gives

(7) $$f^* \leq l^2(\delta M^*)/(\sqrt{3}\pi)$$

for the number of 2-cells f^* of M^*. We complete the proof of a quadratic isoperimetric inequality for the diagram M itself by showing that $l(\delta M^*)$ is bounded by a linear function in terms of $l(\delta M)$ and by transforming (7) into an estimate for f.

The assumption that each generator occurs at least twice in the set of roots $\{s_1, ..., s_m\}$ of the relators implies that the vertices of the star graph S_P have valence ≥ 2. It is then an easy exercise to show that every reduced path w in S_P can be extended to a reduced circuit z containing w as a subpath. This allows us to extend the entire diagram M to a reduced diagram M' over P, that contains M in its interior and is, as a reduced diagram over P, also of type W. This is the point in the proof which prevents us from using "vertex reduced" in the definition of type-W presentations; vertex reduced would not be preserved under the extension of M to M'.

Recall that the combinatorial definition of W lists lower bounds for the valences d_i of the vertices of interior 2-cells of degrees 3, 4, and 5. The maximum of these lower bounds is 42, corresponding to an angle of $2\pi/42 = \pi/21$. If we define the angle of every corner γ at an interior vertex v_i of M' to be $g(\gamma) = \max\{\pi/21, \ 2\pi/d(v_i)\}$ (where $d(v_i)$ is the valence of v_i in M'), the sum of angles at every interior vertex of M' will be $\geq 2\pi$, and condition W guarantees that for every interior 2-cell D of M' the sum of angles of its corners satisfies $\sum_{\gamma \in D} g(\gamma) \leq (d(D) - 2)\pi$. This holds in particular for every 2-cell of M. Under these conditions the curvature lemma for M (with angle function induced from M') reduces to

(8) $$\sum_{v \in \delta M} \kappa(v) = \sum_{v \in \delta M} (\pi - g(v)) \geq 2\pi$$

or simply:

(9) $$\sum_{v \in \delta M} g(v) \leq (l(\delta M) - 2)\pi$$

where the left hand side of (9) is the sum of angles of corners in M that are incident with (vertices in) δM. Using that the number of corners incident with δM equals $l(\delta M^*) + l(\delta M)$ and the fact that the angle of every corner is $\geq \pi/21$, we get from (9) the following estimate for $l(\delta M^*)$:

(10) $$l(\delta M^*) \leq 20l(\delta M) - 42.$$

Taking into account that $V^\circ = V - l(\delta M) = f^*$ (where V, V° is the number of vertices, interior vertices of M, respectively) and that the Euler characteristic of M^* is 2, one can derive from (7) the following estimate for the number f of 2-cells of M:

$$f \leq \frac{2l^2(\delta M^*)}{\sqrt{3}\pi} + l(\delta M) - 2.$$

By substituting (10) for $l(\delta M^*)$ and simplifying we obtain the quadratic isoperimetric inequality for M:

$$f \leq \frac{2(20^2)l^2(\delta M)}{\sqrt{3}\pi}.$$

□

3. Non-Positively Curved PE 2-complexes

Bridson studies in [1] the geometry of metric cell complexes under certain mild restrictions, which are certainly satisfied if the complexes are finite. His "Main Theorem" establishes the equivalence of certain local and global characterizations of non-positive curvature for simply connected metric cell complexes. In this section we will list the basic definitions and the results of [1] that we need in the proof of theorems 2 and 4. We will quote them as they apply to the special case of finite PE 2-complexes, which is a much narrower context than Bridson uses.

In a metric space X, a *geodesic (segment)* is a continuous path α that can be parametrized such that $\alpha : [0, l] \to X$ is an isometry from the interval $[0, l]$ to its image in X with the induced metric. X is called a *geodesic metric space* if each pair of points $x, y \in X$ can be joined by a geodesic. A geodesic metric space X is said to be *convex* if, for any pair of geodesics α, β in X, that are parametrized proportional to arc length on the interval $[0, 1]$, and for any $t \in [0, 1]$, the following inequality holds:

$$d(\alpha(t), \beta(t)) \leq (t - 1)d(\alpha(0), \beta(0)) + td(\alpha(1), \beta(1)).$$

In the following we assume K is a finite connected PE 2-complex. A *pl path* in K is a finite concatenation of straight line segments $[x_i, x_{i+1}]$ where each line segment $[x_i, x_{i+1}]$ is contained in a 2-cell or a 1-cell of K. The length $l(\alpha)$ is the sum of the lengths of the line segments. The distance between two points $x, y \in K$ is defined to be

$$d(x, y) := inf\{l(\alpha) : \alpha \text{ a pl path from } x \text{ to } y\}.$$

Bridson shows that the infimum in the above definition is, in fact, a minimum; i.e. the distance between two points can be realized by a pl (geodesic) path. Hence, d is a metric, called the *intrinsic metric* of the PE complex, and K, with this metric, is a geodesic metric space.

Let the length of an edge of the Whitehead graph $W(K)$ of a PE 2-complex K be the angle at the corresponding corner of a 2-cell in K. A PE 2-complex K is said to satisfy the *link condition* if the length of every non-trivial reduced circuit in $W(K)$ is greater than or equal to 2π. (A reduced circuit is a closed edge path that, considered as a cyclic path, has no backtracking.) This link condition defined by Bridson is a local condition of non-positive curvature. For a diagram M provided with a PE structure, the link condition is equivalent to the sum of angles around every inner vertex of M being $\geq 2\pi$. It also coincides with the definition of non-positive curvature for PE 2-disks which we gave in section 2, above. Therefore, we will use the term "non-positively curved" for a PE 2-complex satisfying the link condition of Bridson.

The following two results from Bridson's article will be used in the proof of theorems 2 and 4. The first is part of the Theorem from section 2. of [1], the second is Lemma 2.3 (from the same section):

1. For a connected and simply connected PE 2-complex K the following statements are equivalent:
   (I) K is non-positively curved, i.e. K satisfies the link condition,
   (II) K has unique geodesics, i.e. for every pair of points $x, y \in K$ there is only one geodesic pl path connecting x to y,
   (III) K is convex (as a geodesic metric space).

2. If the PE 2-complex K has unique geodesics then they vary continuously with their endpoints. More precisely, let $x_i \to x$ and $y_i \to y$ and let α_i, α be the geodesics from x_i to y_i, x to y, respectively, then

$$\|\alpha - \alpha_i\| = \sup\{d(\alpha_i(t), \alpha(t)) : t \in [0, 1]\} \to 0.$$

(Here "$\to$" denotes convergence of a sequence of points in K or of a sequence of real numbers)

4. PROOF OF THEOREMS 2 AND 4

Let u, v be words in F. Then u and v are conjugate in G if and only if there exists an annular diagram M over P whose boundary paths, oriented parallelly, read for suitable choices of start points the words u and v. A *parallel orientation* is an orientation such that the boundary paths are freely homotopic in M. We will always assume that the boundary paths of an annular diagram are oriented parallelly.

Proof of Theorem 2:
Let M be a vertex reduced annular diagram over P whose boundary paths, oriented parallelly, read the words u, v. If the two boundary paths meet in at

least one vertex, then some cyclic permutations of u and v will be equal in G, i.e. $aua^{-1} = bvb^{-1}$ in G, where a, b are subwords of u, v (or their inverses) respectively. This implies $u = a^{-1}bvb^{-1}a$ in G with $|a^{-1}b| \leq 2\max\{|u|, |v|\}$, and hence the conclusion of Theorem 2 holds. Therefore, we may restrict ourselves in the proof of Theorem 2 to the case when the boundary paths of M are disjoint (as subsets of M). In this case M consists of an actual annulus that has, in general, "trees of disks" attached to its boundaries (as sketched in Figure 6).

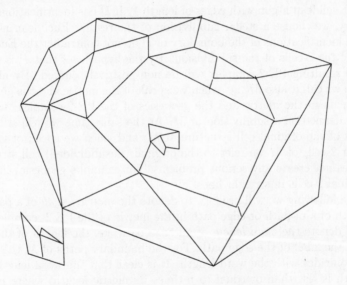

FIGURE 6. An annular diagram

Let M' be the annulus obtained from M by cutting off the trees of disks. To obtain a word w that conjugates v to u we proceed as follows. Choose a shortest edge path w' in M' whose endpoints lie on opposite boundary components, then connect the endpoints of w' along the boundary paths of M to the start points of u and v. This yields a conjugating word $w = xw'y$ where x and y are subwords of u and v (or their inverses), respectively, i.e. $|w| \leq 2\max\{|u|, |v|\} + |w'|$. It remains to show that, upon modifying the diagram M', if necessary, without changing its boundary, we can achieve $|w'| \leq N$ where N is the number of reduced words in F of length $\leq 3 \cdot p \cdot \max\{|u|, |v|\}$. This follows directly from the subsequent Proposition applied to the diagram M', taking into account that the lengths of the boundary paths of M' are less than or equal to the lengths of the corresponding boundary paths of M.

Proposition 7. *Let M be an annular diagram of type W^* over P, that is homeomorphic to an annulus. Let l be the maximum of the lengths of its boundary paths and N the number of words of length $\leq 3 \cdot p \cdot l$ over the alphabet $\{x_i^{\pm 1}\}$. Then there exists a word w in F of length $\leq N$ satisfying the equation $u = wvw^{-1}$ in G, where u and v are the words read from the boundary paths of M for a suitable choice of startpoints.*

Proof: We provide the diagram M with a PE-structure as in the proof of Theorem 1, that is, every 2-cell D of M of degree n is represented by a regular Euclidean n-gon with edges of length 1. If D has identifications on its boundary, we choose a stellar subdivision of the regular Euclidean n-gon so that the identifications of the boundary of D do not contradict the Euclidean metric of the 2-cells of the subdivision. By the hypothesis that P is of type W^*, the resulting PE-2-complex will be non-positively curved. By abuse of notation we call it also M, i.e. when we speak of the metric and the geodesics of M we mean the metric and the geodesics of the PE-2-complex, which is actually defined on a subdivision of M. At the same time we will still think in terms of the original cell structure of M, and when we speak of vertices, edges, or 2-cells of M we refer to the original (unsubdivided) cell structure. This does not create any serious problems. For example, geodesics intersect the original 2-cells in straight lines.

In the following we will use $|\ |_m$ to denote the *metric length* of a path, i.e. the length of a pl-path or edge path in the metric of the PE 2- complex; and $|\ |_w$ will denote the *word length* of an edge path, i.e. the length of the word which is the label of the edge path. For the boundary paths of M the metric length coincides with the word length. It is clear that the word length of an edge path is less than or equal to p times its metric length, where p is the maximal word length of a piece for the presentation P.

Now let $\bar{w}$ be a geodesic path in M whose endpoints are vertices on the two boundary components of M, such that $\bar{w}$ realizes the shortest distance between the sets of vertices of the two boundary components. Let Q be the startpoint and R the endpoint of $\bar{w}$ and let u, v be the closed boundary paths starting at Q, R respectively. (The boundary words read along these paths will also be called u and v.) We cut the annulus M open along $\bar{w}$ to obtain $\overline{M}$. $\overline{M}$ inherits a PE- structure from M. (As mentioned above, $\bar{w}$ will cut through 2-cells of M along straight lines creating pieces that are still convex Euclidean polygons) Since the link of an interior vertex of $\overline{M}$ is the same as the link of the corresponding vertex of M, $\overline{M}$ is a non-positively curved PE-disk. Hence, $\overline{M}$ has unique geodesic segments and is a convex geodesic metric space.

For the following notation compare with Figure 7. Let Q_1 and Q_2, R_1 and R_2, $\bar{w}_1$ and $\bar{w}_2$ be the pairs of vertices or paths in the boundary of $\overline{M}$ that

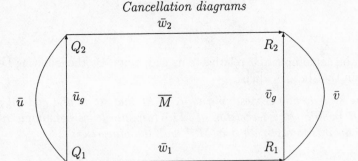

FIGURE 7. the diagram M'

originate from Q, R, $\bar{w}$, respectively, by the splitting of M along $\bar{w}$, and let $\bar{u}$, $\bar{v}$ be the paths in the boundary of $\overline{M}$ that are created by cutting the closed boundary paths u, v of M at Q, R respectively. We will also consider the (unique) geodesics $\bar{u}_g$ from Q_1 to Q_2 and $\bar{v}_g$ from R_1 to R_2 in $\overline{M}$. Together with $\bar{w}_1$ and $\bar{w}_2$ they form a geodesic rectangle in $\overline{M}$. If we parameterize each of the geodesics $\bar{w}_1$ and $\bar{w}_2$ proportional to arc length on the interval $[0,1]$, we obtain by the convexity of $\overline{M}$ for every $t \in [0,t]$:

$$
\begin{aligned}
d(\bar{w}_1(t), \bar{w}_2(t)) &\leq \max\{d(\bar{w}_1(0), \bar{w}_2(0)), d(\bar{w}_1(1), \bar{w}_2(1))\} \\
&= \max\{|\bar{u}_g|_m, |\bar{v}_g|_m\} \\
&\leq \max\{|u|, |v|\} = l,
\end{aligned}
$$

where d denotes the distance in the geodesic metric space $\overline{M}$. For every $t \in [0,1]$ let $\bar{g}_t$ be the unique geodesic in $\overline{M}$ from $\bar{w}_1(t)$ to $\bar{w}_2(t)$ ($\bar{g}_0 = \bar{u}_g$, $\bar{g}_1 = \bar{v}_g$), and let $\overline{M}_1$ be the subspace of $\overline{M}$ bounded by the geodesic rectangle that is formed by $\bar{u}_g, \bar{v}_g, \bar{w}_1, \bar{w}_2$. Since in a simply connected PE-complex of non-positive curvature, geodesic segments vary continuously with their endpoints (see [1]), it follows that the one-parameter family of geodesics $\bar{g}_t$ covers all of $\overline{M}_1$, i.e. the map $\bar{g} : I \times I \to \overline{M}_1$ defined by $\bar{g}(t,s) = \bar{g}_t(s)$ is surjective.

In the annular diagram M let g_t be the family of closed pl-paths that are obtained from $\bar{g}_t$ by identifying the endpoints. Each path g_t is shortest among the paths in M that start and end at $\bar{w}(t)$ and are freely homotopic to u (or v).

Now we replace the pl geodesic path $\bar{w}$ by a shortest edge path in M from Q to R, where "edge path" refers to the original cell structure of M. The label on w reads a word in F that conjugates v to u in G. Note that w may intersect u or v in more than just the endpoints Q or R respectively and w

may not be homotopic to $\bar{w}$ relative to its endpoints. By the following Lemma we obtain that $|w|_m \leq 2|\bar{w}|_m$.

Lemma 8. *A locally geodesic pl-path α in M that starts and ends at points A and B in $M^{(1)}$ (the 1-skeleton of M), respectively, is homotopic relative to its endpoints to a pl-path β in $M^{(1)}$ with the properties:*

1. $|\beta|_m \leq 2|\alpha|_m$,
2. *if A and B are in $M^{(0)}$ then β is an edge path,*
3. *if A or B are not in $M^{(0)}$ then β is an edge path except for the first or last segment.*

We will call β an edge path approximation of α.

Proof: The proof of Lemma 8 is elementary. Simply replace each straight line segment of α that intersects the interior of a 2-cell D of M by the shorter of the two paths on the boundary of D that connect the same endpoints. Elementary geometry shows that the shorter path on the boundary of a regular Euclidean n-gon has at most two times the length of the secant through the interior of the polygon. After reducing any backtracking in the resulting path by a homotopy in the 1-skeleton we obtain a path β that satisfies the conditions of the Lemma. □

Lemma 9. *Let w be the edge path constructed in the proof of Proposition 7 and let $V_0 = Q, V_1, V_2, ..., V_k = R$ be the sequence of vertices (including the interior vertices of valence 2) that are met by the path w. Then, for $i = 1, ..., k$, a shortest closed edge path w_i in M, such that w_i starts and ends at V_i and is homotopic to u (or v), will have metric length $|w_i|_m \leq 3 \cdot l$ and, hence, word length $|w_i|_w \leq 3 \cdot p \cdot l$.*

Proof: Recall that $\overline{M}_1$ is the part of $\overline{M}$ bounded by the geodesic rectangle with sides $\bar{u}_g, \bar{v}_g, \bar{w}_1, \bar{w}_2$. Without loss of generality we can assume that $\bar{u}_g$ and $\bar{v}_g$ do not intersect; if they do, then it is easy to see that the edge path w, defined above, satisfies $|w|_m \leq 2\max\{|u|, |v|\}$ and hence the conclusion of Proposition 7. Let M_1 be the part of M that corresponds to $\overline{M}_1$, i.e. M_1 is $\overline{M}_1$ with $\bar{w}_1$ and $\bar{w}_2$ identified. M_1 is an annulus with possibly one spike on either boundary. (The intersections $\bar{u} \cap \bar{w}_1$ and $\bar{u} \cap \bar{w}_2$, or $\bar{v} \cap \bar{w}_1$ and $\bar{v} \cap \bar{w}_2$ may consist of initial and terminal segments of $\bar{u}$ or $\bar{v}$, respectively, which create a spike)

If a vertex V_i of the path w is in M_1 then at least one of the closed pl-paths g_t, that were constructed above, will pass through V_i. Since $|g_t|_m \leq \max\{|u|, |v|\} = l$, a shortest closed pl-path in M with basepoint V_i that is homotopic to u will have length $\leq l$ and, by Lemma 8, a shortest edge path w_i with the same properties will have length $|w_i|_m \leq 2l$.

If V_i is not in M_1 we proceed as follows. Let u_g, v_g be the closed pl-paths in M that correspond to $\bar{u}_g$, $\bar{v}_g$ in $\overline{M}$; in other words, u_g and v_g are the boundary paths of M_1. Let $[t_a, t_b]$ be the maximal subinterval of $[0,1]$ such that $w([t_a, t_b])$ is contained in M_1. Then $w(t_a)$ and $w(t_b)$ are points in $w \cap u_g$ and $w \cap v_g$ respectively. Since w is a shortest edge path, Lemma 8 implies that the length of $w|_{[0,t_a]}$ is :less than or equal to two times the length of the shorter segment of u_g connecting $w(0) = Q$ to $w(t_a)$. Hence $\left| w|_{[0,t_a]} \right|_m \leq |u_g|_m \leq |u|$. Similarly we obtain $\left| w|_{[t_b,1]} \right|_m \leq |v|$. Assume V_i lies on the segment $w|_{[0,t_a]}$, i.e. $V_i = w(t_i)$ with $t_i < t_a$. Then V_i can be connected to $Q = w(0)$ along the edge path $\lambda_i = (w|_{[0,t_i]})^{-1}$ and $|\lambda_i|_m \leq |u|$. Hence, $\lambda_i u \lambda_i^{-1}$ is a closed edge path homotopic to u with basepoint V_i that has metric length $\leq 3|u| \leq 3l$. If V_i lies on $w|_{[t_b,1]}$ we choose λ_i to be a final segment of w connecting V_i to $R = w(1)$, and the path $\lambda_i v \lambda_i^{-1}$ will have the desired properties and length estimate. □

We are now able to complete the proof of Proposition 7 and thereby the proof of Theorem 2.

Consider the diagram M of Proposition 7 and the edge path w in M from Q to R. Let w_{QV_i}, w_{V_jR} be the segments of w from Q to V_i, V_j to R, respectively, and let w_i be the paths from Lemma 9. Any null homotopic edge path in a diagram represents a word that is equal to 1 in G and hence corresponds to a closed path in the Cayley graph. This applies to the closed edge paths $u w_{QV_i} w_i^{-1} w_{QV_i}^{-1}$ and $w_j w_{V_jR} v w_{V_jR}^{-1}$. We now consider the conjugacy relation $uwv^{-1}w^{-1}$ as a closed path based at 1 in the Cayley graph of G.

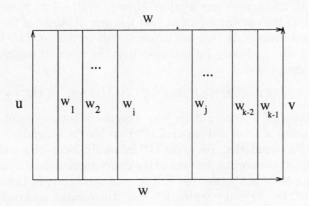

FIGURE 8. A part of the Cayley graph

We then have the situation depicted in Figure 8, and we are able to apply a standard surgery trick: Let N be the number of words of length $\leq 3pl$ over the alphabet $\{x_1^{\pm 1}, \ldots, x_n^{\pm 1}\}$. If $|w|_w > N$ then some of the "vertical words", say w_i and w_j will be equal, and we can do surgery on the diagram of Figure 8 by cutting out the middle part and gluing the left part to the right part along $w_i = w_j$. This is done in the Cayley graph by translating the right part under left multiplication by $w_{V_i V_j}^{-1}$. (Here we use that left multiplication by any group element is an isometry of the Cayley graph that preserves edge labels.) This surgery shortens the conjugating word w by the length of the word $w_{V_i V_j}$. Since the vertical paths in the resulting figure are those in the parts of Figure 8 left of w_i and right of w_j, we can iterate this procedure until the length of the horizontal (conjugating) word becomes less than N. $\square$

Proof of Theorem 4:

We follow the basic idea of the proof of Theorem 2 where the key steps are Proposition 7 and Lemma 9. Given a presentation P that satisfies the hypothesis of Theorem 4, let M be a reduced annular diagram over P that is homeomorphic to an annulus and let M^* be the dual of M. Since M is of type W, M^* is of type W^*.

We first estimate the length of the boundary $l(\delta M^*)$ in terms of $l(\delta M)$ (where the length is measured, as usual, as the number of edges in the boundary paths). We use the same trick as in the proof of Theorem 3 to do so: Extend the diagram M to a diagram M' that contains M in its interior and is also reduced and hence of type W. If we give every corner γ at every interior vertex of M' the angle $g(\gamma) = \max\{\pi/21, 2\pi/d(v_i)\}$ then the sum of angles around every interior vertex is $\geq 2\pi$ and the sum of angles of every interior 2-cell D of M' and hence every 2-cell of M is $\leq (d(D) - 2)\pi$. Since $\chi(M) = 0$ we obtain from the curvature lemma that the sum of angles of the corners of δM, i.e. the corners of M that are incident with vertices in δM, is less than or equal to $l(\delta M)\pi$. Taking into account that the minimal angle of a corner is $\pi/21$, we get:

#of corners of $\delta M = l(\delta M^*) + l(\delta M) \leq 21 \cdot l(\delta M)$ or: $l(\delta M^*) \leq 20 \cdot l(\delta M)$.

As in the proof of Theorem 2 we may assume, without loss of generality, that M^* consists of an actual annulus $M^{*\prime}$ that has (in general) trees of disks attached to its boundaries. We make $M^{*\prime}$ into a PE 2-complex and construct an edge path w^* connecting vertices of the opposite boundary components of $M^{*\prime}$ as in the proof of Proposition 7. Then the conclusion of Lemma 9 holds for w^* in $M^{*\prime}$, i.e. for every vertex V_i^* of w^* there exists a closed edge path w_i^* such that w_i^* starts and ends at V_i^*, is homotopic to a boundary path of $M^{*\prime}$, and $|w_i^*|_m \leq 3l(\delta M^{*\prime}) \leq 3l(\delta M^*)$.

By duality, any edge path α^* in M^* corresponds to a sequence

$$D_0 e_0 D_1 e_1 ... e_{k-1} D_k$$

of 2-cells D_i and 1-cells e_i in M such that $e_i \subset D_i \cap D_{i+1}$ (we call it briefly a *chain of 2-cells*). Clearly, for such a chain of 2-cells in M belonging to an edge path α^* in M^* and for any pair of vertices Q in D_0 and R in D_k one can find an edge path α from Q to R in M that is contained in the subcomplex $\bigcup_{i=1}^k D_i$ and has word-length $|\alpha| \leq r(|\alpha^*| + 1)/2$, where r is the maximal length of the relators of P. We call α an *accompanying path* of α^*. First we apply this to the path w^* and obtain an accompanying edge path w from Q to R in M, where we can choose Q and R to be vertices in opposite boundary components of δM. (This follows since Q^* and R^* are vertices in opposite boundary components of $\delta M^{*\prime} \subset \delta M^*$ and a vertex of M^* belongs to a boundary component of M^* if and only if the corresponding 2-cell of M intersects the corresponding boundary component of M.) By deleting loops or backtracking we may assume that w is a simple path, i.e. without double points. Let V_i be the vertices along w. Then each V_i belongs to a 2-cell D_{j_i} of the chain corresponding to w^* and we can find an accompanying path w_i of $w_{j_i}^*$ with startpoint and endpoint V_i that is freely homotopic to either boundary component of M and has length

$$|w_i|_w \leq \frac{r(|w_{j_i}^*| + 1)}{2} \leq \frac{r(3l(\delta M^*) + 1)}{2} \leq 31 \cdot r \cdot l(\delta M).$$

Now we are ready to do the same surgery trick as in the proof of Proposition 7 and obtain the following result: if u, v are the words read from the boundary of M starting at Q, R, respectively, and N is the number of words in the alphabet $\{x_1^{\pm 1}, \ldots, x_n^{\pm 1}\}$ of length $\leq 31rl(\delta M)$, then there exists a word w in F of length $\leq N$ such that $u = wvw^{-1}$ holds in G.

Finally, if we have an arbitrary reduced annular diagram M with given boundary words u and v and M is not homeomorphic to an annulus, then, by the same argument as in the proof of Theorem 2, there exists a word of length $\leq 2\max\{|u|, |v|\} + N$ conjugating u to v in G. $\qquad\qquad\Box$

5. THE CONDITIONS V AND V^*

A strengthening of the conditions W^* and W leads to classes of presentations V^* and V, defined below. V^* and V, which generalize the classical small cancellation conditions C(7), C(5)T(4), C(4)T(5), and C(3)T(7), imply a linear isoperimetric inequality and, hence word hyperbolicity of the underlying group.

A diagram M is said to be *of type V^** if it is of type W^* with the sum of angles around every interior vertex being strictly greater than 2π. A presentation P is *of type V^** if every vertex reduced diagram M over P is of type V^*. The combinatorial definition of V^* is given in the same way as the combinatorial definition of W^* by the following tables (where d_i are lower bounds for the degrees of 2-cells adjacent to an interior vertex v):

(i) $d(v) = 3$,

d_1	3	3	3	3	3	3	4	4	4	4	5	5	5	6
d_2	7	8	9	10	11	12	5	6	7	8	5	6	7	6
d_3	43	25	19	16	14	13	21	13	10	9	11	8	7	7

(ii) $d(v) = 4$,

d_1	3	3	3	3	3	4
d_2	3	3	3	4	4	4
d_3	4	5	6	4	5	4
d_4	13	8	7	7	5	5

(iii) $d(v) = 5$,

d_1	3	3	3
d_2	3	3	3
d_3	3	3	4
d_4	3	4	4
d_5	7	5	4

(iv) $d(v) = 6$, $d_i \geq 3$ for $i = 1, ..., d(v)$ and $d_i > 3$ for at least one i

(v) $d(v) \geq 7$, $d_i \geq 3$ for $i = 1, ..., d(v)$.

A diagram M is said to be *of type V* if it is of type W with the sum of angles of every interior 2-cell D being strictly less than $d(D) - 2$. A presentation P is *of type V* if every reduced diagram over P is of type V. Combinatorially, type V is characterized by the same tables as above where valences of vertices and degrees of 2-cells are interchanged and the vertices and 2-cells are both from the interior of the diagram.

We obtain the following result using curvature of diagrams and Euler characteristic counts in a way similar to proofs in [3] and [10]:

Theorem 10. *Assume G is a group that has a finite presentation of type V^* or of type V and, in the latter case, assume that each generator occurs at least twice in the set of roots of the relators. Then G satisfies a linear isoperimetric inequality and hence is word-hyperbolic.*

REFERENCES

[1] M. Bridson. Geodesics and curvature in metric simplicial complexes. In *Group theory from a geometrical viewpoint*, E. Ghys, A. Haefliger, A. Verjovsky (eds.), World Scientific, pages 373–463, 1990.

[2] D.J. Collins and J. Huebschmann. *Spherical diagrams and identities among relations*, Math. Ann. 261; pages 155 - 183, 1982.

[3] S. Gersten. *Reducible diagrams and equations over groups*, *Essays in group theory*, S. Gersten (ed.), M.S.R.I. publ. Nr. 8, Springer Verlag, pages 15–73, 1987.

[4] S. Gersten. The isoperimetric inequality and the word problem. unpublished, 1988.

[5] M. Gromov. Hyperbolic groups. In *Essays in group theory*, S. Gersten (ed.), M.S.R.I. publ. Nr. 8, Springer Verlag, pages 75–263, 1987.

[6] G. Huck and S. Rosebrock. Ein verallgemeinerter Gewichtstest mit Anwendungen auf Baumpraesentationen. *Mathematische Zeitschrift*, 211 (3), pages 351–367, 1992.

[7] G. Huck and S. Rosebrock. Applications of diagrams to decision problems. In *Two-dimensional homotopy and combinatiorial group theory*, C. Hog-Angeloni, W. Metzler, A. Sieradski (eds.), LMS Lecture Notes Series 197, Cambridge University Press, pages 189–218, 1993.

[8] A. Juhasz. Small cancellation theory with a unified small cancellation condition I. *J. Lond. Math. Soc.* (2) 40, pages 57–80, 1989.

[9] R. Lyndon and P. Schupp. *Combinatorial group theory*. Springer Verlag, Berlin, 1977.

[10] S. Pride. Star-complexes, and the dependence problems for hyperbolic complexes. *Glasgow Math. J.* 30, pages 155–170, 1988.

[11] Y. G. Reshetnyak. On a special kind of mapping of a cone onto a polyhedral disk. *Math. Sbornik*, V. 53 (95), pages 39–52, 1961.

Stephan Rosebrock
Fakultät für Mathematik
University of Education
Bismarckstr. 10
D-76133 Karlsruhe
Germany
rosebrock@PH-karlsruhe.de

Günther Huck
Department of Mathematics
Northern Arizona University
Flagstaff AZ 86011
USA
gph@math.nau.edu

SOME APPLICATIONS OF PREFIX-REWRITING IN MONOIDS, GROUPS, AND RINGS

KLAUS MADLENER AND FRIEDRICH OTTO

ABSTRACT. Rewriting techniques have been applied successfully to various areas of symbolic computation. Here we consider the notion of *prefix-rewriting* and give a survey on its applications to the *subgroup problem* in combinatorial group theory. We will see that for certain classes of finitely presented groups finitely generated subgroups can be described by convergent prefix-rewriting systems, which can be obtained from a presentation of the group considered and a set of generators for the subgroup by a specialized Knuth-Bendix style completion procedure. In many instances a finite presentation for the subgroup considered can be constructed from such a convergent prefix-rewriting system, thus solving the *subgroup presentation problem*. Finally we will see that the classical procedures for computing Nielsen reduced sets of generators for a finitely generated subgroup of a free group and the Todd-Coxeter coset enumeration can be interpreted as particular instances of prefix-completion. Further, both procedures are closely related to the computation of prefix Gröbner bases for right ideals in free group rings.

1. INTRODUCTION

There is a recent shift in paradigm in mathematics, and in modern algebra in particular, from pure structural considerations back to the notion of computability, that is, one is not merely interested in the structural properties of the mathematical entities under consideration, but one wants to actually perform computations in these structures.

This development has been preceded by that in combinatorial group theory, where algorithmic questions have been of major concern since the beginning of the century. In 1911 Dehn formulated three fundamental decision problems for groups given in terms of generators and defining relations [Deh11], the most famous of which is the *word problem*. Consequently these problems have been shown to be undecidable in general, and much effort has been spent on deriving decidability results for restricted instances. Already in his original paper Dehn gave a solution for the word problem for free groups by establishing that the process of freely reducing strings leads to unique representatives for the elements of the free group. In a subsequent paper

---

Key words and phrases. monoid and group presentations, subgroup problem, subgroup presentation problem, prefix-rewriting, confluence, λ-confluence, coset enumeration, Gröbner bases in monoid and group rings.

he proved that the word problem for the fundamental groups of closed, two dimensional, orientable surfaces of genus larger than or equal to two can be solved by a *monotonic reduction process* [Deh12]. Actually, this algorithm, known as *Dehn's algorithm for the word problem*, extends to the class of all small-cancellation groups, which are groups that satisfy certain combinatorial conditions [LS77].

On the other hand, string-rewriting systems that are *convergent*, that is, noetherian and confluent, yield a unique irreducible string for each element of the group (or monoid) presented [BO93]. Hence, if a group admits a presentation involving a finite convergent string-rewriting system, then its word problem is decidable by reduction. The class of groups that admit such presentations includes for example the finite groups, the free groups of finite rank, the finitely generated polycyclic groups, and the finitely generated virtually free groups. Actually, to solve the word problem for a group by reduction, it suffices that the string-rewriting system presenting the group is finite, noetherian, and λ-confluent, that is, each word presenting the identity of the group considered reduces to the empty string [MO87]. In fact, Dehn's algorithm for the word problem corresponds to the process of reduction with respect to a finite, length-reducing, and λ-confluent string-rewriting system [Büc79, LeC86].

A generalization of the word problem that also has received a lot of attention is the *subgroup problem*, also known in the literature as the *generalized word problem* [MKS76]. Let $\langle \Sigma; S \rangle$ be a finite group presentation of a group G. Then the subgroup problem for $\langle \Sigma; S \rangle$ is the following decision problem:

INSTANCE : A finite set $U \subset \underline{\Sigma}^*$, and a string $w \in \underline{\Sigma}^*$.

QUESTION : Does w represent an element of the subgroup of G that is generated by U?

Since this is a generalization of the word problem, this problem is also undecidable in general. In fact, it is even undecidable for some groups that can be presented by finite convergent string-rewriting systems [Mil71]. On the other hand, if F is the free group given by the free presentation $\langle \Sigma; \emptyset \rangle$, then a finite set $U \subset \underline{\Sigma}^*$ can effectively be transformed into a *Nielsen reduced set* V that is a set of free generators for the subgroup of F generated by U (see, e.g., [MKS76]). Using this set it is easy to decide whether w is a member of the subgroup.

Another classical approach to the subgroup problem is the *Todd-Coxeter coset enumeration method* [TC36], which, given a presentation $\langle \Sigma; S \rangle$ and a set U, enumerates coset representatives of the subgroup $\langle U \rangle$ of G that is generated by U. This procedure succeeds if and only if $\langle U \rangle$ has finite index in G, and in this case it returns a *coset table*, that is, a complete table for the multiplication of cosets of $\langle U \rangle$ with generators of G.

Unfortunately the so-called *index problem*, that is, the problem of deciding whether or not $\langle U \rangle$ has finite index in G, is also undecidable in general. Thus, the Todd-Coxeter coset enumeration only yields a semi-decision procedure for the subgroup problem.

Here we will show that rewriting methods are a powerful tool to solve the subgroup problem and related problems. A set $U \subset \Sigma^*$ induces a left-congruence $\sim_U$ on Σ^* as follows: two strings x and y are congruent modulo $\sim_U$ if and only if x and y belong to the same right coset of $\langle U \rangle$. This congruence can be expressed by a *prefix-rewriting system*, that is, a string-rewriting system where the rules are only used to replace prefixes of strings [Bau81, KM89]. If this prefix-rewriting system is noetherian and confluent, then it yields a set of unique coset representatives for the cosets of $\langle U \rangle$ in G. Thus, if in addition the prefix-rewriting process induced by this system is effective, then prefix-rewriting solves the subgroup problem. Further, if this system is finite or at least left-regular, then the set of representatives is a regular language, and hence, the index of $\langle U \rangle$ in G can be computed.

In this paper we give a survey on the applications of prefix-rewriting to the subgroup problem and related problems in groups, monoids, and rings. After establishing notation in Section 2 we define the problems considered in the following. In addition to the subgroup problem and the index problem they include the *subgroup presentation problem*, which asks to compute a finite presentation for the subgroup $\langle U \rangle$ of the given group G.

In Section 3 we briefly discuss three classical algorithms of combinatorial group theory: Nielsen's method of transforming a set of generators of a subgroup of a free group into a set of free generators, the Todd-Coxeter coset enumeration method, and the method of Reidemeister and Schreier for determining presentations for subgroups [MKS76].

In the following section we introduce prefix-rewriting systems, and show how they are used to represent left-congruences of groups. From a given finite set U a prefix-rewriting system P_U defining $\sim_U$ is easily obtained, but in general this system will not be confluent. However, specially adopted versions of the Knuth-Bendix completion procedure [KB70] have been defined that, given P_U and a reduction ordering as input, try to generate a prefix-rewriting system Q_U that is convergent, and that generates the left-congruence $\sim_U$. Obviously, this procedure will not always terminate, but it has been shown to terminate successfully if and only if there exists a finite system Q_U that generates $\sim_U$ and that is compatible with the given reduction ordering. This section closes by showing that the process of computing Nielsen reduced sets of generators for subgroups of free groups can be interpreted as a particular instance of prefix-completion. In fact there is a one-to-one correspondence between the prefix-interreduced, canonical prefix-rewriting systems on $\langle \Sigma; \emptyset \rangle$ and the generalized Nielsen-reduced subsets of Σ^*.

In Section 5 it is shown that for some classes of presentations of groups, each finitely generated subgroup can be described by a finite set of prefix-rules such that the prefix-rewriting system obtained from this set together with the prefix-rules corresponding to the string-rewriting system of the given presentation is convergent. Here we will follow Cremanns [Cre95], who introduced an abstract condition that guarantees this property, so that then it remains to verify that the classes of presentations considered do indeed satisfy this condition. As it turns out this approach works fine for those finite convergent presentations that involve a special string-rewriting system or a two-monadic string-rewriting system with inverses of length one. In addition it applies to the so-called virtually free presentations, which is a class of finite convergent presentations for the finitely generated virtually free groups. Further, for the class of finite, monadic, and λ-confluent presentations, which also characterize these groups, and for the class of PCP2-presentations, which characterize the finitely presented polycyclic groups, λ-convergent prefix-rewriting systems are obtained.

From a finite convergent presentation of a group G and a convergent prefix-rewriting system for a subgroup H of G, a finite presentation of H can be constructed, if H is indeed finitely presented. In Section 6 this property is characterized with the aid of an infinite graph that describes the induced prefix-rewriting relation. This graph is an extension of the graph that Squier used to define the notion of *finite derivation type* for monoids and groups [SOK94]. In addition a finite presentation for H can be extracted from this graph based on resolutions of the U- and G-critical pairs [Cre95]. For free groups and for groups presented by finite, special, and confluent presentations this yields a finite presentation of the same form for the subgroup H. The corresponding result can also be shown for the virtually free presentations, the PCP2-presentations, and the presentations involving a finite, monadic, and λ-confluent string-rewriting system.

In Section 7 we discuss in short a correspondence between automatic structures and prefix-rewriting systems. Automatic structures for groups and monoids have been found very useful, since they are close to geometrical considerations and use natural automata constructions providing an algorithm for solving the word problem that runs in quadratic time [Eps92]. An automatic structure yields a set of representatives for the monoid or group considered, and although in general these representatives need not be unique, each automatic structure can be replaced by another one that yields unique representatives. Now the solution of the word problem corresponds to the left-to-right computation of the representative that corresponds to a given string, and so it can be interpreted as a kind of prefix-rewriting. And in fact if the set of unique representatives is in addition prefix-closed, then a synchronously regular prefix-rewriting system can be constructed that defines the group or monoid considered, is convergent, and yields the same representatives. Conversely, if such a prefix-rewriting system exists, then based on its set of irreducible strings, an automatic structure can be constructed for the monoid or group considered.

Finally, in Section 8 we point out a close correspondence between the process of computing Nielsen reduced sets of generators in free groups and the Todd-Coxeter coset enumeration method on the one hand and the computation of prefix Gröbner bases in the free group ring on the other hand. It turns out that both these classical procedures of combinatorial group theory have their counterparts in the computation of prefix Gröbner bases in this ring. This section is based on recent work of Reinert et al [RMM98a].

The paper closes with a short summary and some open problems.

2. THE SUBGROUP PROBLEM AND RELATED PROBLEMS

Here we introduce the basic notions and notation concerning presentations of groups, and we state in detail the subgroup problem and some related problems that we are interested in. For additional information on the notions introduced we refer to the literature, where [BO93] serves as our main reference on string-rewriting systems and monoid presentations, while [MKS76, LS77] are our main references on combinatorial group theory.

Let Σ be a finite alphabet, that is, Σ is a finite set of symbols called *letters* or *generators*. Then Σ^* denotes the set of strings over Σ including the empty string λ, and $\Sigma^+ = \Sigma^* \smallsetminus \{\lambda\}$ is the set of all nonempty strings over Σ. The operation of concatenation, which is simply written by juxtaposition, is an associative binary operation on Σ^* with identity λ, and so Σ^* has the algebraic structure of a monoid. It is the *free monoid* generated by Σ.

For $w \in \Sigma^*$, $|w|$ denotes the *length* of w, and $|w|_a$ ($a \in \Sigma$) denotes the *a-length* of w which is simply the number of occurrences of the letter a in w. Analogously, for $\Gamma \subseteq \Sigma$, $|w|_\Gamma := \bigcup_{a \in \Gamma} |w|_a$ is the Γ-*length* of w. Finally, to simplify the notation numerical superscripts will be used to write strings in a more compact and readable form, where $w^0 := \lambda$, $w^1 := w$, and $w^{n+1} := w^n w$ for all $w \in \Sigma^*$ and $n \in \mathbb{N}$.

A *string-rewriting system* on Σ is a subset of $\Sigma^* \times \Sigma^*$. Its elements are called *(rewrite) rules*, and usually they will be written as $(\ell \to r)$. If S is a string-rewriting system on Σ, then $\mathrm{dom}(S) := \{\ell \in \Sigma^* \mid \exists r \in \Sigma^* : (\ell \to r) \in S\}$ and $\mathrm{range}(S) := \{r \in \Sigma^* \mid \exists \ell \in \Sigma^* : (\ell \to r) \in S\}$.

A string-rewriting system S on Σ induces several binary relations on Σ^*, the most basic of which is the *single-step reduction relation* $\to_S :=$ $\{(u\ell v, urv) \mid (\ell \to r) \in S, u, v \in \Sigma^*\}$. Its reflexive transitive closure $\to_S^*$ is the *reduction relation* defined by S, while the reflexive, symmetric, and transitive closure $\leftrightarrow_S^*$ of $\to_S$ is actually a congruence on Σ^*, since $x \leftrightarrow_S^* y$ implies $uxv \leftrightarrow_S^* uyv$ for all $u, v \in \Sigma^*$. It is called the *Thue congruence* generated by S. The set $M_S := \{[w]_S \mid w \in \Sigma^*\}$ of congruence classes forms a monoid with identity $[\lambda]_S$ under the operation $[u]_S \circ [v]_S := [uv]_S$, that is, M_S is the factor monoid $\Sigma^* / \leftrightarrow_S^*$ of the free monoid Σ^* modulo the Thue congruence $\leftrightarrow_S^*$. If M is a monoid that is isomorphic to M_S, then the ordered pair $(\Sigma; S)$ is called a *monoid presentation* of M with generators Σ and *defining relations*

S. The monoid M is called *finitely generated* if it has a presentation with a finite set of generators, and it is called *finitely presented* if it has a finite presentation.

From a finite presentation $(\Sigma; S)$ one can determine a set of strings $\{u_a \mid a \in \Sigma\}$ such that the monoid M_S is a group if and only if $au_a \leftrightarrow^*_S \lambda \leftrightarrow^*_S u_a a$ holds for all $a \in \Sigma$ [Ott86]. This yields a function $^{-1} : \Sigma^* \to \Sigma^*$ that realizes the inverse function for the group M_S simply by defining $\lambda^{-1} := \lambda$ and $(wa)^{-1} := u_a w^{-1}$ for all $w \in \Sigma^*$ and $a \in \Sigma$. However, in combinatorial group theory groups are usually presented by so-called group presentations rather than by monoid presentations.

Let Σ be a finite alphabet, let $\overline{\Sigma}$ be another finite alphabet in one-to-one correspondence with Σ such that $\Sigma \cap \overline{\Sigma} = \emptyset$, let $\underline{\Sigma} := \Sigma \cup \overline{\Sigma}$, and let $S_0 := \{a\bar{a} \to \lambda, \bar{a}a \to \lambda \mid a \in \Sigma\}$, where $^- : \Sigma \to \overline{\Sigma}$ denotes the one-to-one correspondence between Σ and $\overline{\Sigma}$. Then $(\underline{\Sigma}; S_0)$ is a presentation of the *free group* generated by Σ, that is, $M_{S_0} \cong F_n$, where $n = \text{card}(\Sigma)$.

In addition, let S be a string-rewriting system on $\underline{\Sigma}$. Then the monoid presented by $(\underline{\Sigma}; S \cup S_0)$ is a group, and the ordered pair $\langle \Sigma; S \rangle$ is called a *group presentation* of this group. The congruence on $\underline{\Sigma}^*$ that is generated by $S \cup S_0$ will simply be written as $=_S$.

Since group presentations are a special class of monoid presentations we will give the following definitions only in terms of the latter.

Let G be a group that is given by a presentation $(\Sigma; S)$. The *word problem* for $(\Sigma; S)$ is the following decision problem:

INSTANCE: Two strings $u, v \in \Sigma^*$.

QUESTION: Do u and v present the same element of the group G, that is, does $u \leftrightarrow^*_S v$ hold?

Since the function $^{-1} : \Sigma^* \to \Sigma^*$ can be determined effectively from $(\Sigma; S)$, the word problem is equivalent to the *special word problem*:

INSTANCE: A string $w \in \Sigma^*$.

QUESTION: Does w present the identity of the group G, that is, does $w \leftrightarrow^*_S \lambda$ hold?

Actually this is the form in which the word problem is usually stated in combinatorial group theory. A generalization of the word problem is the *subgroup problem*, which is also known as the *generalized word problem*:

INSTANCE: A finite set of strings $U \subseteq \Sigma^*$, and a string $w \in \Sigma^*$.

QUESTION: Does w belong to the subgroup $\langle U \rangle$ of G that is generated by U?

A string w belongs to $\langle U \rangle$ if and only if there exist $u_1, \ldots, u_n \in U$ and $\varepsilon_1, \ldots, \varepsilon_n \in \{1, -1\}$ such that $w \leftrightarrow^*_S u_1^{\varepsilon_1} u_2^{\varepsilon_2} \cdots u_n^{\varepsilon_n}$. Here $u^1 := u$, and u^{-1} denotes the inverse of u. To simplify the notation we will usually assume that the set U is *closed under inverses*, that is, for each $u \in U$ there is an element $v \in U$ such that $v \leftrightarrow_S^* u^{-1}$.

With U we associate a binary relation $\sim_U$ on Σ^* as follows:

$$x \sim_U y \text{ iff } \exists u \in \langle U \rangle : x \leftrightarrow_S^* uy.$$

Then $\sim_U$ is a left-congruence on Σ^*, that is, it is an equivalence relation such that $x \sim_U y$ implies $xz \sim_U yz$ for all $z \in \Sigma^*$.

By $[w]_U$ we denote the congruence class of w modulo $\sim_U$. Obviously, it is simply the left coset of $\langle U \rangle$ in G containing w. The number of left cosets of $\langle U \rangle$ is called the *index* of $\langle U \rangle$ in G and is written as $|G : \langle U \rangle|$. The *index problem* is an important decision problem:

INSTANCE: A finite set of strings $U \subseteq \Sigma^*$.

QUESTION: What is the index of $\langle U \rangle$ in G?

Finally, the subgroup $\langle U \rangle$ of G may or may not be finitely presented as a group. In the former case it would be of interest to actually determine a finite presentation of the group $\langle U \rangle$. This is the *subgroup presentation problem*:

INSTANCE: A finite set of strings $U \subseteq \Sigma^*$.

TASK: Decide whether or not the subgroup $\langle U \rangle$ of G is finitely presented, and in the affirmative determine a finite presentation for $\langle U \rangle$!

There are finitely presented groups with undecidable word problem. Hence, also the subgroup problem is undecidable in general. Actually the subgroup problem is even undecidable for the direct product $F_2 \times F_2$ of the free group F_2 of rank 2 with itself [Mik58], whose word problem is easily decidable. Further there are finitely presented groups for which the index problem is undecidable, and there are finitely presented groups for which it is undecidable whether or not a finitely generated subgroup is itself finitely presented [BBN59]. Thus, all the problems above are undecidable in general.

On the other hand these problems have been solved successfully for some restricted classes of presentations. The restrictions we are interested in concern the properties of the single-step reduction relation and the syntactic form of the rules.

Let S be a string-rewriting system on Σ. A string $u \in \Sigma^*$ is called *reducible* modulo S if there exists a string $v \in \Sigma^*$ such that $u \to_S v$ holds; otherwise, u is called *irreducible* modulo S. By $\text{RED}(S)$ we denote the set of all reducible strings, and by $\text{IRR}(S)$ the set of all irreducible strings. Obviously $\text{RED}(S) = \Sigma^* \cdot \text{dom}(S) \cdot \Sigma^*$ and $\text{IRR}(S) = \Sigma^* \setminus \text{RED}(S)$. Hence, if S is a finite system, then these sets are regular, and in fact in this case deterministic finite-state acceptors (dfsa's) can be constructed for them in polynomial time.

The string-rewriting system S is·called

– *noetherian* if there is no infinite sequence of reduction steps of the form
$$w_0 \to_S w_1 \to_S \ldots \to_S w_i \to_S w_{i+1} \to_S \ldots;$$

- *weight-reducing* if there exists a weight function $g : \Sigma \to \mathbb{N}_+$ such that the extension of g to a morphism $g : \Sigma^* \to \mathbb{N}$ satisfies $g(\ell) > g(r)$ for each rule $(\ell \to r) \in S$;
- *length-reducing* if $|\ell| > |r|$ holds for each rule $(\ell \to r) \in S$.

If S is noetherian, then each string has some irreducible descendants, since the process of performing reduction steps modulo S terminates. Unfortunately, however, it is undecidable in general whether or not a given finite string-rewriting system is noetherian. On the other hand, if $\geq$ is an admissible well-founded partial ordering on Σ^* such that S is *compatible* with $\geq$, that is, $\ell > r$ holds for each rule $(\ell \to r) \in S$, then S is necessarily noetherian. Here a partial ordering $\geq$ on Σ^* is called *admissible* if $x \geq y$ implies that $uxv \geq uyv$ holds for all $u, v \in \Sigma^*$, and it is called *well-founded* if there is no infinite sequence of strings that is strictly decreasing. For example, a weight function $g : \Sigma \to \mathbb{N}_+$ yields the partial ordering $\geq_g$ defined by $x \geq_g y$ iff $g(x) \geq g(y)$, which is admissible and well-founded, but not linear. If we assume that the alphabet Σ is linearly ordered by $\geq_\Sigma$, then we can combine the partial ordering $\geq_g$ and the lexicographical ordering on Σ^* that is induced by $\geq_\Sigma$ into an admissible well-ordering $\geq_{g,\text{lex}}$ as follows:

$$x \geq_{g,\text{lex}} y \quad \text{iff} \quad g(x) > g(y) \quad \text{or}$$
$$g(x) = g(y) \quad \text{and} \quad x \geq_{\text{lex}} y.$$

A special case is the *length-lexicographical ordering* $\geq_{\ell\ell}$, which is obtained by taking the length of a string as its weight.

The string-rewriting system S is called *monadic* if $\text{range}(S) \subseteq \Sigma \cup \{\lambda\}$ and $\ell \geq_{\ell\ell} r$ holds for each rule $(\ell \to r) \in S$, and it is called *special* if it is length-reducing and $\text{range}(S) = \{\lambda\}$. Obviously, monadic systems are noetherian.

Finally, we turn to restrictions that limit the number of irreducible strings that can occur within certain congruence classes. A string-rewriting system S on Σ is called

- *locally confluent* if, for all $u, v, w \in \Sigma^*$, $u \to_S v$ and $u \to_S w$ imply that $\Delta_S^*(v) \cap \Delta_S^*(w) \neq \emptyset$, where $\Delta_S^*(x) := \{z \in \Sigma^* \mid x \to_S^* z\}$ denotes the *set of all descendants* of x modulo S;
- *confluent* if, for all $u, v, w \in \Sigma^*$, $u \to_S^* v$ and $u \to_S^* w$ imply that $\Delta_S^*(v) \cap \Delta_S^*(w) \neq \emptyset$;
- λ-*confluent* if, for all $w \in \Sigma^*$, $w \leftrightarrow_S^* \lambda$ implies that $w \to_S^* \lambda$ holds.

If S is λ-confluent, then $[\lambda]_S$ contains at most a single irreducible string, which has to be λ. Analogously, if S is confluent, then no congruence class can contain more than a single irreducible string. Hence, if S is λ-*convergent*, that is, noetherian and λ-confluent, then $[\lambda]_S \cap \text{IRR}(S) = \{\lambda\}$, and if S is *convergent*, that is, noetherian and confluent, then each congruence class contains a unique irreducible string. Thus, if S is a finite convergent system, then the word problem for S can simply be solved by reduction.

The system S is called *interreduced* if range$(S) \subseteq$ IRR(S) and $\ell \in$ IRR$(S \setminus \{(\ell \to r)\})$ holds for each rule $(\ell \to r) \in S$. A system is called *canonical* if it is convergent and interreduced. In fact, from a convergent system a canonical system can be constructed which generates the same Thue congruence and has the same set of irreducible strings.

In this paper we will mainly be interested in groups that are given by finite presentations involving certain λ-convergent or convergent string rewriting systems. Since the group $F_2 \times F_2$ is given by the presentation $(\Gamma; R)$, where $\Gamma := \{a, b, c, d, \bar{a}, \bar{b}, \bar{c}, \bar{d}\}$ and $R := \{x^\varepsilon \bar{x}^\varepsilon \to \lambda \mid x \in \{a, b, c, d\}, \varepsilon \in \{1, -1\}\} \cup \{y^\varepsilon x^\mu \to x^\mu y^\varepsilon \mid x \in \{a, b\}, y \in \{c, d\}, \varepsilon, \mu \in \{1, -1\}\}$, which is canonical, we see that the subgroup problem is in general even undecidable for groups that are given by finite canonical presentations. Hence, our restrictions will have to be even more specific.

3. CLASSICAL RESULTS

Here we review some classical approaches to the subgroup problem: the Nielsen reduced sets for subgroups of free groups, the Todd-Coxeter method for enumerating cosets of finitely generated subgroups of finitely presented groups, and the Reidemeister-Schreier method for constructing presentations for subgroups of finitely presented groups. For more details we refer to the literature, where [MKS76, LS77] serve as our main references on these topics.

Let $F = \langle \Sigma; \emptyset \rangle$ be the free group generated by Σ, that is, F is given by the monoid presentation $(\Sigma; S_0)$. The system $S_0 = \{a\bar{a} \to \lambda, \bar{a}a \to \lambda \mid a \in \Sigma\}$ is special and convergent, and hence, IRR(S_0) is a set of unique representatives for F. The elements of IRR(S_0) are called *freely reduced* strings, and the process of reduction modulo S_0 is called *free reduction*.

Let $U := \{u_1, \ldots, u_m\}$ be a subset of IRR(S_0), and let $U^{-1} := \{u_i^{-1} \mid u_i \in U\}$ be the corresponding set of irreducible inverses. The *elementary Nielsen transformations* on U are defined as follows:

(NT1) Replace an element $u_i \in U$ by its inverse u_i^{-1}.

(NT2) Replace an element $u_i \in U$ by the irreducible descendant of $u_i u_j$ for some $j \neq i$.

(NT3) Delete some element $u_i \in U$, where $u_i = \lambda$.

In each of these three cases all the other elements of U remain unchanged. A finite sequence of such transformations is called a *Nielsen transformation*.

Proposition 3.1. *Let $U_1 \subseteq$ IRR(S_0), and let U_2 be obtained from U_1 through a Nielsen transformation. Then $\langle U_1 \rangle = \langle U_2 \rangle$, that is, U_1 and U_2 generate the same subgroup of F.*

A subset $U \subseteq$ IRR(S_0) is called *Nielsen reduced* if, for all $v_1, v_2, v_3 \in U \cup U^{-1}$, the following three conditions are satisfied:

(N0) $v_1 \neq \lambda$,

(N1) $v_1v_2 \neq_F \lambda$ implies that $|(v_1v_2)\!\downarrow| \geq \max\{|v_1|, |v_2|\}$, and

(N2) $v_1v_2 \neq_F \lambda$ and $v_2v_3 \neq_F \lambda$ imply that $|(v_1v_2v_3)\!\downarrow| > |v_1| - |v_2| + |v_3|$.

Here $w\!\downarrow$ denotes the (unique) irreducible descendant of w modulo S_0.

Nielsen reduced sets are of importance as they are free generating sets for the subgroups they generate. Actually they satisfy the following strong property.

Proposition 3.2. *Let U be a Nielsen reduced set. Then, for each element $u \in U \cup U^{-1}$, there exist strings a_u and m_u such that $m_u \neq \lambda$ and $u = a_u m_u (a_{u^{-1}})^{-1}$, and if $w = (u_1 u_2 \cdots u_m)\!\downarrow$ for some $u_1, \ldots, u_m \in U \cup U^{-1}$, where $u_i u_{i+1} \neq_F \lambda$ for $i = 1, \ldots, m-1$, then the strings $m_{u_1}, \ldots, m_{u_m}$ remain uncancelled in w.*

Hence, if $w = (u_1 u_2 \cdots u_m)\!\downarrow$ as above, then w has the prefix $a_{u_1} m_{u_1}$. From this prefix u_1 can be determined, and we can consider $w_1 := (u_1^{-1} w)\!\downarrow = (u_2 \cdots u_m)\!\downarrow$. Iterating this process we can reconstruct the sequence $u_1, \ldots, u_m \in U \cup U^{-1}$ from w. Hence, the subgroup problem for F is solved provided the set U given is Nielsen reduced.

Proposition 3.3. *Given a finite set $U \subseteq \mathrm{IRR}(S_0)$ a Nielsen transformation can be found effectively that transforms U into a Nielsen reduced set V.*

Actually this task can be performed in polynomial time [AM84]. This yields the following result.

Corollary 3.4. *For finitely generated free groups the subgroup problem is decidable in polynomial time.*

We will see in the next section how this approach to the subgroup problem of free groups can be described and even extended by using the notion of prefix-rewriting.

Next we turn to the *Todd-Coxeter coset enumeration method* (TC). While it is undecidable in general whether a finitely generated subgroup $\langle U \rangle$ has finite index in a finitely presented group G, TC attempts to verify whether the index is finite by systematically enumerating the cosets of $\langle U \rangle$ in G. It is based on the following observation. Assume that G is given by the finite presentation $\langle \Sigma; S \rangle$. Then G is the quotient of the free group F generated by Σ by the normal subgroup N that is generated by S. This normal subgroup N is the subgroup of F that is generated by the set $N(S) := \{w \cdot \ell r^{-1} \cdot w^{-1} \mid (\ell \to r) \in S, w \in \mathrm{IRR}(S_0)\}$. Thus, N is finitely generated as a *normal* subgroup of F, since S is finite, but N may not be finitely generated as a subgroup of F.

Now let H be the subgroup of G that is generated by $U \subseteq \Sigma^*$. We are interested in the index $|G : H|$ of the subgroup H in the group G. It is easily seen that this index coincides with the index $|F : \langle U \cup N(S) \rangle|$ of the subgroup generated by $U \cup N(S)$ in the free group F. TC now attempts to verify that this index is finite.

For the following considerations we assume that $\langle \Sigma; S \rangle$ is a finite presentation of G, and that $U \subseteq \underline{\Sigma}^*$ is a finite set. Moreover TC requires that each generator $a \in \Sigma$ occurs in at least one of the defining relations S. TC tries to determine the index $|G : H|$ by exploiting the following facts about cosets:

(1.) For each $u \in U$, $Hu = H$.

(2.) For each rule $(\ell \to r) \in S$ and each coset Hu, $H(u \cdot \ell r^{-1} \cdot u^{-1}) = H$.

It proceeds as follows. With each generator $u = a_1 a_2 \cdots a_k \in U$, where $k \geq 1$ and $a_1, \ldots, a_k \in \underline{\Sigma}$, a table of the form below is associated:

a_1	a_2	a_3	$\cdots$	a_{k-1}	a_k
λ					λ

Here λ represents the coset H, and the empty slots are to be filled with representatives for the cosets $Ha_1, Ha_1a_2, \ldots, Ha_1a_2 \cdots a_{k-1}$. Further, with each defining relation $(\ell \to r) \in S$, or more exactly with the freely reduced form $b_1 b_2 \cdots b_m$ of ℓr^{-1}, where $m \geq 1$ and $b_1, \ldots, b_m \in \underline{\Sigma}$, a table of the following form is associated:

b_1	b_2	b_3	$\cdots$	b_{m-1}	b_m
λ					λ
$\vdots$					$\vdots$

These tables will contain a row for each coset encountered. If w is a coset representative, then the slots in the row starting with w will be filled with representatives for the cosets $Hwb_1, Hwb_1b_2, \ldots, Hwb_1b_2 \cdots b_{m-1}$.

Depending on the strategy used for determining the slot to be filled next, different types of equations between coset representatives are deduced. For example, if the representative of $Ha_1a_2 \cdots a_{k-1}$ happens to be the string $w \in \underline{\Sigma}^*$, then we see from the table for $u = a_1 \cdots a_k$ that $(Hw) \cdot a_k = H$, that is, with respect to their operation on cosets we obtain the equation $w \cdot a_k \sim_H \lambda$, which is called a *bonus equation*. On the other hand, if we have the coset representatives w for $Ha_1 \cdots a_i$ and z for $Ha_k^{-1} \cdots a_{i+1}^{-1}$, then we see that $Hw = Ha_1 \cdots a_i = Hu \cdot a_k^{-1} \cdots a_{i+1}^{-1} = Ha_k^{-1} \cdots a_{i+1}^{-1} = Hz$, which implies that w and z represent the same coset. This gives the *collapse equation* $w \sim_H z$. By identifying the cosets represented by w and z we may obtain additional information on other cosets. A detailed presentation of TC can be found in [Sim94].

To illustrate this procedure, and also for future reference, we give a simple example, which is taken from [Joh76], page 71.

Example 3.5. Let G be the Dyck group D(3,3,2), which is given by the presentation $\langle a, b; a^3, b^3, abab \rangle$. Here a^3 stands for the defining relation $a^3 \to \lambda$, and similar for the other strings given. Further, let H be the subgroup that is generated by $U := \{a\}$. Then the table for $a \in U$ yields the equation $a \sim_H \lambda$, and correspondingly $\bar{a} \sim_H \lambda$. In order to choose a unique coset representative we use the length-lexicographical ordering induced by $a < b < \bar{a} < \bar{b}$.

Now filling in the tables for a^3, b^3, and $abab$, and choosing the smaller string as the coset representative whenever an equation is obtained, we get the complete set of coset representatives $\{\lambda, b, \bar{b}, b\bar{a}\}$ and the following coset table for the cosets of H in G:

	a	$\bar{a}$	b	$\bar{b}$
H	H	H	Hb	$H\bar{b}$
Hb	$H\bar{b}$	$Hb\bar{a}$	$H\bar{b}$	H
$H\bar{b}$	$Hb\bar{a}$	Hb	H	Hb
$Hb\bar{a}$	Hb	$H\bar{b}$	$Hb\bar{a}$	$Hb\bar{a}$

The coset representative of the string aba can now be deduced by tracing this table starting from the coset H:

$$H \cdot a = H, \ H \cdot b = Hb, \text{ and } Hb \cdot a =_H H\bar{b}, \text{ that is, } Haba = H\bar{b}.$$

$\square$

Actually the coset table yields a prefix-rewriting system that is convergent, and correspondingly the coset representative for a string w can be obtained by prefix-rewriting (see [RMM98a] for how to deduce this system and for further details).

Concerning the behavior of TC we have the following result.

Proposition 3.6. *Given a finite group presentation $\langle \Sigma; S \rangle$ of a group G such that each generator $a \in \Sigma$ actually occurs in S and a finite set $U \subseteq \Sigma^*$, TC terminates if and only if the index of $\langle U \rangle$ in G is finite. In this case TC determines a set of unique coset representatives for $\langle U \rangle$ in G and the corresponding coset table.*

Finally we restate in short the method of Reidemeister and Schreier for constructing presentations of subgroups. Again G is a group given by a finite group presentation $\langle \Sigma; S \rangle$, and $U \subseteq \Sigma^*$ is a finite set of generators for a subgroup H of G. From $\langle \Sigma; S \rangle$ and U we would like to construct a presentation for H. To do so we assume that in addition to the above we have a complete set C of coset representatives for H in G, and that we have an effective process $\varrho : \Sigma^* \to C$ that maps a string $w \in \Sigma^*$ to its coset representative $\varrho(w) \in C$. Then H is actually generated by the set $V := \{ca\varrho(ca)^{-1} \mid c \in C, a \in \Sigma\}$. Thus, if the index $|G : H|$ is finite, then H is finitely generated.

Let $\Gamma := \{b_{c,a} \mid c \in C, a \in \Sigma\}$ be a new alphabet in one-to-one correspondence to the set V, and let $\varphi : \Gamma^* \to \Sigma^*$ be the morphism that is induced by mapping $b_{c,a}$ to the string $ca\varrho(ca)^{-1}$ ($c \in C, a \in \Sigma$). Then φ induces a homomorphism from the free group F_Γ generated by Γ onto the subgroup H of G. Thus, H is isomorphic to the factor group $F_\Gamma/\ker(\varphi)$ of the free group F_Γ by the kernel of the homomorphism φ.

Since each string $u \in U$ represents an element of H, there exists a string $\tau(u) \in \underline{\Gamma}^*$ such that $\varphi(\tau(u)) =_S u$. In fact, based on C and ϱ, a *rewriting function* $\tau : \langle U \rangle \to \underline{\Gamma}^*$ can be obtained that satisfies $\varphi(\tau(w)) =_S w$ for all $w \in \langle U \rangle$. Such a rewriting function based on coset representatives is called a *Reidemeister rewriting function*.

Proposition 3.7. *If τ is a Reidemeister rewriting function for the subgroup $H := \langle U \rangle$ of the group G presented by $\langle \Sigma; S \rangle$, then H is described by the presentation $\langle \Gamma; \{b_{c,a} \to \tau(ca\varrho(ca)^{-1}) \mid c \in C, a \in \Sigma\} \cup \{\tau(c\ell r^{-1}c^{-1}) \to \lambda \mid c \in C, (\ell \to r) \in S\} \rangle$, where C is the set of coset representatives for H in G underlying τ.*

Thus, if G is finitely presented and H has finite index in G, then H is finitely presented. If the set of coset representatives C is chosen in such a way that it is *prefix-closed*, that is, all the prefixes of a representative $c \in C$ are themselves representatives, then the presentation obtained for H is simplified considerably. A rewriting function that is based on a prefix-closed set of representatives is called a *Reidemeister-Schreier rewriting function*.

Proposition 3.8. *If τ is a Reidemeister-Schreier rewriting function for the subgroup $H := \langle U \rangle$ of the group G presented by $\langle \Sigma; S \rangle$, then H has the presentation*

$$\langle \Gamma; \ \{b_{c,a} \to \lambda \mid c \in C, a \in \Sigma \text{ such that } c \cdot a = \varrho(ca)\} \ \cup \\ \{\tau(c\ell r^{-1}c^{-1}) \to \lambda \mid c \in C, (\ell \to r) \in S\} \rangle.$$

Observe that the presentation obtained for H is infinite whenever the index $|G : H|$ of H in G is infinite. Using prefix-rewriting systems we will see in Section 6 that for certain classes of finite presentations of groups finite presentations can be obtained even for subgroups of infinite index.

4. PREFIX-REWRITING

When using the coset table obtained by the Todd-Coxeter coset enumeration method to determine the representative of a coset $H \cdot w$, this is done by reading w from left to right and replacing each prefix of w by its corresponding representative. Hence, this computation is an application of prefix-rewriting. Here we introduce prefix-rewriting systems in detail and relate them to the subgroup problem.

Let Σ be a finite alphabet. A *prefix-rewriting system* on Σ is a subset of $\Sigma^* \times \Sigma^*$. Its elements are called *prefix-rules*. If P is a prefix-rewriting system on Σ, then $\text{dom}(P)$ and $\text{range}(P)$ are defined as for string-rewriting systems.

The *prefix-reduction relation* $\Rightarrow_P^*$ defined by P is the reflexive transitive closure of the *single-step prefix-reduction relation* $\Rightarrow_P := \{(\ell w, rw) \mid (\ell, r) \in$

$P, w \in \Sigma^*\}$, and by $\Leftrightarrow_P^*$ we denote the reflexive, symmetric, and transitive closure of $\Rightarrow_P$. Obviously $\Leftrightarrow_P^*$ is a left-congruence on Σ^*.

A string $u \in \Sigma^*$ is called *reducible* modulo P if $u \Rightarrow_P v$ holds for some $v \in \Sigma^*$; otherwise, u is *irreducible* modulo P. By $\mathrm{RED}(P)$ we denote the set of all reducible strings, and $\mathrm{IRR}(P)$ denotes the set of irreducible strings. Obviously, $\mathrm{RED}(P) = \mathrm{dom}(P) \cdot \Sigma^*$ and $\mathrm{IRR}(P) = \Sigma^* \setminus \mathrm{RED}(P)$. Hence, if $\mathrm{dom}(P)$ is a regular language, then $\mathrm{RED}(P)$ and $\mathrm{IRR}(P)$ are regular languages as well. In this situation the prefix-rewriting system P is called *left-regular*.

The prefix-rewriting system P is called *noetherian, confluent, convergent, λ-confluent, λ-convergent, interreduced,* or *canonical* if the corresponding condition is satisfied by $\Rightarrow_P$. It is interesting to observe that a prefix-rewriting system is convergent whenever it is interreduced [Sny89, Ott98b], that is, it is canonical if and only if it is interreduced.

Next we will show how prefix-rewriting systems are related to the subgroup problem. Let G be a group that is given by a finite presentation $(\Sigma; S)$, and let $^{-1} : \Sigma^* \to \Sigma^*$ denote a function realizing the inverse function of G. Further, let $U \subseteq \Sigma^*$ be a finite set. Without loss of generality we can assume that U is closed under inverses. Hence, a string $w \in \Sigma^*$ presents an element of the subgroup $\langle U \rangle$ of G if and only if there exist $u_1, \ldots, u_k \in U$ such that $w \leftrightarrow_S^* u_1 u_2 \cdots u_k$.

With $(\Sigma; S)$ and U we now associate a prefix-rewriting system $P := P_U \cup P_S$, where

$$P_U := \{(u, \lambda) \mid u \in U\}$$

and

$$P_S := \{(x\ell, xr) \mid x \in \Sigma^* \text{ and } (\ell \to r) \in S\}.$$

Then P is a left-regular system, and the following property is easily verified.

Proposition 4.1. *The left-congruences $\sim_U$ and $\Leftrightarrow_P^*$ coincide.*

Hence, if P is λ-confluent, then a string $w \in \Sigma^*$ belongs to $\langle U \rangle$ if and only if $w \Rightarrow_P^* \lambda$, and if P is convergent, then $\mathrm{IRR}(P)$ is a complete set of coset representatives for $\langle U \rangle$ in G.

If S is noetherian, then P is noetherian, but in general P will not be convergent even in case S is. For future reference we consider the following simple example.

Example 4.2. Let $F_2 = \langle a, b; \emptyset \rangle$, and let $U_0 := \{ab, ba, aa\}$. Then $U := U_0 \cup \{\bar{b}\bar{a}, \bar{a}\bar{b}, \bar{a}\bar{a}\}$ is closed under inverses, and $\langle U_0 \rangle = \langle U \rangle$. Let $w := \bar{b}\bar{a}aaa\bar{a}\bar{b}$. Then $w \to_{S_0}^* \bar{b}\bar{b} \in \mathrm{IRR}(P)$, and $w \Rightarrow_{P_U}^* \lambda$. Hence, P is not even λ-confluent, although S_0 is a canonical string-rewriting system. $\qquad \Box$

In order to solve the subgroup problem by prefix-rewriting we need a procedure that transforms the prefix-rewriting system P defined above into an equivalent prefix-rewriting system P_1 that is convergent or at least λ-convergent.

Let $(\Sigma; S)$ be a finite presentation of (a group) G, let P_U be a finite prefix-rewriting system on Σ, and let $P := P_U \cup P_S$, where P_S is the infinite prefix-rewriting system corresponding to the string-rewriting system S as above. If P is noetherian, for example it may be compatible with some admissible well-founded partial ordering on Σ^*, then P is confluent if and only if it is locally confluent. In order to determine the points of local divergence we introduce various forms of critical pairs.

If there are two rules $(u_1, v_1), (u_2, v_2) \in P_U$ such that $u_1 = u_2 y$ for some $y \in \Sigma^*$, where either $y \neq \lambda$ or $v_1 \neq v_2$, then $v_1 \Leftarrow_{P_U} u_1 = u_2 y \Rightarrow_{P_U} v_2 y$, and $(v_1, v_2 y)$ is called a *U-critical pair* of P. If there are rules $(u, v) \in P_U$ and $(\ell \to r) \in S$ such that $uy = x\ell z$ for some $x, y, z \in \Sigma^*$ satisfying $|x| < |u|$ and at least one of y or z is λ, then $vy \Leftarrow_{P_U} uy = x\ell z \Rightarrow_{P_S} xrz$, and (vy, xrz) is called a *G-critical pair* of P. Finally, if there are rules $(\ell_1 \to r_1), (\ell_2 \to r_2) \in S$ such that $\ell_1 = x\ell_2 y$ or $\ell_1 y = x\ell_2$, where $|x| < |\ell_1|$, then $(r_1, xr_2 y)$ or $(r_1 y, xr_2)$ is a critical pair of the string-rewriting system S, and correspondingly $(wr_1, wxr_2 y)$ or $(wr_1 y, wxr_2)$ is an *S-critical pair* of P for each string $w \in \Sigma^*$.

Proposition 4.3. *The prefix-rewriting system $P = P_U \cup P_S$ is locally confluent iff all U-, G-, and S-critical pairs of P resolve, that is, if (p, q) is one of these critical pairs, then p and q have a common descendant modulo P.*

The string-rewriting system S is locally confluent if and only if each of its critical pairs resolves. Thus, if S is locally confluent, then all S-critical pairs of P resolve, that is, in this situation it suffices to consider the U- and G-critical pairs of P. If P_U and S are both finite, then there are only finitely many of these pairs, and they can be computed in polynomial time. Hence, we have the following decidability result.

Proposition 4.4. *Let $(\Sigma; S)$ be a finite presentation such that S is convergent, and let P_U be a finite prefix-rewriting system on Σ. If $P := P_U \cup P_S$ is noetherian, then it is decidable whether or not P is confluent.*

If P is not confluent, since some U- or G-critical pairs do not resolve, then by introducing additional prefix-rules, these critical pairs can be resolved. This is the basic idea of the following procedure which is an adapted version of the Knuth-Bendix completion procedure [KB70, KN85]. Here we present only the most basic form of this procedure in order to illustrate it.

Procedure 4.5. Knuth-Bendix completion procedure for prefix rewriting systems.

INPUT: A finite convergent presentation $(\Sigma; S)$, a finite prefix-rewriting system P_U on Σ, and an admissible well-founded partial ordering $\geq$ on Σ^* such that S is compatible with $\geq$.

begin $Q_0 \leftarrow \emptyset$;
     while $P_U \neq \emptyset$ **do**

begin choose $(u, v) \in P_U$;
    if u and v are incomparable under $\geq$ **then failure**;
    if $u > v$ **then** $Q_0 \leftarrow Q_0 \cup \{(u, v)\}$;
    if $v > u$ **then** $Q_0 \leftarrow Q_0 \cup \{(v, u)\}$;
    $P_U \leftarrow P_U \smallsetminus \{(u, v)\}$
end;
(Comment: Q_0 is obtained from P_U by orienting all prefix-rules with
        respect to $\geq$)
$i \leftarrow -1$;
repeat $i \leftarrow i + 1$; $Q_{i+1} \leftarrow \emptyset$;
    $CP \leftarrow$ set of U- and G-critical pairs of $Q_i \cup P_S$;
    while $CP \neq \emptyset$ **do**
    begin choose a minimal pair $(p, q) \in CP$;
        compute normal forms $\hat{p}$ and $\hat{q}$ of p and q modulo
        $Q_i \cup P_S$;
        if $\hat{p}$ and $\hat{q}$ are incomparable under $\geq$ **then failure**;
        if $\hat{p} > \hat{q}$ **then** $Q_{i+1} \leftarrow Q_{i+1} \cup \{(\hat{p}, \hat{q})\}$;
        if $\hat{q} > \hat{p}$ **then** $Q_{i+1} \leftarrow Q_{i+1} \cup \{(\hat{q}, \hat{p})\}$;
        $CP \leftarrow CP \smallsetminus \{(p, q)\}$
    end;
    (Comment: all critical pairs of $Q_i \cup P_S$ have been resolved)
    if $Q_{i+1} \neq \emptyset$ **then** $Q_{i+1} \leftarrow Q_i \cup Q_{i+1}$
until $Q_{i+1} = \emptyset$;
$Q_U \leftarrow \bigcup_{i \geq 0} Q_i$
end.

Concerning the behavior of this procedure the following results have been obtained.

Proposition 4.6. [KM89, Kuh91] *Let $(\Sigma; S)$ be a finite convergent presentation, let P_U be a finite prefix-rewriting system on Σ, and let $\geq$ be an admissible well-founded partial ordering on Σ^* such that S is compatible with $\geq$. If the Knuth-Bendix completion procedure does not stop with failure given $(\Sigma; S)$, P_U and $\geq$ as input, then the system Q_U generated has the following properties:*

(1.) $Q_U \cup P_S$ is equivalent to $P_U \cup P_S$.
(2.) $Q_U \cup P_S$ is compatible with $\geq$.
(3.) $Q_U \cup P_S$ is convergent.

Proposition 4.7. [KM89, Kuh91] *Let $(\Sigma; S)$ and P_U be as above, and let $\geq$ be an admissible well-ordering on Σ^* such that S is compatible with $\geq$. Then given $(\Sigma; S)$, P_U and $\geq$ as input, the Knuth-Bendix completion procedure will enumerate a prefix-rewriting system Q_U that has the properties (1.) to (3.) of the previous proposition. It will terminate after finitely many steps, thus generating a finite system Q_U, if and only if there exists some finite prefix-rewriting system on Σ that has these properties.*

In particular, if the induced left-congruence has finite index, then termination of the procedure is guaranteed.

If P_U has the property that $\text{dom}(P_U) \cup \text{range}(P_U) \subseteq \text{IRR}(S)$, then also the resulting system Q_U has this property. In this case we say that Q_U is *S-reduced*. Further, we will call a prefix-rewriting system $P = P_U \cup P_S$ *prefix-interreduced*, *p-interreduced* for short, if $v \in \text{IRR}(P)$ and $u \in \text{IRR}(P \smallsetminus \{(u,v)\})$ hold for each prefix-rule $(u,v) \in P_U$, and we call it *p-canonical* if it is convergent and p-interreduced.

Example 4.2. (continued). Let $\geq$ be the length-lexicographical ordering on $\underline{\Sigma}_2^*$ that is induced by $\bar{b} > b > \bar{a} > a$, and let

$$P_U := \{(ab, \lambda), (ba, \lambda), (aa, \lambda), (\bar{b}\bar{a}, \lambda), (\bar{a}\bar{b}, \lambda), (\bar{a}\bar{a}, \lambda)\}.$$

Then $Q_0 = P_U$, there are no U-critical pairs, but there are three G-critical pairs for Q_0 : $CP_0 = \{(\bar{b}, a), (\bar{a}, b), (\bar{a}, a)\}$. Hence, we get $Q_1 := Q_0 \cup \{(\bar{b}, a), (b, \bar{a}), (\bar{a}, a)\}$.

The G-critical pairs of Q_1 all resolve, but there is an unresolved U-critical pair: $(a\bar{b}, \lambda)$. This yields $Q_2 := Q_1 \cup \{(a\bar{b}, \lambda)\}$, which in turn gives the unresolved G-critical pair (b, a). Finally, we obtain $Q_3 := Q_2 \cup \{(b, a)\}$, and $Q_3 \cup P_{S_0}$ is convergent.

By interreduction Q_3 is transformed into the system

$$Q_U := \{(ab, \lambda), (\bar{b}, a), (a\bar{b}, \lambda), (b, a), (aa, \lambda), (\bar{a}, a)\}.$$

The prefix-rewriting system $P := Q_U \cup P_{S_0}$ is p-canonical, satisfying $\Leftrightarrow_P^* = \sim_U$. Thus, $\text{IRR}(P) = \{\lambda, a\}$ is a complete set of coset representatives for $\langle U \rangle$ in F_2, showing that $|F_2 : \langle U \rangle| = 2$. $\qquad \square$

For the following considerations we fix a group presentation $\langle \Sigma; \emptyset \rangle$ of a free group F. Recall that S_0 denotes the set of trivial rules $S_0 = \{a\bar{a} \to \lambda, \bar{a}a \to \lambda \mid a \in \Sigma\}$.

Proposition 4.8. [Cre95] *Let P_U be a prefix-rewriting system on $\langle \Sigma; \emptyset \rangle$, and let $P := P_U \cup P_{S_0}$. Then the system P is p-canonical if and only if the following two conditions are satisfied:*

(1.) *P is p-interreduced, and*
(2.) *for each rule $(ua, v) \in P_U$, where $u, v \in \underline{\Sigma}^*$ and $a \in \underline{\Sigma}$, (va^{-1}, u) is also a rule of P_U.*

Let $P = P_U \cup P_{S_0}$ be a p-canonical system on $\langle \Sigma; \emptyset \rangle$. If the left-congruence $\Leftrightarrow_P^*$ on $\underline{\Sigma}^*$ is finitely generated, then P_U contains a finite subsystem P_{U_1} such that $P_1 := P_{U_1} \cup P_{S_0}$ also generates this left-congruence. Let $P_{U_2} := P_{U_1} \cup \{(va^{-1}, u) \mid (ua, v) \in P_{U_1}\}$. Then P_{U_2} is still a finite subsystem of P_U, and $P_2 := P_{U_2} \cup P_{S_0}$ generates the left-congruence $\Leftrightarrow_P^*$. However, by Proposition 4.8 P_2 is itself p-canonical, which implies that P_2 coincides with P_U. This gives the following result.

Proposition 4.9. [Cre95] *Let P_U be a prefix-rewriting system on $\langle \Sigma; \emptyset \rangle$ such that $P := P_U \cup P_{S_0}$ is p-canonical. If the left-congruence $\Leftrightarrow_P^*$ is finitely generated, then P_U is itself finite.*

Now let U be a finite subset of Σ^* that is closed under inverses, and let $P_U := \{(u, \lambda) \mid u \in U\}$. Then $\Leftrightarrow_P^* = \sim_U$, where $P := P_U \cup P_{S_0}$ and hence, $\Leftrightarrow_P^*$ is finitely generated. Thus, if $\geq$ is any admissible well-ordering on Σ^*, then the p-canonical prefix-rewriting system Q_U that is compatible with $\geq$ and that satisfies $\Leftrightarrow_{Q_U \cup S_0}^* = \sim_U$ is finite. Hence, given $(\Sigma; S_0)$, P_U and $\geq$ as input, the Knuth-Bendix completion procedure for prefix-rewriting systems is guaranteed to terminate successfully.

Actually we can characterize the p-canonical prefix-rewriting systems on $\langle \Sigma; \emptyset \rangle$ as follows.

Definition 4.10. Let $U \subseteq \Sigma^*$ be a set of freely reduced strings, that is, $U \subseteq \text{IRR}(S_0)$.

(a) The set U is called *reduced* if $\lambda \notin U$ and if, for each string $w \in U$, $w^{-1} \notin U$. By $\underline{U}$ we denote the set $\underline{U} := U \cup U^{-1}$. A prefix of a string in $\underline{U}$ is called *isolated* if it does not occur as a prefix of any other string of $\underline{U}$.

(b) The set U is called *marked* if $\lambda \notin U$ and one letter of each string $w \in U$ is marked as the *central factor* of w. For a string $w = uav \in U$, where $u, v \in \Sigma^*$ and $a \in \Sigma$ is the marked letter, ua is called the *major prefix* of w, and $v^{-1}a^{-1}$ is called the *major suffix* of w.

(c) The set U is called *generalized Nielsen reduced* if it is reduced and marked, and if each major prefix and each major suffix is isolated.

It can be checked fairly easily that a Nielsen reduced set is generalized Nielsen reduced. For a string of odd length, the letter in the middle can be marked, and for a string of even length one of the letters next to the middle of the string can be marked (see, e.g., [MKS76], p.123).

Generalized Nielsen reduced sets can be used to characterize p-canonical prefix-rewriting systems of the form $P := P_U \cup P_{S_0}$. Let P_U be a prefix-rewriting system on $\langle \Sigma; \emptyset \rangle$ such that P is p-canonical. Then by Proposition 4.8 P_U is the union of two-rule systems of the form $\{(ua, v), (va^{-1}, u)\}$. With a two-rule system of this form we associate the freely reduced string uav^{-1}, where we mark the letter a at position $|u| + 1$. Let V be the set consisting of all these strings. Since P is p-reduced, it is easily verified that V is a generalized Nielsen reduced set, and that $\Leftrightarrow_P^* = \sim_V$. Conversely, if $V \subseteq \Sigma^+$ is a generalized Nielsen reduced set, then with each string $w = uav \in V$, where $a \in \Sigma$ is the marked letter, we associate the two prefix-rules (ua, v^{-1}) and $(v^{-1}a^{-1}, u)$. Let P_U be the prefix-rewriting system consisting of all these rules, and let $P := P_U \cup P_{S_0}$. Since all elements of V are freely reduced, we see that $ua, v^{-1}, v^{-1}a^{-1}, u \in \text{IRR}(S_0)$. Further, since each major prefix ua and each major suffix $v^{-1}a^{-1}$ is isolated, P is in fact p-interreduced. Thus, by

Proposition 4.8 P is p-canonical. Hence, we have the following result which extends a result of Bauer [Bau81].

Proposition 4.11. [Cre95] *There is a one-to-one correspondence between the set of p-canonical prefix-rewriting systems of the form* $P = P_U \cup P_{S_0}$ *on* $\langle \Sigma; \emptyset \rangle$ *and the set of generalized Nielsen reduced subsets of* Σ^+.

In particular, it follows that the computation of a (generalized) Nielsen reduced set V from a given finite set U can be interpreted as completing the prefix-rewriting system $P_U \cup P_{S_0}$.

We close this section by taking another look at the Todd-Coxeter coset enumeration method. If $\langle \Sigma; S \rangle$ is a finite presentation, $U \subseteq \Sigma^*$ a finite set, and $\geq$ an admissible well-ordering on Σ^*, then in TC we can always choose those coset representatives that are minimal with respect to $\geq$. By this we mean the following. If w is currently a coset representative, and the slot for $w \cdot a$ is not yet filled in one of the tables, then wa is taken as a representative. If later a collapse equation $wa \sim z$ or a bonus equation $wa \cdot b \sim z$ is discovered, then wa and z, respectively wa and zb^{-1}, are compared with respect to $\geq$, and the smaller of the two strings will be chosen as the new representative for the coset of wa. If TC terminates, then the set of coset representatives is prefix-closed, and the prefix-rewriting system $\{(ua, v) \mid u, v$ are representatives, $a \in \Sigma, ua \neq v\}$, which is essentially just a description of the non-trivial part of the coset multiplication table determined by TC, is canonical and it defines the left-congruence $\sim_U$. Thus, TC can also be seen as a method that determines finite canonical prefix-rewriting systems for subgroups of finite index. The correspondence between TC and standard string-rewriting completion has first been observed by Benninghofen et al [BKR87] in the case that the subgroup considered is trivial.

5. Uniform solvability of the subgroup problem

As seen in the previous section for free groups the subgroup problem can be solved by determining p-canonical prefix-rewriting systems for the left-congruences generated by the subgroups considered. This approach exploited the fact that the free groups considered are given by standard presentations of the form $\langle \Sigma; \emptyset \rangle$, that is, by finite, special, and confluent presentations of the form $(\Sigma; S_0)$. Here we will see that these results can be carried over to certain classes of more general presentations. We follow the development given by Cremanns [Cre95].

Definition 5.1. Let $(\Sigma; S)$ be a finite convergent presentation. This presentation is said to satisfy the *local finiteness condition for p-canonical prefix-rewriting systems* if the following condition holds for each p-canonical prefix-rewriting system $P := P_U \cup P_S$ on Σ^*:

(lfc) for each rule $(u, v) \in P_U$, there exists a finite subsystem $P_{U'}$ of P_U such that $(u, v) \in P_{U'}$ and $P' := P_{U'} \cup P_S$ is p-canonical.

Proposition 4.8 shows essentially that the presentation $\langle \Sigma; \emptyset \rangle$ satisfies the above condition. Accordingly the arguments leading to Proposition 4.9 can be generalized, giving the following result.

Proposition 5.2. *Let $(\Sigma; S)$ be a finite convergent presentation that satisfies the local finiteness condition for p-canonical prefix-rewriting systems, and let $P := P_U \cup P_S$ be a p-canonical prefix-rewriting system on Σ. If the left-congruence $\Leftrightarrow_P^*$ is finitely generated, then the subsystem P_U is itself finite.*

Thus, if $(\Sigma; S)$ is a finite presentation of a group G such that S is convergent and $(\Sigma; S)$ satisfies the condition (lfc), then the Knuth-Bendix completion procedure for prefix-rewriting systems is guaranteed to terminate with success, when $(\Sigma; S)$, a finite set $P_U = \{(u, \lambda) \mid u \in U \cup U^{-1}\}$, and an arbitrary admissible well-ordering $\geq$ are given as input. The resulting finite set Q_U has the property that $Q_U \cup P_S$ is convergent and generates the left-congruence $\sim_U$.

Corollary 5.3. *For each finite convergent presentation $(\Sigma; S)$ satisfying the condition (lfc) the subgroup problem and the index problem can be solved by effectively determining a finite set of prefix-rules P_Q such that $P := P_Q \cup P_S$ is convergent, and $\Leftrightarrow_P^* = \sim_U$.*

Which classes of finite convergent presentations of groups do satisfy the condition (lfc)? In order to verify the condition (lfc) for certain classes of presentations, Cremanns introduced the following technical notion.

Definition 5.4. Let $(\Sigma; S)$ be a finite convergent presentation of a group, and let $^{-1} : \Sigma^* \to \Sigma^*$ be a function realizing the inverse function of the group presented such that $a^{-1} \neq \lambda$ for all $a \in \Sigma$. The presentation $(\Sigma; S)$ is called *compact* if there exists a set of strings $C \subseteq \Sigma^*$ that satisfies all of the following conditions:

(1.) C only contains elements of finitely many congruence classes modulo S, that is, there is a finite subset $C' \subseteq C$ such that $\bigcup_{u \in C} [u]_S = \bigcup_{u \in C'} [u]_S$;

(2.) $C \supseteq \mathrm{dom}(S)$;

(3.) C is closed under the following operations:

  (3.1) substrings, that is, if $xuy \in C$, then $u \in C$,
  (3.2) S-reductions, that is, if $u \in C$, then $\Delta_S^*(u) \subseteq C$,
  (3.3) overlaps, that is, if $xy \in C$ and $yz \in C$ for some $y \neq \lambda$, then $xyz \in C$,
  (3.4) inverses, that is, if $u \in C$, then $u^{-1} \in C$;

(4.) for all $a \in \Sigma$, $aa^{-1} \in C$ and $a^{-1}a \in C$.

Assume that $(\Sigma; S)$ is compact, and that $C \subseteq \Sigma^*$ is the corresponding compact subset of Σ^*. By (4.) and (3.1) we have $\Sigma \subseteq C$. Further (4.) and (3.3), (3.4) imply that $uu^{-1}, u^{-1}u \in C$ for all $u \in C$. Compact presentations are of interest because of the following result.

Proposition 5.5. *If a finite convergent presentation of a group is compact, then it satisfies the local finiteness condition for p-canonical prefix-rewriting systems.*

Hence, Corollary 5.3 applies to finite convergent presentations of groups that are compact.

Proposition 5.6. (a) *If $(\Sigma; S)$ is a finite presentation of a group such that S is a special and canonical string-rewriting system, then $(\Sigma; S)$ is compact.*
(b) *If $(\Sigma; S)$ is a finite presentation of a group such that S is a monadic and canonical string-rewriting system satisfying $\mathrm{dom}(S) \subseteq \Sigma^2$ and $|a^{-1}| = 1$ for all $a \in \Sigma$, then $(\Sigma; S)$ is compact.*
(c) *If $(\Sigma; S)$ is a finite convergent presentation of a finite group, then $(\Sigma; S)$ is compact.*

The presentations in (b) are called *two-monadic* with *inverses of length* 1. Thus, we have the following result.

Corollary 5.7. [Kuh91, Cre95] *Let a group G be given by a finite canonical presentation $(\Sigma; S)$ such that S is special or two-monadic with inverses of length 1, and let $U \subseteq \Sigma^*$ be a finite set. For any admissible well-ordering that is compatible with S, a finite set of prefix-rules Q_U can be determined such that $P := Q_U \cup P_S$ is p-canonical and $\Leftrightarrow_P^* = \sim_U$, thus solving the subgroup problem and the index problem for $(\Sigma; S)$.*

If $(\Sigma; S)$ is a finite presentation of a group such that the string-rewriting system S is two-monadic and confluent, then a finite subset Σ_1 of Σ exists such that each letter $a \in \Sigma_1$ has an inverse of length 1, the subsystem $S_1 := S \cap (\Sigma_1^2 \times (\Sigma_1 \cup \{\lambda\}))$ is confluent, and $(\Sigma_1; S_1)$ presents the same group as $(\Sigma; S)$ [AMO86]. Hence, $(\Sigma_1; S_1)$ satisfies the hypothesis of Corollary 5.7. Since Σ_1 can be determined effectively from $(\Sigma; S)$, this implies that Corollary 5.7 extends to two-monadic confluent presentations of groups.

There is one other class of finite convergent presentations for which the subgroup problem has been solved by computing p-canonical prefix-rewriting systems. This is a very special class of presentations for the finitely generated virtually free groups, which by a result of Muller and Schupp coincide with the context-free groups [MS83].

Let F_Σ be the free group presented by $\langle \Sigma; \emptyset \rangle$, and let E be a finite group. Further, let D be a finite alphabet in one-to-one correspondence to $E \smallsetminus \{1\}$, where we assume without loss of generality that $D \cap \underline{\Sigma} = \emptyset$. For each $a \in D$, let $\varphi_a : \underline{\Sigma} \to \mathrm{IRR}(S_0)$ be a function, and for all $a, b \in D \cup \{\lambda\}$, let $z_{a,b} \in \mathrm{IRR}(S_0)$ be a freely reduced string. By φ_λ we denote the inclusion $\underline{\Sigma} \to \mathrm{IRR}(S_0)$. Now let $\Gamma := \underline{\Sigma} \cup D$, and let S be the string-rewriting system on Γ that consists of the following three groups of rules:

(F) $s\bar{s} \to \lambda$, $\bar{s}s \to \lambda$ for all $s \in \Sigma$;

(E) $ab \to cz_{a,b}$ for all $a, b \in D, c \in D \cup \{\lambda\}$ satisfying $ab =_E c$;

(K) $sa \to a\varphi_a(s)$ for all $s \in \Sigma$ and $a \in D$.

It is easily seen that S is noetherian, but in general S will not be confluent. The presentation $(\Gamma; S)$ is called *virtually free* if S is confluent, that is, S is even canonical.

Lemma 5.8. [CO94] *The presentation $(\Gamma; S)$ described above is canonical if and only if the following three conditions are satisfied:*

(1.) *For all $a \in D$ and all $s \in \Sigma$, $\varphi_a(s)\varphi_a(s^{-1}) \hookrightarrow^*_{S_0} \lambda$.*

(2.) *For all $a, b, c \in D \cup \{\lambda\}$, $\varphi_c(z_{a,b}) \hookrightarrow^*_{S_0} z^{-1}_{ab,c} z_{a,bc} z_{b,c}$.*

(3.) *For all $a, b \in D \cup \{\lambda\}$ and all $s \in \Sigma$, $\varphi_b(\varphi_a(s)) \hookrightarrow^*_{S_0} z^{-1}_{a,b}\varphi_{ab}(s)z_{a,b}$.*

If $(\Gamma; S)$ is a virtually free presentation, then the monoid presented by $(\Gamma; S)$ is obviously a group. In fact, it is an extension of the free group F_Σ by the finite group E. Conversely, each group of this form admits a virtually free presentation. Hence, we have the following characterization.

Proposition 5.9. *A monoid admits a virtually free presentation if and only if it is a finitely generated virtually free group, that is, a context-free group.*

Because of Lemma 5.8 it is easily decidable whether or not a finite presentation is virtually free. Hence, the virtually free presentations form an easily recognizable subclass of the class of all finite canonical presentations of groups. Notice that context-free groups may also admit finite canonical presentations of a different form, and that the virtually free presentations are not at all succinct, since they contain the complete multiplication table of the finite extension group.

Let $(\Gamma; S)$ be a virtually free presentation of a group G. Then G is represented by the set of normal forms $\text{IRR}(S) = (D \cup \{\lambda\}) \cdot \text{IRR}(S_0)$. Accordingly in order to describe subgroups of G we can restrict our attention to sets P_U of prefix-rules satisfying $P_U \subseteq \text{IRR}(S) \times \text{IRR}(S)$.

Proposition 5.10. [CO94] *Let $(\Gamma; S)$ be a virtually free presentation, and let $P_U \subseteq \text{IRR}(S) \times \text{IRR}(S)$ be a set of prefix-rules. Then the prefix-rewriting system $P := P_U \cup P_S$ is p-canonical if and only if the following two conditions are satisfied:*

(1.) *P is p-interreduced, and*

(2.) *for each rule $(aus, bv) \in P_U$, where $a, b \in D \cup \{\lambda\}$, $u, v \in \Sigma^*$, and $s \in \Sigma$, (bvs^{-1}, au) is also a rule of P_U.*

This yields as a consequence the fact that each virtually free presentation satisfies the local finiteness condition for p-canonical prefix-rewriting systems, which in turn implies that the Knuth-Bendix completion procedure for prefix-rewriting systems terminates with success whenever it is given a virtually free presentation $(\Gamma; S)$, a finite set of prefix-rules P_U and an arbitrary admissible well-ordering on Γ^* (Proposition 5.2). Actually by exploiting the particular properties of virtually free presentations the following stronger result can be derived.

Proposition 5.11. [CO94] *There exists an algorithm that solves the following task in polynomial time:*

INPUT: *A virtually free presentation* $(\Gamma; S)$ *and a finite set of generators* $U \subseteq \Gamma^*$.

OUTPUT: *A finite set* P_U *of prefix-rules such that the prefix-rewriting system* $P := P_U \cup P_S$ *is p-canonical, and* P *generates the left-congruence* $\sim_U$ *on* Γ^*. *Further,* P_U *has a partition* $P_U = P_1 \cup P_2$ *of the form* $P_1 \subseteq D \times (D \cup \{\lambda\}) \cdot \underline{\Sigma}^*$ *and* $P_2 \subseteq \bigcup_{a \in D \cup \{\lambda\}} (a \cdot \underline{\Sigma}^* \times a \cdot \underline{\Sigma}^*)$.

In order to solve the subgroup problem for a presentation $(\Sigma; S)$ by prefix-rewriting, it suffices to construct a finite set P_U of prefix rules such that the system $P := P_U \cup P_S$ generates the left-congruence $\sim_U$, and P is noetherian and λ-confluent. Unfortunately λ-confluence is in general much harder to decide than confluence. In the following we restrict our attention to finite canonical presentations $(\Sigma; S)$ of groups.

Let P_U be a finite set of prefix-rules on Σ, and let $P := P_U \cup P_S$. For $w \in \Sigma^*$ let $RC(w) := \{z \in \mathrm{IRR}(S) \mid wz \Rightarrow_P^* \lambda\}$ be the set of *right contexts* of w. Using sets of this form we obtain the following characterization for λ-confluence of P.

Proposition 5.12. [KMO90] *Let* $(\Sigma; S)$ *be a finite convergent presentation, and let* P_U *be a set of prefix-rules on* Σ *such that* $P := P_U \cup P_S$ *is noetherian. Then the following two statements are equivalent:*

(1.) *The prefix-rewriting system* P *is* λ-*confluent.*

(2.) *For each U- or G-critical pair* (p, q) *of* P, p *and* q *are joinable modulo* P *or* $RC(p) = RC(q)$.

Although the sets of U- and G-critical pairs are finite for each finite system P_U, the characterization above has the serious drawback that the sets $RC(w)$ of right contexts need not even be recursive. However, in certain restricted instances these sets are quite simple.

Example 5.13. Let $G = \mathbb{Z} \times \mathbb{Z}_2$ be given by the presentation $(\Sigma; S)$, where $\Sigma := \{a, \bar{a}, b\}$ and $S := \{ba \to ab, b\bar{a} \to \bar{a}b, b^2 \to \lambda, a\bar{a} \to \lambda, \bar{a}a \to \lambda\}$, and let $P_U := \{(b, \lambda)\}$, that is, we consider the subgroup of G that is generated by $U := \{b\}$.

Then it can be shown that there does not exist a finite set of prefix-rules P_1 such that the prefix-rewriting system $P_1 \cup P_S$ is convergent and equivalent to $P := P_U \cup P_S$. On the other hand, there are no U-critical pairs for P_U and only the following three G-critical pairs: $(a, ab), (\bar{a}, \bar{a}b), (b, \lambda)$.

The pair (b, λ) is joinable by $b \Rightarrow_P \lambda$. Further,
$$RC(a) = \{a^i b^\varepsilon \mid i \in \mathbb{Z}, \varepsilon \in \{0, 1\} \text{ such that } a^{i+1} b^\varepsilon \Rightarrow_P^* \lambda\} = \{\bar{a}, \bar{a}b\},$$
$$RC(ab) = \{a^i b^\varepsilon \mid i \in \mathbb{Z}, \varepsilon \in \{0, 1\} \text{ such that } aba^i b^\varepsilon \Rightarrow_P^* \lambda\} = \{\bar{a}, \bar{a}b\},$$
$$RC(\bar{a}) = \{a, ab\} = RC(\bar{a}b).$$
Hence, P is λ-confluent, that is, $w \in \langle U \rangle$ if and only if $w \Rightarrow_P^* \lambda$. □

Proposition 5.12 yields a test for deciding λ-confluence of P whenever the presentation $(\Sigma; S)$ has the following properties:

(a) The languages of the form $RC(w)$ belong to a family of languages for which equality is decidable.

(b) From $(\Sigma; S)$, P_U and $w \in \Sigma^*$, a specification of the language $RC(w)$ can be computed effectively such that the equality test of (a) is applicable to this specification.

For example, if $(\Sigma; S)$ is a finite, confluent, and weight-reducing presentation of a group, then the languages of the form $RC(w)$ are regular, and finite-state acceptors can effectively be constructed for them [KMO90]. Thus, λ-confluence of prefix-rewriting systems is decidable for this class of presentations. Actually, since the equivalence problem is decidable for multi-tape deterministic finite-state acceptors [HK91], this result carries over to finite direct products of groups presented by finite, confluent, and weight-reducing presentations. Observe, however, that the finite, confluent, and weight-reducing presentations only describe a proper subclass of the context-free groups [MO89].

In addition to Proposition 5.12 there is still another criterion for deciding λ-confluence. Let $(\Sigma; S)$ be a finite convergent presentation of a group, let P_U be a set of prefix-rules on Σ, and let $P := P_U \cup P_S$, where we assume that the set $U := \{uv^{-1} \mid (u, v) \in P_U\}$ is closed under taking inverses. Then $[\lambda]_P = \langle U \rangle$, and hence, $w \in [\lambda]_P$ if and only if $w \Leftrightarrow_P^* z$ for some $z \in U^*$. Now P is λ-confluent if and only if each string $w \in \langle U \rangle \smallsetminus \{\lambda\}$ is reducible by P, that is, if and only if $[\lambda]_P \cap \text{IRR}(P) = \{\lambda\}$. However, since S is convergent, the latter equality is equivalent to the equality $(\Delta_S^*(U^*) \cap \text{IRR}(S)) \cap \text{IRR}(P) = \{\lambda\}$.

If S and P_U are both finite, then the sets $\text{IRR}(S)$ amd $\text{IRR}(P)$ are both regular, and finite-state acceptors for them can be constructed effectively. Also U^* is a regular set in this situation. Hence, this criterion becomes decidable whenever the set $\Delta_S^*(U^*)$ (or the set $\Delta_S^*(U^*) \cap \text{IRR}(S)$) allows an effective specification for which the intersection with the regular set $\text{IRR}(P)$ can be determined effectively.

If $(\Sigma; S)$ is a finite, weight-reducing, and confluent presentation of a group and $U \subseteq \Sigma^*$ is a finite set, then it is still an open problem whether or not the set $\Delta_S^*(U^*)$ is necessarily regular. However, if we restrict the set $\Delta_S^*(U^*)$ to only those strings that are obtained by left-most reductions, then this subset $\Delta_{L,S}^*(U^*)$ can be shown to always be regular [Kuh91]. In fact, a finite-state acceptor for this language can be constructed effectively. Since $\Delta_{L,S}^*(U^*) \cap$

$\mathrm{IRR}(S) = \Delta_S^*(U^*) \cap \mathrm{IRR}(S)$, we obtain a finite-state acceptor for the set $\Delta_S^*(U^*) \cap \mathrm{IRR}(P)$. This gives the following decidability result.

Proposition 5.14. [Kuh91] *Let $(\Sigma; S)$ be a finite, weight-reducing, and confluent presentation of a group, and let P_U be a finite set of prefix-rules on Σ such that the set $U := \{uv^{-1} \mid (u,v) \in P_U\}$ is closed under taking inverses. Then it is decidable whether the prefix-rewriting system $P := P_U \cup P_S$ is λ-confluent.*

Now assume that $(\Sigma; S)$ is a finite, weight-reducing, and confluent presentation of a group G, let $U \subseteq \Sigma^+$ be a finite set that is closed under taking inverses, and let $P_U := \{(u, \lambda) \mid u \in U\}$. From $(\Sigma; S)$ and U we can construct a finite-state acceptor $A = (Q, \Sigma, q_0, F, \delta)$ for the language $\Delta_S^*(U^*) \cap \mathrm{IRR}(S)$. From A we extract a finite set of prefix-rules P_U' as follows, where we identify A with its state graph in order to simplify the notation:

(i) For every simple path in A leading from the initial state q_0 to a final state $q_f \in F$, which does not pass through any final state, we put the rule (x, λ) into P_U', where x is the label along the path considered.

(ii) For every path p in A from q_0 to a final state $q_f \in F$, which does not pass through any final state, and which can be partitioned into three parts $p = p_1, p_2, p_3$ such that p_1 is a simple path, and p_2 is a simple loop, we put the rule $(x_1 x_2, x_1)$ into P_U', where x_i is the label along the subpath p_i, $i = 1, 2$.

Obviously, P_U' is a finite set of prefix-rules that can effectively be obtained from A. For $w \in \langle U \rangle$ there exists a unique string $w_0 \in \mathrm{IRR}(S)$ such that $w \to_S^* w_0$. Since $w \in \langle U \rangle$, $w_0 \in \Delta_S^*(U^*) \cap \mathrm{IRR}(S)$, and hence, w is accepted by A. From the construction of P_U' it follows that $w \Rightarrow_{P'} \lambda$ holds, where $P' := P_U' \cup P_S$. Since $u \sim_U v$ holds for each rule $(u, v) \in P_U'$, it follows that $\Leftrightarrow_{P'}^* = \sim_U$, and P' is a λ-convergent prefix-rewriting system for $\sim_U$.

Proposition 5.15. [Kuh91] *Let $(\Sigma; S)$ be a finite, weight-reducing, and confluent presentation of a group G, and let $U \subseteq \Sigma^+$ be a finite set. Then a finite set of length-reducing prefix-rules P_U' can be determined effectively such that the prefix-rewriting system $P := P_U' \cup P_S$ is λ-convergent and $\Leftrightarrow_P^* = \sim_U$.*

If $(\Sigma; S)$ is a finite convergent presentation of a group, and $U \subseteq \Sigma^*$ is a finite set that is closed under taking inverses, then $\Delta_S^*(U^*) \cap \mathrm{IRR}(S)$ is a set of unique representatives for the subgroup $\langle U \rangle$. However, if the string-rewriting system S is not confluent, then there may exist irreducible strings $w \in \mathrm{IRR}(S)$ such that $w \in \langle U \rangle$, but $w \notin \Delta_S^*(U^*)$. In this situation the construction above will not yield a λ-convergent prefix-rewriting system presenting $\sim_U$. The situation can be saved if the string-rewriting system S is finite, monadic, and λ-confluent, because of the following fact.

Proposition 5.16. [MO91] *Let S be a finite, monadic, and λ-confluent string-rewriting system on Σ such that the monoid presented by $(\Sigma; S)$ is a* group. Then, for each regular language $L \subseteq \Sigma^*$, the set $I_S(L) := \{w \in \text{IRR}(S) \mid \exists u \in L : u \leftrightarrow^*_S w\}$ is regular as well. In addition, a finite-state acceptor for $I_S(L)$ can be constructed in polynomial time from S and a finite-state acceptor for L.

Observe that a group can be given by a finite, monadic, and λ-confluent string-rewriting system if and only if it is a context-free group [ABS87]. By applying the construction of a prefix-rewriting system described above to a finite-state acceptor for $I_S(U^*)$ we obtain the following result.

Proposition 5.17. [KMO94] *Let $(\Sigma; S)$ be a finite, monadic, and λ-confluent presentation of a group, and let $U \subseteq \Sigma^+$ be a finite set. Then a finite set of length-reducing prefix-rules P'_U can be constructed effectively such that the prefix-rewriting system $P := P'_U \cup P_S$ is λ-convergent and $\leftrightarrow^*_P = \sim_U$.*

Even though the prefix-rewriting system P is not convergent, it can be used to determine the index $|G : \langle U \rangle|$, where G is the group presented by $(\Sigma; S)$. This is due to the fact that in the situation described in the proposition above, the intersection $[y]_P \cap \text{IRR}(P)$ is finite for each string $y \in \Sigma^*$. Thus, the index $|G : \langle U \rangle|$ is infinite if and only if $\text{IRR}(P)$ is infinite, and in case $\text{IRR}(P)$ is finite, we can actually determine the index $|G : \langle U \rangle|$ by checking which of the finitely many strings of $\text{IRR}(P)$ belong to the same coset.

We finally turn to presentations of polycyclic groups.

Let $\Sigma = \{a_1, \bar{a}_1, \ldots, a_n, \bar{a}_n\}$, let $\Sigma_i := \{a_i, \bar{a}_i, \ldots, a_n, \bar{a}_n\}$ for $i = 1, \ldots, n$, and let $\Sigma_{n+1} := \emptyset$. We need several particular classes of rules on Σ. A rule $(\ell \to r)$ is called

- a *CP2-rule* if $\ell = a_j^\delta a_i^\varepsilon$ and $r = a_i^\varepsilon z$ for some $j > i$, $\delta, \varepsilon \in \{1, -1\}$, and $z \in \Sigma_{i+1}^*$,
- a *positive P-rule* if $\ell = a_i^k$ and $r \in \Sigma_{i+1}^*$ for some $i \in \{1, \ldots, n\}$ and $k > 0$,
- a *negative P-rule* if $\ell = \bar{a}_i$ and $r = a_i^k z$ for some $i \in \{1, \ldots, n\}$, $k \geq 0$, and $z \in \Sigma_{i+1}^*$.

A set S of rules is called

- a *P-system* if it contains P-rules only, and for each $i \in \{1, \ldots, n\}$, S either contains exactly one rule with left-hand side a_i^k for some $k > 0$ and exactly one rule with left-hand side $\bar{a}_i$, or S contains no rule with left-hand side from $\{a_i, \bar{a}_i\}^+$,
- a *CP2-system* if it contains CP2-rules only, and for each $i, j \in \{1, \ldots, n\}$, $j > i$, and each $\delta, \varepsilon \in \{1, -1\}$, S contains exactly one rule with left-hand side $a_j^\delta a_i^\varepsilon$.

A presentation $(\Sigma; S)$ is called a *PCP2-presentation* if S can be written as $S = S_0 \cup R \cup C$, where $S_0 = \{a_i \bar{a}_i \to \lambda, \bar{a}_i a_i \to \lambda \mid i = 1, \ldots, n\}$ is the set of trivial rules, R is a P-system, and C is a CP2-system.

Using a particular ordering Wißmann shows that a string-rewriting system of this form is noetherian. Further, he proves that a group G can be described by a finite PCP2-presentation if and only if G is a finitely presented polycyclic group [Wiß89]. Actually the PCP2-presentations are closely related to the polycyclic presentations of [BCM77]. In fact, based on a specialized version of the Knuth-Bendix completion procedure Wißmann proves that each finite PCP2-presentation of a polycyclic group G can be transformed into a finite convergent PCP2-presentation of G. Accordingly in the following we consider polycyclic groups that are given by finite convergent PCP2-presentations.

Let $(\Sigma; S)$ be a finite convergent PCP2-presentation of a polycyclic group G, let U be a finite subset of Σ^+, and let $H := \langle U \rangle$. As shown by Wißmann a canonical base Ω for the subgoup H can be constructed effectively by employing a specialized completion procedure [Wiß89]. With Ω a finite set of prefix-rules P_U can be associated in a straightforward manner. The resulting prefix-rewriting system $P := P_U \cup P_S$ is λ-convergent, and $\Leftrightarrow_P^* = \sim_U$. Thus, we have the following result.

Proposition 5.18. [Wiß89] *Let $(\Sigma; S)$ be a finite (convergent) PCP2-presentation of a polycyclic group G, and let U be a finite subset of Σ^+. Then a finite set of prefix-rules P_U can be constructed effectively such that the prefix-rewriting system $P := P_U \cup P_S$ is λ-convergent, and $\Leftrightarrow_P^* = \sim_U$.*

For a more detailed presentation of this result see for example [KMO94]. Using a different ordering on the letters and slightly different convergent presentations for polycyclic groups this result can be strengthened to obtain a finite set of prefix-rules such that P is even convergent [MR98c].

6. The subgroup presentation problem

Using prefix-rewriting systems that are λ-convergent or even convergent we have been able to solve the subgroup problem for certain classes of finite presentations of groups. The groups presented by these classes of presentations are the finitely generated polycyclic groups, the context-free groups, and certain subclasses thereof. However, the polycyclic groups and the context-free groups are both subgroup closed, that is, each finitely generated subgroup of a finitely presented polycyclic, respectively context-free, group is itself a finitely presented polycyclic, respectively context-free, group. Hence, each subgroup of this form has a finite presentation, and accordingly, from the given presentation of the group and the given set of generators for the subgroup we want to determine a finite presentation of the subgroup effectively. Preferably we would like to obtain a presentation for the subgroup that is of the same type as the presentation of the group given.

For the class of free groups given by free presentations of the form $\langle \Sigma; \emptyset \rangle$ this problem has already been solved implicitly in Section 4. For a presentation $\langle \Sigma; \emptyset \rangle$ and a finite set of strings $U \subseteq \Sigma^*$, a finite set P_U of prefix-rules can be constructed effectively such that the prefix-rewriting system $P := P_U \cup P_{S_0}$ is convergent, and $\Leftrightarrow_P^* = \sim_U$. From P_U a finite set V can be obtained that is generalized Nielsen reduced, and that generates the same subgroup as U.

Thus, V is a set of free generators for $\langle U \rangle$, that is, $\langle \Gamma_V; \emptyset \rangle$ is a free presentation of $\langle U \rangle$, where Γ_V is an alphabet in one-to-one correspondence to V.

Actually this process is a special case of a procedure that applies to finite convergent presentations of groups and convergent prefix-rewriting systems presenting subgroups. In order to describe this procedure we need some additional notions.

Let $(\Sigma; S)$ be a finite canonical presentation of a group G, let P_U be a finite set of prefix-rules on Σ such that the prefix-rewriting system $P := P_U \cup P_S$ is convergent, and let H denote the subgroup of G that is generated by P_U. Our aim is to present a construction that yields a finite presentation for H from $(\Sigma; S)$ and P_U.

With $(\Sigma; S)$ and P_U we associate an infinite graph $\Gamma := (V, E, \sigma, \tau, ^{-1})$, where

(a) $V := \Sigma^*$ is the set of vertices,

(b) $E := \{(x, (\ell, r), y, \varepsilon) \mid (\ell, r) \in S, x, y \in \Sigma^*, \varepsilon \in \{1, -1\}\} \cup$
$\{((u, v), y, \varepsilon) \mid (u, v) \in P_U, y \in \Sigma^*, \varepsilon \in \{1, -1\}\}$

　　is the set of edges,

(c) $\sigma, \tau : E \to V$ are mappings that associate with each edge $e \in E$ its initial vertex $\sigma(e)$ and its terminal vertex $\tau(e)$:

$$\sigma(x, (\ell, r), y, \varepsilon) := \left\{ \begin{array}{ll} x\ell y, & \text{if } \varepsilon = 1 \\ xry, & \text{if } \varepsilon = -1 \end{array} \right\},$$

$$\tau(x, (\ell, r), y, \varepsilon) := \left\{ \begin{array}{ll} xry, & \text{if } \varepsilon = 1 \\ x\ell y, & \text{if } \varepsilon = -1 \end{array} \right\},$$

and

$$\sigma((u, v), y, \varepsilon) := \left\{ \begin{array}{ll} uy, & \text{if } \varepsilon = 1 \\ vy, & \text{if } \varepsilon = -1 \end{array} \right\}, \quad \tau((u, v), y, \varepsilon) := \left\{ \begin{array}{ll} vy, & \text{if } \varepsilon = 1 \\ uy, & \text{if } \varepsilon = -1 \end{array} \right\},$$

(d) $^{-1} : E \to E$ is a mapping that associates with each edge $e \in E$ its inverse edge e^{-1}:

$(x, (\ell, r), y, \varepsilon)^{-1} := (x, (\ell, r), y, -\varepsilon)$, and $((u, v), y, \varepsilon)^{-1} := ((u, v), y, -\varepsilon)$.

Thus, this graph represents the relation $\Rightarrow_P$ on Σ^*. It is obtained from the graph Γ_S considered by Squier [SOK94] for the (finite) monoid presentation $(\Sigma; S)$ by adding the edges corresponding to the prefix-rules P_U.

This graph has some additional structure in the form of a right action of Σ^* on Γ. Let $z \in \Sigma^*$. Then $w \cdot z := wz$ for each $w \in V$, $(x, (\ell, r), y, \varepsilon) \cdot z := (x, (\ell, r), yz, \varepsilon)$ and $((u, v), y, \varepsilon) \cdot z := ((u, v), yz, \varepsilon)$ for all edges $(x, (\ell, r), y, \varepsilon)$ and $((u, v), y, \varepsilon) \in E$.

By $P(\Gamma)$ we denote the set of paths in Γ, where we include a path (w) of length zero from w to w for each $w \in V$. The mappings $\sigma, \tau, ^{-1}$ and the right action of Σ^* easily extend to paths. By $P^{(2)}(\Gamma)$ we denote the following set of pairs of paths in Γ:

$$P^{(2)}(\Gamma) := \{(p, q) \mid p, q \in P(\Gamma), \sigma(p) = \sigma(q), \text{ and } \tau(p) = \tau(q)\}.$$

We are interested in certain subsets of $P^{(2)}(\Gamma)$. The *set of disjoint derivations* D_Γ is defined as

$$D_\Gamma := \{(p \cdot \sigma(q) \circ \tau(p) \cdot q, \sigma(p) \cdot q \circ p \cdot \tau(q)) \mid p \in P(\Gamma), \text{ and } q \in P(\Gamma)$$
$$\text{such that } q \text{ does not contain any } P_U\text{−edges}\},$$

where $\circ$ denotes the concatenation of paths. The *set of inverse derivations* I_Γ is defined as

$$I_\Gamma := \{(p \circ p^{-1}, (\sigma(p))) \mid p \in P(\Gamma)\}.$$

Clearly, D_Γ and I_Γ are subsets of $P^{(2)}(\Gamma)$. Further, by $P^{(2)}(\Gamma_S)$ we denote the set of pairs $(p, q) \in P^{(2)}(\Gamma)$ of paths such that neither p nor q does contain any P_U-edges.

Definition 6.1. An equivalence relation $\simeq$ on $P(\Gamma)$ is called a *homotopy relation* if it satisfies all of the following conditions:

(1.) $D_\Gamma \cup I_\Gamma \cup P^{(2)}(\Gamma_S) \subseteq \simeq \subseteq P^{(2)}(\Gamma)$,

(2.) if $p \simeq q$, then $p \cdot z \simeq q \cdot z$ for all $z \in \Sigma^*$, and

(3.) if $p, q_1, q_2, r \in P(\Gamma)$ satisfy $\tau(p) = \sigma(q_1) = \sigma(q_2)$, $\tau(q_1) = \tau(q_2) = \sigma(r)$, and $q_1 \simeq q_2$, then $p \circ q_1 \circ r \simeq p \circ q_2 \circ r$.

For each subset $B \subseteq P^{(2)}(\Gamma)$ there exists a smallest homotopy relation $\simeq_B$ on $P(\Gamma)$ that contains B. Accordingly $\simeq_B$ is called the homotopy relation *generated* by B.

Definition 6.2. The prefix-rewriting system $P = P_U \cup P_S$ is said to have *finite derivation type* if there exists a finite set $B \subseteq P^{(2)}(\Gamma)$ that generates the homotopy relation $P^{(2)}(\Gamma)$, that is, $P^{(2)}(\Gamma)$ is the only homotopy relation on $P(\Gamma)$ containing B.

Squier introduced this very notion for monoid presentations proving that a finite canonical monoid presentation has finite derivation type [SOK94]. Although this property has been investigated thoroughly [OK97], no algebraic characterization has been obtained so far for the class of finitely presented monoids (or groups) that have finite derivation type. For the case of prefix-rewriting systems presenting subgroups of groups the situation is more positive.

Proposition 6.3. [Cre95] *Let $(\Sigma; S)$ be a finite presentation of a group G, let P_U be a finite set of prefix-rules on Σ, let $P := P_U \cup P_S$, and let H be the subgroup of G that is generated by P_U. Then H is itself finitely presented if and only if P has finite derivation type.*

If $(\Sigma; S)$ is convergent, and P is convergent as well, then it can be shown that P does have finite derivation type. This yields the following consequence.

Corollary 6.4. [Cre95] *Let $(\Sigma; S)$ be a finite convergent presentation of a group G, and let P_U be a finite set of prefix-rules on Σ such that the prefix-rewriting system $P := P_U \cup P_S$ is convergent. Then the subgroup H of G that is generated by P_U is finitely presented.*

Actually a finite presentation for H can be constructed from $(\Sigma; S)$ and P_U as follows. Let $(\Sigma; S)$ be a finite canonical presentation of a group G, let P_U be a finite set of prefix-rules on Σ such that the prefix-rewriting system $P := P_U \cup P_S$ is p-canonical, let H be the subgroup of G that is generated by P_U, and let Γ be the graph associated with $(\Sigma; S)$ and P_U. By $P_+(\Gamma)$ we denote the set of all paths in Γ that only contain positive edges, that is, edges of the form $(x, (\ell, r), y, 1)$ and $((u, v), y, 1)$. In analogy to the definition of critical pairs we obtain critical pairs of edges.

An ordered pair (e_1, e_2) of edges of Γ is called a *G-critical pair of edges*, if $e_1 = ((uw, v), z, 1)$ and $e_2 = (u, (wz, r), \lambda, 1)$, where $(uw, v) \in P_U$, $(wz, r) \in S$, and $w \neq \lambda$. Then (e_1, e_2) corresponds to the G-critical pair (vz, ur) of P.

The ordered pair (e_1, e_2) of edges of Γ is called a *U-critical pair of edges*, if $e_1 = ((uw, v), \lambda, 1)$ and $e_2 = ((u, z), w, 1)$, where $(uw, v), (u, z) \in P_U$ and $w \neq \lambda$ or $v \neq z$. Then (e_1, e_2) corresponds to the U-critical pair (v, zw) of P.

Since the prefix-rewriting system P is convergent, there exists a pair (p_1, p_2) of paths $p_1, p_2 \in P_+(\Gamma)$ for each critical pair of edges (e_1, e_2) such that $\sigma(p_1) = \tau(e_1)$, $\sigma(p_2) = \tau(e_2)$, and $\tau(p_1) = \tau(p_2)$, that is, $(e_1 \circ p_1, e_2 \circ p_2) \in P_+^{(2)}(\Gamma)$. Such a pair (p_1, p_2) is called a *resolution* of (e_1, e_2).

Let B denote the set consisting of all pairs $(e_1 \circ p_1, e_2 \circ p_2)$, where (e_1, e_2) is a G- or a U-critical pair of edges, and (p_1, p_2) is a fixed resolution chosen for (e_1, e_2). Then $B \subseteq P_+^{(2)}(\Gamma)$ is a finite set of pairs of positive paths. Further, let Γ_P be a new alphabet in one-to-one correspondence to P_U, and let f denote the function $f : P_+(\Gamma) \to \Gamma_P^*$ that is defined as follows:

$$f(p) \quad := \lambda, \text{ if } p \text{ is a path of length } 0;$$

$$f(p \circ e) \quad := \begin{cases} f(p), & \text{if } p \in P_+(\Gamma), e = (x, (\ell, r), y, 1), \text{ and } \tau(p) = x\ell y, \\ f(p) \cdot a_{(u,v)}, & \text{if } p \in P_+(\Gamma), e = ((u, v), y, 1), \text{ and } \tau(p) = uy. \end{cases}$$

Here $a_{(u,v)} \in \Gamma_P$ is the letter corresponding to the prefix-rule $(u, v) \in P_U$. Thus, for $p \in P_+(\Gamma)$, $f(p)$ yields (an encoding of) the sequence of prefix-rules from P_U that are applied along the path p. Then the following result holds.

Proposition 6.5. [CO94] *Let $(\Sigma; S)$ be a finite canonical presentation of a group G, let P_U be a finite set of prefix-rules on Σ such that the prefix-rewriting system $P := P_U \cup P_S$ is p-canonical, and let H be the subgroup of G that is generated by P_U. Then $(\Gamma_P; f(B))$ is a finite presentation of H.*

Observe that $(\Gamma_P; f(B))$ can be constructed effectively from $(\Sigma; S)$ and P_U.

Example 6.6. Let $\Sigma = \{a, \bar{a}, b, \bar{b}\}$, and let $S = \{b^\varepsilon a^\mu \to a^\mu b^\varepsilon \mid \varepsilon, \mu \in \{1, -1\}\} \cup S_0$. Then $(\Sigma; S)$ is a finite canonical presentation of the group $G = F_1 \times F_1$. Let H be the subgroup of G that is generated by $\{ab\}$, and let $P_U := \{(a, \bar{b}), (\bar{a}, b)\}$. Then $P := P_U \cup P_S$ is a p-canonical prefix-rewriting system such that $\overset{*}{\Leftrightarrow}_P = \sim_H$.

P has no U-critical pairs, but it has the following two G-critical pairs: $\{(\bar{b}\bar{a}, \lambda), (ba, \lambda)\}$. Since $a\bar{a} \Rightarrow_{P_U} \bar{b}\bar{a} \rightarrow_S \bar{a}\bar{b} \Rightarrow_{P_U} b\bar{b} \rightarrow_{S_0} \lambda \leftarrow_{S_0} a\bar{a}$, and $\bar{a}a \Rightarrow_{P_U} ba \rightarrow_S ab \Rightarrow_{P_U} b\bar{b} \rightarrow_{S_0} \lambda \leftarrow_{S_0} \bar{a}a$, we obtain the following presentation for H, where α corresponds to the prefix-rule $(a, \bar{b})$, and β corresponds to the prefix-rule $(\bar{a}, b)$:

$$(\{\alpha, \beta\}; \{\alpha\beta \rightarrow \lambda, \beta\alpha \rightarrow \lambda\}).$$

In particular, this shows that H is the free group of rank one.   □

The general result above has several consequences. First of all when applied to the free presentation $\langle \Sigma; \emptyset \rangle$ of a free group and a finite set of prefix-rules P_U such that $P := P_U \cup P_{S_0}$ is p-canonical, the above construction yields a free presentation for the subgroup H generated by P_U.

The class of groups that admit a finite presentation of the form $(\Sigma; S)$, where S is a special and canonical string-rewriting system on Σ, is exactly the class of groups that are free products of finitely many finite and infinite cyclic groups [Coc76]. Since this class of groups is closed under the operation of taking finitely generated subgroups, each such subgroup can be given by a finite presentation involving a special canonical string-rewriting system. By Corollary 5.7 for each subgroup of this form a finite set of prefix-rules can be constructed such that the resulting prefix-rewriting system is p-canonical. Hence, Proposition 6.5 yields a finite presentation for this subgroup. Actually, the following result holds.

Proposition 6.7. [Cre95] *Let $(\Sigma; S)$ be a finite, special, and canonical presentation of a group G, let U be a finite set of strings from Σ^+, and let $P := P_U \cup P_S$ be a p-canonical prefix-rewriting system for $\sim_U$. Then the construction of Proposition 6.5 yields a finite presentation $(\Gamma; T)$ for the subgroup $\langle U \rangle$ of G such that the string-rewriting system T is special and canonical.*

Using a slightly different approach N. Kuhn shows that the presentation $(\Gamma; T)$ of $\langle U \rangle$ can be obtained in polynomial time from $(\Sigma; S)$ and U [Kuh91].

The construction of Proposition 6.5 also applies to virtually free presentations because of Proposition 5.11. By analyzing it in detail the following result has been obtained.

Proposition 6.8. [CO94] *Given a virtually free presentation $(\Sigma; S)$ of a group G and a finite set of generators $U \subseteq \Sigma^+$, a virtually free presentation $(\Gamma; T)$ for the subgroup $\langle U \rangle$ of G can be constructed in polynomial time.*

For a finitely generated subgroup H of a polycyclic group G that is given by a finite convergent PCP2-presentation $(\Sigma; S)$, Proposition 5.18 yields a finite set of prefix-rules P_U such that the prefix-rewriting system $P := P_U \cup P_S$ is noetherian, p-interreduced, and λ-confluent. Since P is in general not confluent, Proposition 6.5 does not apply. However, since P_U corresponds to a canonical base Ω of H, a finite PCP2-presentation $(\Gamma; T)$ for H can be constructed from P_U. This PCP2-presentation is even confluent, that is, we have the following result.

Proposition 6.9. [KMO94] *Given a finite PCP2-presentation of a polycyclic group G and a finite set U of generators, a finite convergent PCP2-presentation for the subgroup $\langle U \rangle$ of G can be constructed effectively.*

In the remaining part of this section we briefly review the results that have been obtained for various classes of finite presentations of groups that involve monadic string-rewriting systems. For these classes of presentations a different approach has been shown to be advantageous. This approach is based on automata-theoretical constructions.

From a finite, monadic, and confluent presentation $(\Sigma; S)$ of a group G and a finite set $U \subseteq \Sigma^+$ that is closed under inverses a deterministic finite-state acceptor A is constructed for $\Delta_S^*(U^*)$. From A a finite set REP is extracted that forms a partial set of coset representatives for $\langle U \rangle$ in G. By applying the Reidemeister-Schreier rewriting process to $(\Sigma; S)$ and $\langle U \rangle$ using REP, a finite monadic presentation $(\Gamma; T)$ is obtained for $\langle U \rangle$. If the given presentation $(\Sigma; S)$ is two-monadic, then the resulting presentation $(\Gamma; T)$ can be transformed into a two-monadic confluent presentation of $\langle U \rangle$ [Kuh91].

In [KMO94] this procedure is carried over to the class of all finite monadic and λ-confluent presentations of groups that have inverses of length one. Here a monadic and λ-confluent presentation is obtained for each finitely generated subgroup. In fact, for both these cases the presentation for the subgroup is obtained in polynomial time.

7. AUTOMATIC MONOIDS

An automatic structure for a monoid presentation $(\Sigma; S)$ can be interpreted as a finite description of the multiplication table of the monoid M_S. We will see that under certain restrictions there exists a close correspondence between automatic structures and infinite canonical prefix-rewriting systems for $(\Sigma; S)$. In order to define these structures we need the following definition as we will be dealing with infinite sets of pairs of strings that are to be recognized by finite-state acceptors (fsa's).

Let Σ be a finite alphabet, and let $\# \notin \Sigma$ be an additional "padding" symbol. Then by $\Sigma_\#$ we denote the following finite alphabet:

$$\Sigma_\# := ((\Sigma \cup \{\#\}) \times (\Sigma \cup \{\#\})) \smallsetminus \{(\#, \#)\}.$$

This alphabet is called the *padded extension* of Σ. A mapping $\nu : \Sigma^* \times \Sigma^* \to \Sigma_\#^*$ is now defined as follows:
if $u := a_1 a_2 \cdots a_n$ and $v := b_1 b_2 \cdots b_m$, where $a_1, \ldots, a_n, b_1, \ldots, b_m \in \Sigma$, then

$$\nu(u, v) := \begin{cases} (a_1, b_1)(a_2, b_2) \cdots (a_m, b_m)(a_{m+1}, \#) \cdots (a_n, \#), & \text{if } m < n, \\ (a_1, b_1)(a_2, b_2) \cdots (a_m, b_m), & \text{if } m = n, \\ (a_1, b_1)(a_2, b_2) \cdots (a_n, b_n)(\#, b_{n+1}) \cdots (\#, b_m) & \text{if } m > n. \end{cases}$$

A *prefix-rewriting system* P on Σ is called *synchronously regular, s-regular* for short, if $\nu(P)$ is accepted by some fsa over $\Sigma_\#$. Obviously, if P is s-regular, then $\mathrm{dom}(P)$ and $\mathrm{range}(P)$, and therewith also $\mathrm{RED}(P)$ and $\mathrm{IRR}(P)$, are regular languages.

An *automatic structure* for a finitely generated monoid presentation $(\Sigma; S)$ consists of a fsa W over Σ, a fsa $M_=$ over $\Sigma_\#$, and fsa's M_a $(a \in \Sigma)$ over $\Sigma_\#$ satisfying the following conditions:

(0.) $L(W) \subseteq \Sigma^*$ is a complete set of (not necessarily unique) representatives for the monoid M_S, that is, $L(W) \cap [u]_S \neq \emptyset$ holds for each $u \in \Sigma^*$,

(1.) $L(M_=) = \{\nu(u,v) \mid u,v \in L(W) \text{ and } u \leftrightarrow_S^* v\}$, and

(2.) for all $a \in \Sigma$, $L(M_a) = \{\nu(u,v) \mid u,v \in L(W) \text{ and } ua \leftrightarrow_S^* v\}$.

Actually, one may require that the set $L(W)$ is a cross-section for M_S, in which case we say that we have an *automatic structure with uniqueness* [Eps92]. In this situation the fsa $M_=$ is trivial, and hence, it will not be mentioned explicitly.

A monoid presentation is called *automatic* if it has an automatic structure, and a monoid is called *automatic* if it has an automatic presentation. Automatic monoids have word problems that are decidable in quadratic time based on the automatic structure. For automatic groups many additional nice properties have been obtained, while for automatic monoids in general the situation is not quite as nice [CRRT97, OSKM98, Ott98a]. Here we are interested in automatic structures with uniqueness, for which the set of representatives considered is in addition prefix-closed. It is an open problem whether or not every automatic group does have an automatic structure with this additional property.

Recall that the set $\mathrm{IRR}(P)$ of strings that are irreducible with respect to a prefix-rewriting system P is prefix-closed. Finally, we say that the prefix-rewriting system P on Σ is equivalent to a string-rewriting system S on the same alphabet Σ, if the relations $\Leftrightarrow_P^*$ and $\leftrightarrow_S^*$ coincide.

The following observation is rather straightforward.

Lemma 7.1. *If there exists an s-regular canonical prefix-rewriting system P on Σ that is equivalent to the string-rewriting system S, then $C := \mathrm{IRR}(P)$ is part of an automatic structure with uniqueness for $(\Sigma; S)$.*

On the other hand, the following result can be proved by taking P to be the prefix-rewriting system $P := \{(ua, v) \mid u,v \in L(W), a \in \Sigma, ua \leftrightarrow_S^* v, \text{ but } ua \neq v\}$.

Proposition 7.2. *Let $(W, A_a(a \in A))$ be an automatic structure with uniqueness for $(\Sigma; S)$ such that the set $L(W)$ is in addition prefix-closed. Then there exists an s-regular canonical prefix-rewriting system P on Σ that is equivalent to S such that $\mathrm{IRR}(P) = L(W)$.*

Together these two facts yield the following characterization.

Corollary 7.3. *Let* $(\Sigma; S)$ *be a finitely generated monoid presentation. Then the following two statements are equivalent:*

(a) *There exists an automatic structure* $(W, A_a(a \in \Sigma))$ *with uniqueness for* $(\Sigma; S)$ *such that the set* $L(W)$ *is prefix-closed.*

(b) *There exists an s-regular canonical prefix-rewriting system* P *on* Σ *that is equivalent to* S.

Is there a corresponding characterization for automatic structures (with uniqueness), where $L(W)$ is **not** prefix-closed?

There exists a group with a finite convergent presentation, which does not admit an automatic structure [Ger92]. Hence, no finitely generated presentation of this group has an s-regular canonical prefix-rewriting system that defines the corresponding Thue congruence.

The monoid N of [OSKM98] has an automatic structure that is based on a regular cross-section that is the set of irreducible strings modulo some infinite left-regular convergent string-rewriting system. Hence, this set is certainly prefix-closed and so Proposition 7.2 shows that this presentation of N admits an s-regular canonical prefix-rewriting system. However, N does not admit any finite convergent presentation. These observations yield the following result.

Corollary 7.4. *The class of finitely presented monoids that admit a finite convergent presentation and the class of finitely presented monoids that admit an s-regular canonical prefix-rewriting system are incomparable under set inclusion.*

8. COSET ENUMERATION AND GRÖBNER BASES

In this section we will point out a close correspondence between the Todd-Coxeter coset enumeration method for finitely generated subgroups of finitely presented groups as presented in Section 3 and the computation of prefix Gröbner bases of finitely generated binomial right ideals in the corresponding free group ring. It is based on a recent paper by Reinert, Madlener, and Mora [RMM98a].

Let $\langle \Sigma; S \rangle$ be a finite group presentation of a group G, where we assume without loss of generality that S is a special string-rewriting system, that is, $\text{dom}(S)$ is the defining set of relators for the group. Let $U \subseteq \Sigma^+$ be a finite set generating a subgroup H of G, and let F denote the free group generated by Σ, that is, $F = \langle \Sigma; \emptyset \rangle \cong (\Sigma; S_0)$. The elements of F are represented by the set $\text{IRR}(S_0)$ of freely reduced strings.

By $\mathbb{K}[F]$ we denote the *free group ring* of F, where $\mathbb{K}$ is a field. The elements of $\mathbb{K}[F]$ are the formal polynomials of the form $\sum_{i=1}^{k} \alpha_i \cdot w_i$, where $\alpha_i \in \mathbb{K} \setminus \{0\}$ and $w_i \in F$. Here $\cdot$ denotes the scalar multiplication, while $*$ will be used to denote the multiplication in the ring $\mathbb{K}[F]$. On Σ we fix the precedence $a_1 < a_2 < \cdots < a_n < \bar{a}_1 < \cdots < \bar{a}_n$, where $\Sigma = \{a_1, a_2, \ldots, a_n\}$, which induces the length-lexicographical ordering $\geq_{\ell\ell}$ on F. This ordering is linear and well-founded, but it is not compatible with the group structure of F. Nevertheless, this ordering can be lifted to $\mathbb{K}[F]$ and used to distinguish the largest term of a polynomial $f \in \mathbb{K}[F]$ as its *head term* $HT(f)$, this term's coefficient as its *head coefficient* $HC(f)$, and the *head monomial* $HM(f) = HC(f) \cdot HT(f)$. Further, for a subset $Q \subseteq \mathbb{K}[F]$, $HT(Q) := \{HT(f) \mid f \in Q\}$.

As each element of F is identified with its freely reduced representative, we can define the following concept of reduction on $\mathbb{K}[F]$ based on prefixes of strings: For two non-zero polynomials $p, f \in \mathbb{K}[F]$ we say that f *prefix-reduces* p to q at a monomial $\alpha \cdot w$ of p, where $\alpha \in \mathbb{K} \smallsetminus \{0\}$ and $w \in F$, in a single step, denoted by $p \Rightarrow_f q$, if there exists a string $z \in \mathrm{IRR}(S_0)$ such that $w = HT(f)z$, and $q = p - \alpha \cdot HC(f)^{-1} \cdot f * z$. If $Q \subseteq \mathbb{K}[F]$, then $p \Rightarrow_Q q$ denotes the fact that $p \Rightarrow_f q$ holds for some $f \in Q$. In the reduction process the monomial $\alpha \cdot w$ is replaced by a sum of smaller monomials, which means that the prefix-reduction relation $\Rightarrow_Q$ on $\mathbb{K}[F]$ is noetherian.

A basis G of a right ideal i of $\mathbb{K}[F]$ is called a *prefix Gröbner basis* of i, if $HT(i) = \{uz \mid u \in HT(G), z \in F\}$. In this case every non-zero polynomial $p \in i$ is prefix-reducible by G, which implies immediately that $p \Rightarrow_G^* 0$ holds for each $p \in i$. Since it is even true that $p \Rightarrow_G^* 0$ if and only if $p \in i$, prefix Gröbner bases can be used to decide the membership problem for i by reduction. The set G is called *reduced* if no polynomial of G is prefix-reducible by any other polynomial of G, that is, the elements of $HT(G)$ are incomparable under the prefix relation. More on prefix Gröbner bases in monoid and group rings can be found in [MR93, MR98a, MR98b]. Here we only need that they are finite and computable for finitely generated right ideals in $\mathbb{K}[F]$ (see also [Ros93]).

As in the commutative case congruences on the free group F can be modelled by certain ideals of $\mathbb{K}[F]$. A subset $B \subseteq \mathbb{K}[F]$ is called a *binomial basis* of an ideal $i \subseteq \mathbb{K}[F]$ if it is a basis that consists only of polynomials of the form $u - v$, where $u, v \in F$ and $u >_{\ell\ell} v$. An ideal will be called *binomial* if it admits a binomial basis. These ideals are closely related to the word problem in groups (see [MR98a, Rei95]). Prefix Gröbner bases of finitely generated binomial right ideals can be computed by a procedure prefix_Gröbner_basis in polynomial time when starting with an arbitrary binomial basis.

In the following we will show how prefix Gröbner bases of binomial ideals can be used to do coset enumeration in a way directly related to the Todd-Coxeter coset enumeration method. In order to describe the input $\langle \Sigma; S \rangle$ and U of the Todd-Coxeter enumeration method by binomials we choose the sets $F_S := \{s - 1 \mid s \in \mathrm{dom}(S)\}$ and $F_U := \{u - 1 \mid u \in U\}$. The initial goal is to check whether the subgroup of the free group F generated by $U \cup N(S)$ is finitely generated. This is done in an incremental fashion. Using prefix Gröbner bases we can solve the membership problem for a finitely generated subgroup of the free group F, and by adding polynomials that are obtained by multiplying the generating elements of the subgroup considered with suitably chosen group elements from the left we approximate the normal subgroup $N(S)$ of F.

Next we want to provide the theoretical foundation of these ideas.

Proposition 8.1. [MR93] *Let U be a finite subset of $\underline{\Sigma}^+$, and let $w \in \underline{\Sigma}^+$. Then w defines an element of the subgroup $\langle U \rangle$ of the free group $F = \langle \Sigma; \emptyset \rangle$ if and only if $w \Rightarrow_G^* 1$, where G is the monic prefix Gröbner basis for the right ideal of $\mathbb{K}[F]$ that is generated by the set $\{u - 1 \mid u \in U\}$.*

As the prefix-rewriting systems in Section 4, the monic prefix Gröbner bases are related to Nielsen reduced sets.

Proposition 8.2. [Rei95] *Let U be a finite subset of $\underline{\Sigma}^+$, and let G be the monic prefix Gröbner basis for the right ideal of $\mathbb{K}[F]$ that is generated by the set $\{u - 1 \mid u \in U\}$. Then the set $X_G := \{uv^{-1} \mid (u - v) \in G\}$ is a Nielsen reduced set of generators for the subgroup $\langle U \rangle$ of F.*

However, here we are interested in the general case that the subgroup H is generated by the set $U \cup N(S)$, where the set S of defining relations is not empty. Later we will see how this possibly infinite set can be approximated in a finitary manner.

What we are aiming at is a procedure with the following specification: Given a finite group presentation $\langle \Sigma; S \rangle$, where S is a special system, and a finite set $U \subseteq \underline{\Sigma}^+$, our procedure produces the following output:

(i) if $S = \emptyset$, that is, the group G presented by $\langle \Sigma; S \rangle$ is the free group F generated by Σ, then our procedure terminates, and it produces the monic prefix Gröbner basis G for the right ideal of $\mathbb{K}[F]$ generated by the set $\{u - 1 \mid u \in U\}$;

(ii) if $S \neq \emptyset$, then our procedure enumerates cosets of the subgroup of the free group F generated by the set $U \cup N(S)$, and if it terminates, it returns a complete set of coset representatives and the non-trivial part of the coset table encoded in the prefix Gröbner basis.

In contrast to the Todd-Coxeter enumeration method we need not require that each generator does actually occur in at least one defining relation.

Informally the procedure works as follows. The input $\langle \Sigma; S \rangle$ and U are encoded as sets of binomials $F_S := \{s-1 \mid s \in \mathrm{dom}(S)\}$ and $F_U := \{u-1 \mid u \in U\}$, and computations are performed in the group ring $\mathbb{K}[F]$. The following additional sets will be used:

(1.) A set $N \subseteq F$ of potential coset representatives of the subgroup $H :=\langle U \cup N(S) \rangle$ in F.

(2.) A set $B \subseteq F$ that serves as a test set for possible coset representatives.

(3.) A set $D \subseteq \mathbb{K}[F]$ that is used to increment the generating set considered in order to obtain a generating set for H.

(4.) A monic, reduced prefix Gröbner basis $G \subseteq \mathbb{K}[F]$ that is used to decide whether or not the elements in B are indeed coset representatives of H.

First the procedure checks whether or not the set of defining relations S is empty. If it is, then a monic prefix Gröbner basis is computed for the right ideal of $\mathbb{K}[F]$ that is generated by F_U. Based on Proposition 8.2 this yields a Nielsen reduced set of generators for the subgroup H. If the set S is not empty, then N is initialized as the set containing the empty string only, which is the coset representative of the subgroup itself. During the computation N will always be prefix-closed. The set B is initialized as $B := \{a \mid a \in \underline{\Sigma}\}$, and then G is computed as the monic prefix Gröbner basis for the right ideal of $\mathbb{K}[F]$ that is generated by $F_S \cup F_U$. Hence, G corresponds to the subgroup of F that is generated by $\mathrm{dom}(S) \cup U$. This completes the initialization phase.

Now as long as there still are elements in the set B the following actions are performed. The smallest element is chosen from B (with respect to the length-lexicographical ordering $\geq_{\ell\ell}$). Call it τ. It is removed from B, and if τ is not prefix-reducible by G, then it is added to N and all freely reduced elements of the form τa are added to B, where $a \in \underline{\Sigma}$. Next $D := \{\tau * (s - 1) \mid s \in \mathrm{dom}(S)\}$ is determined, which in TC corresponds to the process of marking the first and the last slot of each relator table with the newly found coset representative τ. Finally the monic prefix Gröbner basis of the set $G \cup D$ is computed, which corresponds to the subgroup of F that is generated by $X_G \cup \{\tau \cdot s \cdot \tau^{-1} \mid s \in \mathrm{dom}(S)\}$. Hence, we approximate the potentially infinite generating set of the subgroup $\langle U \cup N(S) \rangle$ of F. Based on the new prefix Gröbner basis some elements of N may become prefix-reducible. These have to be removed from N, as they are no longer coset representatives. This corresponds to the collapse of cosets in TC. Here is the complete description of the procedure.

Procedure 8.3. Extended TC simulation [RMM98a].

INPUT: $F_S = \{s - 1 \mid s \in \mathrm{dom}(S)\}$, and $F_U = \{u - 1 \mid u \in U\}$.

> **begin** $N \leftarrow \emptyset$;
>   **if** $S = \emptyset$ **then** $G \leftarrow$ prefix_Gröbner_basis (F_U)
>   **else begin** $N \leftarrow \{\lambda\}$;
> $B \leftarrow \underline{\Sigma}$;
> $G \leftarrow$ prefix_Gröbner_basis $(F_S \cup F_U)$;
>   **while** $B \neq \emptyset$ **do**
>   **begin** $\tau \leftarrow \min_{\leq_{\ell\ell}}(B)$, $B \leftarrow B \smallsetminus \{\tau\}$;
>   **if** τ is not prefix-reducible by G **then**
>   **begin** $N \leftarrow N \cup \{\tau\}$;
> $B \leftarrow B \cup \{\tau a \mid a \in \underline{\Sigma}, \tau a \in \mathrm{IRR}(S_0)\}$;
> $D \leftarrow \{\tau * (s - 1) \mid s \in \mathrm{dom}(S)\}$;
> $G \leftarrow$ prefix_Gröbner_basis $(G \cup D)$;
> $M \leftarrow \{w \in N \mid w \text{ is prefix-reducible by } G\}$;

$$N \leftarrow N \smallsetminus M$$
$$\text{end}$$
$$\text{end}$$
$$\text{end}$$
end.

The correctness of the procedure follows from the following proposition.

Proposition 8.4. [RMM98a] *Let* $\langle \Sigma; S \rangle$ *be a finite group presentation, and let* $U \subseteq \Sigma^+$ *be a finite set.*

(a) *If Procedure 8.3 terminates on input* (F_S, F_U), *then the subgroup of the free group* F *that is generated by the set* $U \cup N(S)$ *is finitely generated.*

(b) *If the subgroup of* F *generated by* $U \cup N(S)$ *has finite index in* F, *then Procedure 8.3 terminates on input* (F_S, F_U). *In this case the set* N *computed is a complete set of coset representatives.*

The procedure has been implemented in the MRC package for computing Gröbner bases in monoid and group rings [ReZe98]. Since its description differs very much from the original table based method of Todd and Coxeter, there are new possibilities for creating new cosets. Right now new strategies involving different algorithms for the computation of the prefix Gröbner bases, different orderings on $\mathbb{K}[F]$ which have an effect on the selection of the next τ, and new data structures for representing the cosets are studied. We are aware that representing cosets by words is very space consuming, but we are looking for *knowledge* from prefix string-rewriting theory which then could be integrated into other Todd-Coxeter coset enumeration procedures.

In [RMM98a] it is further shown how this result leads to a completion-based procedure for prefix-rewriting systems that emulates the Todd-Coxeter coset enumeration. This procedure turned out to be similar to one given in [La90]. Another approach by Sims in [Sim94] to simulate Todd-Coxeter coset enumeration by a modification of the Knuth-Bendix algorithm for string-rewriting systems can also be compared to this approach, although in some cases additional efforts are necessary to make it terminating for subgroups of finite index. A thorough comparison of Todd-Coxeter enumeration methods can be found in [RMM98b].

9. Concluding remarks

As we have seen prefix-rewriting is a general method that is well suited to dealing with subgroup problems. Nielsen's process of transforming a set of generators of a subgroup of a free group into a free generating set can easily be interpreted as a special instance of prefix-completion, and in fact by prefix-completion a generalized notion of Nielsen reduced sets has been obtained (Section 4). On the other hand, as prefix-rewriting yields Gröbner bases for right ideals in the free group ring, Nielsen reduced sets are also obtained from prefix Gröbner bases of right ideals (Section 8). In Section 8 we have seen that also the Todd-Coxeter coset enumeration method can be simulated by a procedure that is based on computing prefix Gröbner bases in the free group ring. Further, prefix-rewriting methods provide unique coset representatives for finitely generated subgroups of some classes of groups even if the index of the subgroup considered is not finite (Section 5). Finally, for some classes of groups finite presentations for finitely generated subgroups can be obtained from a prefix-rewriting system for the subgroup (Section 6).

The question arises of whether these results extend to some other classes of groups. In particular for the class of automatic groups it is not yet clear how far the methods based on prefix-rewriting will lead. Recently in [HH98] the concept of an automatic group has been generalized to a group that is automatic with respect to a specified subgroup. Automatic coset systems are then used to solve the subgroup problem and to compute subgroup presentations. The relation between prefix-rewriting and these methods has not yet been studied. Further it remains to investigate the extend to which these methods apply to the submonoid problem, that is, in how far can they be adopted to the problem of deciding whether a given element belongs to a given finitely generated submonoid of a given monoid. Observe that in [MR98a] a correspondence between the submonoid problem in monoids and the subalgebra problem in the corresponding monoid ring is described.

REFERENCES

[ABS87]   J.-M. Autebert, L. Boasson, and G. Senizergues. Groups and NTS languages. *J. Computer System Sciences*, 35:243–267, 1987.

[AM84]   J. Avenhaus and K. Madlener. The Nielsen reduction and *p*-complete problems in free groups. *Theoretical Computer Science*, 32:61–76, 1984.

[AMO86]   J. Avenhaus, K. Madlener, and F. Otto. Groups presented by finite two-monadic Church-Rosser Thue systems. *Transactions American Mathematical Society*, 297:427–443, 1986.

[Bau81]   G. Bauer. *Zur Darstellung von Monoiden durch konfluente Regelsysteme*. Doctoral diss., Fachbereich Informatik, Universität Kaiserslautern, 1981.

[BBN59]   G. Baumslag, W.W. Boone, and B.H. Neumann. Some unsolvable problems about elements and subgroups of groups. *Math. Scand.*, 7:191–201, 1959.

[BCM77]   G. Baumslag, F.B. Cannonito, and C.F. Miller III. Infinitely generated subgroups of finitely presented groups. *Mathematische Zeitschrift*, 153:117–134, 1977.

[BKR87]   B. Benninghofen, S. Kemmerich, and M.M. Richter. *Systems of Reductions*. Lecture Notes in Computer Science 277. Springer-Verlag, Berlin, 1987.

[BO93]   R.V. Book and F. Otto. *String-Rewriting Systems*. Springer-Verlag, New York, 1993.

[Büc79]   H. Bücken. Reduction-systems and small cancellation theory. In *Proceedings 4th Workshop on Automated Deduction*, pages 53–59, 1979.

[CO94]   R. Cremanns and F. Otto. Constructing canonical presentations for subgroups of context-free groups in polynomial time. In J. von zur Gathen and M. Giesbrecht, editors, *Proceedings ISSAC'94*, pages 147–153. ACM, New York, 1994.

[Coc76]   Y. Cochet. Church-Rosser congruences on free semigroups. *Colloquia Mathematica Societatis János Bolyai*, 20:51–60, 1976.

[Cre95]   R. Cremanns. *Finiteness conditions for rewriting systems*. Doctoral diss., Fachbereich Mathematik/Informatik, Universität Kassel, 1995.

[CRRT97]   C.M. Campbell, E.F. Robertson, N. Ruškuc, and R.M. Thomas. *Automatic semigroups*. Technical Report No. 1997/29, Dep. of Mathematics and Computer Science, University of Leicester, 1997.

[Deh11]   M. Dehn. Über unendliche diskontinuierliche Gruppen. *Math. Ann.*, 71:116–144, 1911.

[Deh12]   M. Dehn. Transformation der Kurven auf zweiseitigen Flächen. *Math. Ann.*, 72:413–421, 1912.

[Eps92]   D.B.A. Epstein. *Word Processing In Groups*. Jones and Bartlett Publishers, Boston, 1992.

[Ger92]   S.M. Gersten. Dehn functions and l_1-norms of finite presentations. In G. Baumslag and C.F. Miller III, editors, *Algorithms and Classification in Combinatorial Group Theory*, Math. Sciences Research Institute Publ. 23, pages 195–224. Springer-Verlag, New York, 1992.

[HH98]   D.F. Holt and D.F. Hurt. Computing automatic coset systems and subgroup presentations. *J. Symbolic Computation*, 25:1–19, 1998.

[HK91]   T. Harju and J. Karhumäki. The equivalence problem of multitape finite automata. *Theoretical Computer Science*, 78:347–355, 1991.

[Joh76]   D.L. Johnson. *Presentation of Groups*, volume 22 of *London Mathematical Society Lecture Notes Series*. Cambridge University Press, Cambridge, 1976.

[KB70]   D. Knuth and P. Bendix. Simple word problems in universal algebras. In J. Leech, editor, *Computational Problems in Abstract Algebra*, pages 263–297. Pergamon Press, New York, 1970.

[KM89]   N. Kuhn and K. Madlener. A method for enumerating cosets of a group presented by a canonical system. In *Proc. ISSAC'89*, pages 338–350. ACM Press, New York, 1989.

[KMO90]   N. Kuhn, K. Madlener, and F. Otto. A test for λ-confluence for certain prefix rewriting systems with applications to the generalized word problem. In S. Watanabe and M. Nagata, editors, *Proceedings ISSAC'90*, pages 8–15. ACM, New York, 1990.

[KMO94]   N. Kuhn, K. Madlener, and F. Otto. Computing presentations for subgroups of polycyclic groups and of context-free groups. *Applicable Algebra in Engineering, Communication and Computing*, 5:287–316, 1994.

[KN85]   D. Kapur and P. Narendran. The Knuth-Bendix completion procedure and Thue systems. *SIAM J. Computing*, 14:1052–1072, 1985.

[KRW90]   A. Kandri-Rody and V. Weispfennig. Non-commutative Gröbner bases in algebras of solvable type. *J. Symbolic Computation*, 9:1–26, 1990.

[Kuh91]   N. Kuhn. *Zur Entscheidbarkeit des Untergruppenproblems für Gruppen mit kanonischen Darstellungen*. Doctoral diss., Fachbereich Informatik, Universität Kaiserslautern, 1991.

[La90]   G. Labonté. An algorithm for the construction of matrix representations for finitely presented non-comutative algebras. *J. Symbolic Computation*, 9:27–38, 1990.

[LeC86]   P. LeChenadec. *Canonical Forms in Finitely Presented Algebras*. Research Notes in Theoretical Computer Science. Pitman, London, 1986.

[LS77]   R.C. Lyndon and P.E. Schupp. *Combinatorial Group Theory*. Springer-Verlag, Berlin, 1977.

[Mik58]   K.A. Mikhailova. The occurrence problem for direct products of groups. *Dokl. Akad. Nauk SSSR*, 119:1103–1105, 1958.

[Mil71]   C.F. Miller. *On group-theoretic decision problems and their classification*, volume 68 of *Annals of Mathematical Studies*. Princeton University Press, Princeton, 1971.

[MKS76]   W. Magnus, A. Karrass, and D. Solitar. *Combinatorial Group Theory*. Second revised edition. Dover, New York, 1976.

[MO87]   K. Madlener and F. Otto. Using string-rewriting for solving the word problem for finitely presented groups. *Information Processing Letters*, 24:281–284, 1987.

[MO89]   K. Madlener and F. Otto. About the descriptive power of certain classes of finite string-rewriting systems. *Theoretical Computer Science*, 67:143–172, 1989.

[MO91]   K. Madlener and F. Otto. Decidable sentences for context-free groups. In C. Choffrut and M. Jantzen, editors, *Proceedings STACS'91*, Lecture Notes in Computer Science 480, pages 160–171. Springer-Verlag, Berlin, 1991.

[MR93]   K. Madlener and B. Reinert. Computing Gröbner bases in monoid and group rings. In M. Bronstein, editor, *Proc. ISSAC'93*, pages 254–263. ACM, New York, 1993.

[MR98a]   K. Madlener and B. Reinert. Relating rewriting techniques on monoids and rings: Congruences on monoids and ideals in monoid rings. *Theoretical Computer Science*, 208:3–31, 1998.

[MR98b]   K. Madlener and B. Reinert. String rewriting and Gröbner bases – a general approach to monoid and group rings. In *Proceedings of the Workshop on Symbolic Rewriting Techniques, Monte Verita, 1995*, pages 127–180. Birkhäuser, 1998.

[MR98c]   K. Madlener and B. Reinert. A generalization of Gröbner basis algorithms to polycyclic group rings. *J. Symbolic Computation*, 25:23–45, 1998.

[MS83]   D.E. Muller and P.E. Schupp. Groups, the theory of ends, and context-free languages. *J. Computer System Sciences*, 26:295–310, 1983.

[OK97]   F. Otto and Y. Kobayashi. Properties of monoids that are presented by finite convergent string-rewriting systems - a survey. In D.Z. Du and K. Ko, editors, *Advances in Algorithms, Languages and Complexity*, pages 226–266. Kluwer Academic Publ., Dordrecht, 1997.

[OSKM98]   F. Otto, A. Sattler-Klein, and K. Madlener. Automatic monoids versus monoids with finite convergent presentations. In T. Nipkow, editor, *Rewriting Techniques and Applications, Proceedings RTA'98*, Lecture Notes in Computer Science 1379, pages 32–46. Springer-Verlag, Berlin, 1998.

[Ott86]   F. Otto. On deciding whether a monoid is a free monoid or is a group. *Acta Informatica*, 23:99–110, 1986.

[Ott98a]   F. Otto. *On Dehn functions of finitely presented bi-automatic monoids*. Mathematische Schriften Kassel 8/98, Universität Kassel, July 1998.

[Ott98b]   F. Otto. *On s-regular prefix-rewriting systems and automatic structures*. Mathematische Schriften Kassel 9/98, Universität Kassel, September 1998.

[Rei95]   B. Reinert. *On Gröbner Bases in Monoid and Group Rings*. Doctoral diss., Fachbereich Informatik, Universität Kaiserslautern, 1995.

[RMM98a]   B. Reinert, K. Madlener, and T. Mora. A note on Nielsen reduction and coset enumeration. In *Proceedings ISSAC'98*, pages 171–178. ACM, New York, 1998.

[RMM98b]   B. Reinert, K. Madlener, and T. Mora. *Coset enumeration - a comparison of methods*. Technical report, Universität Kaiserslautern, 1998.

[ReZe98]   B. Reinert and D. Zeckzer. MRC: A System for Computing Gröbner Bases in Monoid and Group Rings, In *6th Rhine Workshop on Computer Algebra*, Sankt Augustin, 1998.

[Ros93]   A. Rosenmann. An algorithm for constructing Gröbner and free Schreier bases in free group algebras. *J. Symbolic Computation*, 16:523–549, 1993.

[Sim94]   C.C. Sims. *Computation With Finitely Presented Groups*, volume 48 of *Encyclopedia of Mathematics and its Applications*. Cambridge University Press, New York, 1994.

[Sny89]   W. Snyder. Efficient ground completion: an $O(n \log n)$ algorithm for generating reduced sets of ground rewrite rules equivalent to a set of ground equations E. In N. Deshowitz, editor, *Rewriting Techniques and Applications, Proceedings RTA '89*, Lecture Notes in Computer Science 355, pages 419–433. Springer-Verlag, Berlin, 1989.

[SOK94]   C.C. Squier, F. Otto, and Y. Kobayashi. A finiteness condition for rewriting systems. *Theoretical Computer Science*, 131:271–294, 1994.

[TC36]   J.A. Todd and H.S.M. Coxeter. A practical method for enumerating cosets of a finite abstract group. *Proc. Edinburgh Math. Soc.*, 5:26–34, 1936.

[Wiß89]   D. Wißmann. *Anwendung von Rewrite-Techniken in polyzyklischen Gruppen*. Doctoral diss., Fachbereich Informatik, Universität Kaiserslautern, 1989.

Klaus Madlener
Fachbereich Informatik,
Universität Kaiserslautern,
D-67653 Kaiserslautern
madlener@informatik.uni-kl.de
http://www.uni-kl.de/AG-AvenhausMadlener/AG-Madlener.html

Friedrich Otto
Fachbereich Mathematik/Informatik,
Universität Kassel,
D-34109 Kassel
otto@theory.informatik.uni-kassel.de
http://www.db.informatik.uni-kassel.de/FG_TH/otto/

VERALLGEMEINERTE BIASINVARIANTEN UND IHRE BERECHNUNG

WOLFGANG METZLER

ABSTRACT. The bias-invariant was derived from congruences of the second homology of a 2-complex modulo spherical elements and lead to distinctions of homotopy types in cases where the (abelian) fundamental group and the Euler characteristic coincide.

But, in general, the determination of the second homotopy group and of its image under the Hurewicz map are undecidable. We hence generalize the bias construction to congruences modulo H_2-images of coverings which correspond to characteristic subgroups of the fundamental group.

In the case of the commutator subgroup, we get computable invariants for distinctions of homotopy types, where it may be impossible to calculate the classical bias moduli.

1. EINLEITUNG

Die Unterscheidung verschiedener Homotopietypen von 2-Komplexen K, L mit gleichem, endlich abelschem π_1 und minimaler Eulerscher Charakteristik durch den Unterzeichneten [15] und Sieradski [20] geschah durch ein Kongruenzargument modulo sphärischer Elemente in den 2. Homologiegruppen: Nicht immer gibt es eine (stetige) Abbildung $K \to L$, die in π_1 und H_2 Isomorphismen induziert (*Homologieäquivalenz*). Allgemein kann man fragen, wann sich ein Isomorphismus von $H_2(\pi_1)$ zu einem geometrisch induzierten Isomorphismus der 2. Homologie der Komplexe hochheben läßt und erhält die sogenannte *Biasinvariante* (§2), siehe Dyer [5] und Latiolais [13]. Für deren praktische Berechnung entsteht jedoch die Aufgabe, aus den Komplexen, beziehungsweise aus von ihnen abgelesenen π_1-Präsentationen, die Untergruppe Σ_2 der sphärischen Zyklen in $H_2(K)$, das heißt, das π_2-Bild zu bestimmen. Tatsächlich liegt hier ein im allgemeinen unentscheidbares Problem vor[1] [7,18].

Um zu vereinfachten, aber berechenbaren Invarianten zu gelangen, betrachten wir Überlagerungen $\overline{K}$ (zwischen K und der universellen Überlagerung $\tilde{K}$), für welche die zugehörige Untergruppe U charakteristisch in $\pi_1(K)$ liegt. Das Bild $\overline{\Sigma_2}$ von $H_2(\overline{K})$ in $H_2(K)$ liefert verallgemeinerte Biasinvarianten,

---

[1]Andererseits hat Brown [3] Bedingungen an Präsentationen angegeben, aus denen sich explizit Eilenberg-MacLane Räume gewinnen lassen, die in jeder Dimension nur endlich viele Zellen besitzen. Meines Wissens ist die Frage noch offen, ob sich eine algorithmische $H_2(\pi_1)$-Bestimmung bereits aus der Lösbarkeit des Wortproblems für π_1 ergibt.

welche im allgemeinen nur noch notwendige Bedingungen für Homologie- und Homotopieäquivalenz ergeben (§3).

Wenn insbesondere U die Kommutatorgruppe von $\pi_1(K)$ ist und die von K nach dem Standardverfahren abgelesene Präsentation sich durch Hinzunahme weiterer Relationen zu einer Präsentation der Abelschmachung $\pi_1(K)/U$ auf Niveau $\chi_{\min}$ ergänzen läßt, erhalten wir in einem zweiten Reduktionsschritt berechenbare Invarianten. Das Berechnungsverfahren nutzt (nach Reidemeister [19]) den Kalkül aus, der sich ergibt, wenn man die Erzeugenden einer Präsentation mit den Kommutatoren vertauschbar macht (§4). Einen Teil der Ergebnisse von §4 hat Michalik [17] schon vor einigen Jahren auf meine Anregung hin auf algebraischem Weg erzielt. Mühevoll war dabei insbesondere der Nachweis der Unabhängigkeit von der Auswahl spezieller Erzeugender und definierender Relationen. Dieser kann durch die hier angegebene topologische Deutung entfallen.

In §5 A) geben wir Beispiele an von 2-Komplexen mit gleichem, nichtabelschem π_1 und minimaler Eulerscher Charakteristik, welche sich durch die in §3 und §4 gewonnenen Ergebnisse (homolog und) homotop unterscheiden lassen, obwohl die Berechnung der klassischen Biasinvariante dabei im allgemeinen vermutlich unmöglich ist. Bei einer 2. Beispielserie wird ein Resultat aus der kürzlich fertiggestellten Diplomarbeit von Grabo [8] verwendet, in welche umgekehrt (Vorlesungs-) Notizen von mir zum Gegenstand dieser Arbeit eingeflossen sind. Wertvolle Anregungen verdanke ich auch insbesondere Cynthia Hog-Angeloni und Jens Harlander.

Alle im folgenden vorkommenden Grundkomplexe sind endlich und zusammenhängend, jedoch treten unendliche Überlagerungskomplexe auf.

2. Ein Überblick über den gewöhnlichen Bias

Ist $\alpha : \pi_1(K^2) \to \pi_1(L^2)$ ein Isomorphismus, so gibt es eine stetige Abbildung $f : K^2 \to L^2$, die α induziert. Sind zwei solche Abbildungen gegeben, so können sie zur Übereinstimmung gebracht werden, indem man auf eine von ihnen

(1) eine Homotopie anwendet und anschließend das Bild jeder 2-Zelle um ein 2-Sphärenbild modifiziert.

Jede solche sphärische Modifikation führt umgekehrt zu einer stetigen Abbildung, welche α realisiert. α bestimmt also eine Klasse geometrisch induzierter Abbildungen $H_2(K^2) \to H_2(L^2)$, die bezüglich einer beliebigen Basis[2] von $H_2(K^2)$

(2) durch Modifikation des Bildes jedes Basiselements um (je) ein Element von $\Sigma_2(L^2)$ hervorgehen.

---

[2] $H_2(K^2)$ ist direkter Summand von $C_2(K^2)$.

Wenn diese Klasse einen Isomorphismus $H_2(K^2) \to H_2(L^2)$ enthält, bestimmt α eine Homologieäquivalenz. Auf jeden Fall induziert α einen Isomorphismus

$$H_2(\pi_1(K)) = H_2(K)/\Sigma_2(K) \to H_2(L)/\Sigma_2(L) = H_2(\pi_1(L)),$$

und es entsteht die Frage, ob dieser sich zu einem geometrisch induzierten Isomorphismus $H_2(K) \to H_2(L)$ hochheben läßt.

Zu diesem Zweck betrachten wir die Lage von Σ_2 in H_2. Bezüglich einer geeigneten (kanonischen) Basis wird diese durch die Matrix

$$(3) \quad \Sigma_2 \left\{ \left(\begin{array}{ccccc|c} 1 & & & & & \\ & \ddots & & & 0 & \\ & & 1 & & & \\ & & & m_1 & & \quad 0 \\ & 0 & & & \ddots & \\ & & & & & m_k \end{array} \right) \right\} \underbrace{}_{H_2} \text{mit } 1 \neq m_1 | m_2 | \ldots | m_k$$

beschrieben, und

(4) die Teilerkette $(1, \ldots, 1, m_1, \ldots, m_k)$ hängt nur von π_1 und der Eulerschen Charakteristik χ ab,[3]

nicht aber von dem speziellen Komplex (und der kanonischen Basis).

Wenn die Teilerkette keine Einsen enthält, heißt m_1 (oder 0) *der Biasmodul für π_1 und χ*; jeder Teiler $m \neq 1$ des Biasmoduls heißt *ein* Biasmodul. Ein Biasmodul m ist also eine nichtnegative ganze Zahl $m \neq 1$ derart, daß

$$(5) \quad \Sigma_2 \subseteq m \cdot H_2$$

gilt, und m_1 ist die größte solche Zahl (oder 0). Mit (4) sind diese Begriffe nur von π_1 und χ abhängig. Wie man durch Anhängen von 2-Sphären an Komplexe minimaler Eulercharakteristik $\chi_{\min}(\pi_1)$ sieht, sind bei (3) für $\chi > \chi_{\min}$ stets Einsen vorhanden. Die Existenz eines Biasmoduls für π_1, χ impliziert also

$$(6) \quad \chi = \chi_{\min}(\pi_1).$$

Seien jetzt K^2, L^2 Komplexe mit isomorphen Fundamentalgruppen, $\alpha : \pi_1(K^2) \to \pi_1(L^2)$ ein gegebener Isomorphismus und $\chi(K^2) = \chi(L^2) = \chi_{\min}(\pi_1)$. Ferner existiere für π_1, χ der Biasmodul m_1. Sind f und f' geometrische Abbildungen $K^2 \to L^2$, die α induzieren, so gilt nach (2) und (5), daß die Matrizen $\mathfrak{A}$ und $\mathfrak{A}'$, welche die zugehörigen H_2-Abbildungen beschreiben (bezüglich fester Basen für $H_2(K)$ und $H_2(L)$)

$$(7) \quad \mathfrak{A} \equiv \mathfrak{A}' \mod m_1$$

---

[3]Dies ist ein Spezialfall von Hilfssatz 1 in §3.

erfüllen.[4] Für beliebige Basen in der 2. Homologie folgt daraus

$$(8) \qquad \overline{\mathrm{Det}\ \mathfrak{A}} \equiv \pm \mathrm{Det}\ \mathfrak{A}' \quad \mathrm{mod}\ m_1.$$

Durch (8) ist also ein Paar $b(\alpha, K, L) = \pm \overline{\mathrm{Det}\ \mathfrak{A}}$ von Restklassen in $\mathbb{Z}_{m_1}$ gegeben, der durch K, L und α bestimmte *Bias*. Damit α zu einer Homologie- beziehungsweise Homotopieäquivalenz gehört, ist

$$(9) \qquad b(\alpha, K, L) = \pm 1$$

notwendig. Wenn (9) für keinen Isomorphismus $\alpha : \pi_1(K) \to \pi_1(L)$ gilt, sind K und L von verschiedenem Homologie- und Homotopietyp. Nach [5] (vergleiche [13]) läßt sich umgekehrt zeigen, daß α für $b(\alpha, K, L) = \pm 1$ zu einer Homologieäquivalenz gehört.[5] Über repräsentierende Abbildungen und den Determinantenmultiplikationssatz erhält man für Komplexe K^2, L^2, M^2 zu $\chi_{\min}$ und komponierte Isomorphismen

$$\pi_1(K^2) \xrightarrow{\alpha} \pi_1(L^2) \xrightarrow{\beta} \pi_1(M^2)$$

ferner die Produktregel

$$(10) \qquad b(\beta\alpha, K, M) = b(\beta, L, M) \cdot b(\alpha, K, L).$$

Zusammen mit der Tatsache, daß $b(\mathrm{id}, K, K) = \pm 1$ gilt, folgt hieraus, daß

(11) $\quad b(\alpha, K, L)$ multiplikativ invertierbare Restklassen in $\mathbb{Z}_{m_1}$ sind.[6]

Aus (10) ergibt sich auch, daß man alle Biaswerte für $K \to L$ erhält, in- dem man den Bias für einen festen Isomorphismus $\pi_1(K) \to \pi_1(L)$ mit allen multipliziert, die sich für Automorphismen der Fundamentalgruppe eines der Komplexe K, L ergeben.

3. Verallgemeinerung

$p : \overline{K}^2 \to K^2$ sei eine Überlagerung derart, daß die zugehörige Untergruppe U charakteristisch in $\pi_1(K)$ ist. Insbesondere ist p regulär. Für $\overline{\Sigma}_2(K) = p_*(H_2(\overline{K})) \subseteq H_2(K)$ gilt

$$(12) \qquad \overline{\Sigma}_2(K) \supseteq \Sigma_2(K).$$

Ist $\alpha : \pi_1(K^2) \to \pi_1(L^2)$ ein Isomorphismus, so wird dabei U isomorph in eine (von α unabhängige) Untergruppe $V \subseteq \pi_1(L^2)$ überführt, und für L gilt die zu (12) analoge Beziehung. Ausgehend von einer stetigen Abbildung $K^2 \to L^2$, welche α induziert, bestimmt α wiederum eine Klasse von Homomorphismen $H_2(K^2) \to H_2(L^2)$, die bezüglich einer beliebigen Basis von $H_2(K^2)$

(13) durch Modifikation des Bildes eines Basiselements um (je) ein Element von $\overline{\Sigma}_2(L^2)$ hervorgehen, und die wegen (12) alle zu α gehörenden geo- metrisch induzierten $H_2(K^2) \to H_2(L^2)$ Abbildungen umfaßt.

---

[4]Wegen $\chi(K) = \chi(L)$ sind $\mathfrak{A}$ und $\mathfrak{A}'$ quadratisch.

[5]Dies gilt auch stets, wenn bei (3) Einsen vorkommen.

[6]Der Bias läßt sich also auch als ein Element in der multiplikativen Gruppe $\mathbb{Z}^*_{m_1}/ \pm 1$ auffassen.

Aber im allgemeinen sind nicht mehr alle Homomorphismen dieser Klasse geometrisch.

Wir betrachten wiederum die Inklusion von $\overline{\Sigma}_2$ in H_2. Bezüglich einer geeigneten (kanonischen) Basis wird diese durch eine Matrix

$$(14) \quad \overline{\Sigma}_2 \left\{ \underbrace{\left(\begin{array}{c|c} \begin{matrix} 1 & & & & \\ & \ddots & & & 0 \\ & & 1 & & \\ & & & \overline{m}_1 & \\ 0 & & & & \ddots \\ & & & & & \overline{m}_{\overline{k}} \end{matrix} & \mbox{\LARGE 0} \end{array} \right)}_{H_2} \right. \quad \text{mit } 1 \neq \overline{m}_1 | \ldots | \overline{m}_{\overline{k}}$$

beschrieben. Hierfür gilt der

Hilfsatz 1. *Die Teilerkette* $(1, \ldots, 1, \overline{m}_1, \ldots \overline{m}_{\overline{k}})$ *hängt nur von* π_1, χ *und* U *ab.*

Beweis. Nach dem Satz von Tietze gehen 2-Komplexe mit denselben Daten durch eine endliche Folge von (i) Homotopieäquivalenzen und (ii) Einpunktvereinigung mit 2-Sphären beziehungsweise dem Schritt (ii)$^{-1}$ auseinander hervor.[7] Bei einem Schritt (i) bleibt die Teilerkette zugleich mit dem Isomorphietyp des Paares $(H_2, \overline{\Sigma}_2)$ ungeändert, denn U ist charakteristisch; durch (ii) erhält die Teilerkette eine zusätzliche 1, beziehungsweise (ii)$^{-1}$ entfernt eine solche. Bei gleicher Eulerscher Charakteristik der Ausgangskomplexe kommt der Prozeß (ii) ebensooft vor wie (ii)$^{-1}$. Da die Teilerketten insgesamt ineinander überführt werden, sind sie also gleich. q.e.d.

Die folgenden Begriffe und Fakten ergeben sich nun analog zu §2:

Wenn die Teilerkette keine Einsen enthält, heißt $\overline{m}_1$ (oder 0) *der Biasmodul für* π_1, χ *und* U; jeder Teiler $\overline{m} \neq 1$ des Biasmoduls heißt *ein* Biasmodul. Ein solcher ist also eine nichtnegative ganze Zahl $\overline{m} \neq 1$ derart, daß

$$(15) \quad \overline{\Sigma}_2 \subseteq \overline{m} \cdot H_2$$

gilt, und $\overline{m}_1$ ist die größte solche Zahl (oder 0). Alle diese Begriffe sind nur von π_1, χ und U abhängig. Die Existenz eines Biasmoduls für π_1, χ, U impliziert wiederum (6).

Seien jetzt K^2, L^2 Komplexe mit isomorphen Fundamentalgruppen, $\alpha : \pi_1(K^2) \to \pi_1(L^2)$ ein gegebener Isomorphismus, $\chi(K^2) = \chi(L^2) = \chi_{\min}(\pi_1)$ und $\overline{m}_1$ zugehöriger Biasmodul bezüglich entsprechender charakteristischer Untergruppen. Sind f und f' geometrische Abbildungen, die α induzieren, so erfüllen die Matrizen $\mathfrak{A}$ und $\mathfrak{A}'$ zugehöriger H_2-Abbildungen aufgrund von (13) und (15)

$$(16) \quad \text{Det } \mathfrak{A} \equiv \pm \text{ Det } \mathfrak{A}' \quad \mod \overline{m}_1.$$

---

[7]Siehe die Fassung des Satzes von Tietze in [10], S. 28.

Das durch (16) gegebene Restklassenpaar in $\mathbb{Z}_{\overline{m}_1}$ werde als *verallgemeinerter Bias* $\overline{b}(\alpha, K, L)$ bezeichnet.

Nun folgt aber aus $\Sigma_2 \subseteq \overline{\Sigma}_2$ und (15), daß $\Sigma_2 \subseteq \overline{m} \cdot H_2$ gilt.

(17) Jeder Biasmodul $\overline{m}$ für π_1, χ und U, insbesondere $\overline{m}_1$, ist also auch ein gewöhnlicher Biasmodul für π_1 und χ.

Wenn die $\overline{\Sigma}_2$-Teilerkette keine Einsen besitzt, gilt das demnach ebenfalls für diejenige von Σ_2; die Umkehrung ist aber im allgemeinen falsch. Aus analogen Schlüssen zu §2 und dem Gesagten folgt nun

Satz 1. *Mit den obigen Bezeichnungen und Voraussetzungen gilt:*

(i) *Damit α zu einer Homologie- beziehungsweise Homotopieäquivalenz gehört, ist $\overline{b}(\alpha, K, L) = \pm 1$ notwendig.*

(ii) *Für Komplexe K^2, L^2, M^2 zu $\chi_{\min}$ und komponierte Isomorphismen*
$$\pi_1(K^2) \xrightarrow{\alpha} \pi_1(L^2) \xrightarrow{\beta} \pi_1(M^2) \text{ gilt } \overline{b}(\beta\alpha, K, M) = \overline{b}(\beta, L, M) \cdot \overline{b}(\alpha, K, L).$$

(iii) *$\overline{b}(\alpha, K, L)$ sind multiplikativ invertierbare[8] Restklassen in $\mathbb{Z}_{\overline{m}_1}$.*

(iv) *$\overline{b}(\alpha, K, L)$ ist der modulo $\overline{m}_1$ reduzierte gewöhnliche Bias $b(\alpha, K, L)$.*

Der Vorteil dieser Verallgemeinerung beziehungsweise Reduktion liegt darin, daß sich $\overline{\Sigma}_2$ unter Umständen algorithmisch bestimmen läßt, auch wenn dies für Σ_2 mühsam wäre oder nicht gelingt. Wenn zum Beispiel die Deckbewegungsgruppe $\pi_1(K)/U$ endlich ist und zu einem explizit gegebenen Kettenkomplex $C(\overline{K})$ führt, lassen sich die $H_i(\overline{K})$ und daher auch $\overline{\Sigma}_2$ nach der üblichen Methode für Kettenkomplexe aus freien abelschen Gruppen endlichen Ranges berechnen. Eine weitere spezielle Reduktion werden wir im folgenden Abschnitt ausführlich behandeln.

4. Der Fall der Abelschmachung

A) Von jetzt an sei $p : \overline{K}^2 \to K^2$ die Überlagerung bezüglich der (charakteristischen) Kommutatorgruppe $U = [\pi_1, \pi_1]$. Sie hat die Abelschmachung $\overline{\pi}_1 = \pi_1(K)/U$ als Deckbewegungsgruppe. Wir machen ferner im folgenden die Zusatzvoraussetzung, daß sich

(18) K^2 durch Hinzufügen endlich vieler 2-Zellen e_j^2 zu einem Komplex $K^2 \cup (\bigcup e_j^2)$ mit $\overline{\pi}_1$ als Fundamentalgruppe und $\chi(K^2 \cup (\bigcup e_j^2)) = \chi_{\min}(\overline{\pi}_1)$ ergänzen läßt.

Es gilt dann

Hilfsatz 2. *Die zusätzlichen 2-Zellen e_j^2 lassen sich so wählen, daß ihre Randworte S_j (bezüglich kanonischer Ablesungen der Fundamentalgruppe) Kommutatorprodukte sind ("gute" Ergänzung).*

---

[8]Vergleiche Fußnote 6.

Beweis. Für gegebene (endlich viele) e_j^2 lassen sich deren Randworte als $S_j = V_j \cdot W_j$ schreiben, wobei die V_j Folgen definierender Relationen von $\pi_1(K^2)$ sind und die W_j Kommutatorprodukte. Die definierenden Relationen von $\pi_1(K^2)$ und die W_j haben dann zusammen denselben Normalenabschluß (in der freien Gruppe der Erzeugenden) wie die definierenden $\pi_1(K^2)$-Relationen und die S_j. Also können wir die 2-Zellen e_j^2 durch ebensoviele ersetzen, welche von vorneherein die W_j als Randworte besitzen.   q.e.d.

Wir werden stets solche "guten" Ergänzungen $\{e_j^2\}$ betrachten.

Bei $p : \overline{K} \to K$ und der universellen Überlagerung

$$\tilde{p} : \widetilde{K \cup (\bigcup e_j^2)} \to K \cup (\bigcup e_j^2)$$

liegt dieselbe Deckbewegungsgruppe $\overline{\pi}_1$ vor, und die Inklusion $K \hookrightarrow K \cup (\bigcup e_j^2)$ läßt sich zu einer $\overline{\pi}_1$-äquivarianten Injektion der Überlagerungskomplexe hochheben. Wir erhalten also ein kommutatives Diagramm

$$(19) \qquad \begin{array}{ccc} \overline{K} & \hookrightarrow & \widetilde{K \cup (\bigcup e_j^2)} \\ \downarrow p & & \downarrow \tilde{p} \\ K & \hookrightarrow & K \cup (\bigcup e_j^2). \end{array}$$

Wenn $\overline{\pi}_1$ die Normalformdarstellung $\mathbb{Z}_{e_1} \times \ldots \times \mathbb{Z}_{e_r} \times \mathbb{Z}^s$ mit $1 \neq e_1 | \ldots | e_r$ oder $r = 0$ hat, so gilt

Hilfsatz 3. *Der Biasmodul m' für $\overline{\pi}_1$ und $\chi_{\min}(\overline{\pi}_1)$ existiert und hat den Wert $m' = e_1$ oder 0. Genau dann gilt $m' = e_1$, wenn einer der Fälle $r \geq 2$, oder $r = 1$ und $s \geq 1$ vorliegt.*

Der Beweis kann nach [11] (siehe auch [15,20]) geführt werden, indem man einen Standardkomplex [$\mathfrak{P}$] zu einer Präsentation $\mathfrak{P}$ nimmt, welche für jeden Faktor $\mathbb{Z}_{e_i}$ respektive $\mathbb{Z}$ eine Erzeugende besitzt und als definierende Relationen diejenigen, welche a) die Ordnungen e_i ausdrücken und b) für jedes Paar von Erzeugenden einen Kommutator, vergleiche unten (30). $H_2([\mathfrak{P}])$ hat die 2-Zellen vom Typ b) als Basis; und Erzeugende von π_2 respektive Σ_2 erhält man, indem man [$\mathfrak{P}$] als 2-Gerüst eines Produktes aus dreidimensionalen Linsenräumen und 1-Sphären auffaßt und die Ränder der 3-Zellen betrachtet, siehe unten (37),(38). Insbesondere hat [$\mathfrak{P}$] minimale Eulersche Charakteristik vom Wert

$$(20) \qquad \chi_{\min}(\overline{\pi}_1) = 1 - s + \binom{r+s}{2}$$

für $\overline{\pi}_1$. Aus Hilfssatz 3 folgt

$$(21) \qquad \Sigma_2(K \cup (\bigcup e_j^2)) \subseteq m' \cdot H_2(K \cup (\bigcup e_j^2)) \text{ mit } m' = e_1 \text{ oder } 0,$$

und ein System "guter" Ergänzungen $\{e_j^2\}$ gemäß Hilfssatz 2 ergibt

$$(22) \qquad H_2(K \cup (\bigcup(e_j^2))) = H_2(K) \oplus \mathbb{Z}^\ell,$$

wobei ℓ die Anzahl der e_j^2 bezeichnet. Das Diagramm (19) impliziert $\overline{\Sigma}_2(K) \subseteq \Sigma_2(K \cup (\bigcup e_j^2))$ und, daß $\overline{\Sigma}_2(K)$ im Summanden $H_2(K)$ der rechten Seite von (22) liegt. Daher ergibt (21)

$$(23) \qquad \overline{\Sigma}_2(K) \subseteq m' \cdot H_2(K) \text{ mit } m' = e_1 \text{ oder } 0.$$

Mit den Bezeichnungen und Voraussetzungen dieses Paragraphen gilt also nach (15) und (17):

Hilfsatz 4. *Es existieren der Biasmodul $\overline{m}_1$ für $\pi_1(K), \chi(K), U = [\pi_1, \pi_1]$ sowie der Biasmodul m_1 für $\pi_1(K), \chi(K)$, und diese Moduln erfüllen*

$$(24) \qquad m' | \overline{m}_1 | m_1.$$

Insbesondere ist $\chi(K)$ minimal für π_1.

Seien jetzt K^2, L^2 Komplexe mit $\chi(K) = \chi(L) = \chi$, die den Voraussetzungen dieses Paragraphen genügen; $\alpha : \pi_1(K^2) \to \pi_1(L^2)$ sei ein gegebener Isomorphismus. Durch α wird ein Isomorphismus $\overline{\alpha}$ zwischen den Abelschmachungen induziert, und mit $1 \leq j \leq \ell = \chi_{\min}(\overline{\pi}_1) - \chi$ lassen sich K^2 und L^2 durch gleichviele 2-Zellen zu Komplexen mit $\overline{\pi}_1$ als Fundamentalgruppe auf Niveau $\chi_{\min}(\overline{\pi}_1)$ ergänzen. Zu gegebenen Ergänzungen $\{e_{jK}^2\}$ und $\{e_{jL}^2\}$ gibt es dann einen Bias $b(\overline{\alpha}, K \cup (\bigcup e_{jK}^2), L \cup (\bigcup e_{jL}^2))$.

Satz 2. *Werden die e_{jL}^2 so gewählt, daß ihre Randworte durch eine α induzierende stetige Abbildung $f : K^2 \to L^2$ aus denen eines "guten" Systems $\{e_{jK}^2\}$ hervorgehen, so gilt:*

$$(25) \qquad b(\alpha, K, L) \equiv \overline{b}(\alpha, K, L) \equiv b(\overline{\alpha}, K \cup \bigcup(e_{jK}^2), L \cup \bigcup(e_{jL}^2)) \mod m'.$$

Beweis. Die erste Kongruenzbeziehung folgt aus Satz 1,(iv) durch Reduktion des Moduls. Für die zweite seien Basen in $H_2(K)$ und $H_2(L)$ gewählt sowie f eine α induzierende stetige Abbildung $K \to L$. Die zugehörige $H_2(K) \to H_2(L)$-Abbildung werde bezüglich der gewählten Basen durch $\mathfrak{A}$ beschrieben, so daß

$$(26) \qquad \overline{b}(\alpha, K, L) = \pm \overline{\text{Det } \mathfrak{A}} \text{ in } \mathbb{Z}_{\overline{m}_1}$$

gilt. In $H_2(K \cup \bigcup(e_{jK}^2)) = H_2(K) \oplus \mathbb{Z}^\ell$ wählen wir die Basis, welche aus der von $H_2(K)$ und der von den e_{jK}^2 bestimmten Ergänzung für $\mathbb{Z}^\ell$ zusammengesetzt ist; analog verfahren wir bezüglich L. f setzt sich zu einer stetigen Abbildung $K \cup (\bigcup e_{jK}^2) \to L \cup (\bigcup e_{jL}^2)$ fort, bei welcher die e_{jK}^2 und ihre Randworte in die e_{jL}^2 und deren Randworte überführt werden. Bezüglich der gewählten Basen wird dann die zugehörige Abbildung $H_2(K \cup \bigcup(e_{jK}^2)) \to H_2(L \cup \bigcup(e_{jL}^2))$ durch eine Matrix

$$(27) \qquad \begin{pmatrix} \mathfrak{A} & 0 \\ 0 & \mathfrak{C} \end{pmatrix}$$

beschrieben.

Die zweite Kongruenz folgt nun aus (26),(27) und (24). q.e.d.

Damit α zu einer Homologie- beziehungsweise Homotopieäquivalenz $K \to L$ gehört, ist bei (25) wiederum das Restklassenpaar ± 1 in $\mathbb{Z}_{m'}$ vonnöten. Für konkrete Unterscheidungen von Komplexen (siehe insbesondere die Beispiele in §5, B)) kann es wichtig sein, sich auf diejenigen Isomorphismen $\overline{\pi}_1(K) \to \overline{\pi}_1(L)$ zu beschränken, welche von einem Isomorphismus $\pi_1(K) \to \pi_1(L)$ herrühren.[9] Ist $b(\alpha, K, L)$ in (25) kongruent zu $\pm 1 \mod m'$, so gehört – wie im Anschluß an (9) zitiert – $\overline{\alpha}$ zu einer Homologieäquivalenz der vergrößerten Komplexe in Satz 2, obwohl dies für α und die ursprünglichen Komplexe nicht der Fall zu sein braucht. Der geometrische Grund hierfür ist, daß auch nichtsphärische Elemente von $\overline{\Sigma}_2(L)$ bei der Vergrößerung von L sphärisch werden und dann zur sphärischen Modifikation stetiger Abbildungen gemäß (1),(2) zur Verfügung stehen.

B) Sind 2-Komplexe - etwa als Standardkomplexe endlicher Präsentationen - gegeben, so lassen sich $\overline{\pi}_1$ und m' berechnen, auch wenn $\overline{m}_1$ oder m_1 nicht algorithmisch ermittelbar sein sollten (siehe die Beispielserie in §5, A)). Für K^2 und L^2 mit $\chi(K) = \chi(L)$ nehmen wir ferner durch Bildworte für Erzeugende einen Isomorphismus $\alpha : \pi_1(K) \to \pi_1(L)$ sowie gemäß (18) eine gute Ergänzung $\{e_{jK}^2\}$ als explizit gegeben an[10]. *Aus diesen Daten wollen wir die reduzierte Biasinvariante in (25) bestimmen.* Alle Schritte sind dabei algorithmisch durchführbar, auch wenn wir dies nicht jedesmal notieren.

Ist für $\overline{\pi}_1 = \pi_1(K^2 \cup (\bigcup e_{jK}^2)) = H_1(K^2 \cup (\bigcup e_{jK}^2))$ eine Präsentation abgelesen, so können wir nach der Methode, Kettenkomplexe aus freien abelschen Gruppen endlichen Ranges in Homologienormalform zu überführen, ihre Erzeugenden und definierenden Relationen durch Nielsentransformationen so abändern, daß die Form

$$\mathfrak{P}_K = \langle u_\lambda, v_\mu, w_\nu | R_\lambda, R_\mu', R_j'' \rangle$$

---

[9]Sämtliche Isomorphismen zu bestimmen ist bekanntlich im allgemeinen wiederum ein algorithmisches Problem, welches wir in dieser Arbeit jedoch nicht behandeln. Ein weiteres liegt bei (18) vor.

[10]Eine entsprechende Ergänzung $\{e_{jL}^2\}$ für Satz 2 läßt sich daraus bestimmen.

entsteht, wobei in $\overline{\pi}_1 \approx \mathbb{Z}_{e_1} \times \ldots \times \mathbb{Z}_{e_r} \times \mathbb{Z}^s$ (mit $1 \neq e_1 | \ldots | e_r$ oder $r = 0$)

(28) (i)   die u_λ trivial werden,
      (ii) $\mathbb{Z}_{e_\mu}$ von v_μ erzeugt wird,
      (iii) jeder Faktor von $\mathbb{Z}^s$ von genau einem w_ν erzeugt wird,
            sowie als Worte in der freien Gruppe $F = F(u_\lambda, v_\mu, w_\nu)$,
      (iv) $R_\lambda = u_\lambda \cdot S_\lambda$, $R'_\mu = v_\mu^{e_\mu} \cdot S'_\mu$ gilt und
      (v)   die S_λ, S'_μ und R''_j in $[F, F]$ liegen.

Bei dem Übergang von $K^2 \cup (\bigcup e_{jK}^2)$ zum Standardkomplex $[\mathfrak{P}_K]$ handelt es sich um eine 3-Deformation im Sinne des einfachen Homotopietyps, siehe [10]. Wir erhalten also bezüglich der durch den Erzeugendenwechsel gegebenen Identifikation i_K der Fundamentalgruppen

$$(29) \qquad b(i_K, K \cup (\bigcup e_{jK}^2), [\mathfrak{P}_K]) = \pm 1 \text{ in } \mathbb{Z}_{m'}.$$

Andererseits besitzt $\overline{\pi}_1$ in den Erzeugenden v_μ, w_ν auch die kanonische Präsentation

$$(30) \qquad \mathfrak{Q}_K = \langle v_\mu, w_\nu | v_\mu^{e_\mu}, [v_\mu, v_{\mu'}]_{\mu < \mu'}, [w_\nu, w_{\nu'}]_{\nu < \nu'}, [v_\mu, w_\nu] \rangle;$$

dabei hat $\chi([\mathfrak{Q}_K])$ den Wert $1 - s + \dbinom{r + s}{2}$, ebenso wie $\chi(K \cup \{e_{jK}^2\})$ und $\chi([\mathfrak{P}_K])$.

Als nächstes berechnen wir $b(\mathrm{id}, [\mathfrak{P}_K], [\mathfrak{Q}_K])$, wobei wir die Fundamentalgruppen vermöge ihrer nichttrivialen Erzeugenden identifizieren:

Die R''_j bestimmen aufgrund von (28) eine Basis von $H_2([\mathfrak{P}_K])$, die Elementarkommutatoren in (30) eine von gleicher Anzahl $\dbinom{r + s}{2}$ für $H_2([\mathfrak{Q}_K])$.

R''_j läßt sich als Konjugiertenprodukt Π_j in den Elementarkommutatoren der u_λ, v_μ, w_ν schreiben. $\mathfrak{A}_K$ sei die Matrix der Exponentensummen, wie oft (mit Vorzeichen) ein gegebener Elementarkommutator

$$[v_\mu, v_{\mu'}]_{\mu < \mu'}, \ [w_\nu, w_{\nu'}]_{\nu < \nu'}, \ [v_\mu, w_\nu]$$

in Π_j vorkommt, wobei die u_λ-haltigen Kommutatoren vernachlässigt werden und die Konjugatoren unterdrückt. Diese Matrix $\mathfrak{A}_K$ beschreibt die Abbildung $H_2([\mathfrak{P}_K]) \to H_2([\mathfrak{Q}_K])$ (bezüglich der gewählten Basen) für ein stetiges $f : [\mathfrak{P}_K] \to [\mathfrak{Q}_K]$, welches die identische Abbildung des 1-Gerüsts fortsetzt. Wir erhalten also

$$(31) \qquad b(\mathrm{id}, [\mathfrak{P}_K], [\mathfrak{Q}_K]) = \pm \overline{\mathrm{Det}\,\mathfrak{A}_K} \text{ in } \mathbb{Z}_{m'}.$$

Da es bei der Ablesung von $\mathfrak{A}_K$ aus den Π_j auf die Konjugatoren nicht ankommt, können wir zuvor nach [19] bei F die Erzeugenden mit den Kommutatoren vertauschbar machen. Im Quotient $F/[F, [F, F]]$ wird aus R''_j (vermöge Π_j) ein eindeutiges Potenzprodukt aus miteinander kommutierenden Elementarkommutatoren, und die Exponenten der aus den v_μ und w_ν gebildeten ergeben $\mathfrak{A}_K$. $\mathfrak{A}_K$ kann also bestimmt werden, indem man die

Erzeugenden in R''_j aneinander "vorbeischiebt", bis R''_j in das triviale Wort verwandelt worden ist, und die dabei verwendeten Kommutierschritte "notiert"[11].

Dieselben Prozesse wie für K führen wir nun bezüglich L durch mit analogen Bezeichnungen und Ergebnissen, die (29) und (30) entsprechen.

Es verbleibt die Aufgabe, den Bias $b(\overline{\alpha}, [\mathfrak{Q}_K], [\mathfrak{Q}_L])$ zu bestimmen, wobei $\overline{\overline{\alpha}}$ derjenige Isomorphismus ist, der sich aus $\overline{\alpha}$ bei den Übergängen zu den kanonischen Präsentationen $\mathfrak{Q}_K, \mathfrak{Q}_L$ ergibt. $\overline{\overline{\alpha}}$ können wir bezüglich der Erzeugenden von $\mathfrak{Q}_K, \mathfrak{Q}_L$ durch eine Matrix $\mathfrak{B}$ geben, weil die Gruppen abelsch sind.

Nach Sieradski, siehe Anhang B von [4], läßt sich $\overline{\overline{\alpha}}$ zerlegen in

(32) (i) endlich viele elementare freie Transformationen[12] der Erzeugenden von $\mathfrak{Q}_K$,

(ii) einen Diagonalautomorphismus $v_\mu \rightarrow v_\mu^{d_\mu}$ mit $(d_\mu, e_\mu) = 1$, und anschließende "natürliche" Identifizierung der Fundamentalgruppen von $[\mathfrak{Q}_K]$ und $[\mathfrak{Q}_L]$ durch diejenige entsprechender Erzeugenden von $\mathfrak{Q}_K$ und $\mathfrak{Q}_L$.

Analog zu $\mathfrak{A}_K$ für (31) kann der Bias für jeden Schritt berechnet werden, indem man in $F/[F, [F, F]]$ das Bild der Basiskommutatoren (unter dem induzierten Isomorphismus) als Potenzprodukt der Basiskommutatoren ausdrückt:

Für einen Schritt (i) führt das zu einer Matrix, welche einen Automorphismus von $H_2([\mathfrak{Q}_K])$ beschreibt, also die Determinante ± 1 hat. Gleiches gilt für die abschließende natürliche Identifizierung. (ii) können wir in Schritte zerlegen, bei denen jeweils eine Erzeugende potenziert wird. Bei jedem solchen Teilschritt gilt $[v_\mu, *] \rightarrow [v_\mu^{d_\mu}, *] \sim [v_\mu, *]^{d_\mu}$ mod $[F, [F, F]]$ bezüglich der $r + s - 1$ Kommutatoren, welche v_μ enthalten; alle übrigen werden identisch abgebildet. Die Determinante der $H_2([\mathfrak{Q}_K])$-Teilabbildung hat also den Wert d_μ^{r+s-1}.

Wir erhalten daher nach der Produktregel in Satz 1:

$$(33) \qquad b(\overline{\alpha}, [\mathfrak{Q}_K], [\mathfrak{Q}_L]) = \pm \overline{(\,\mathrm{Det}\,\mathfrak{B})}^{r+s-1} \text{ in } \mathbb{Z}_{m'}.$$

Ebenfalls mit Hilfe der Produktregel folgt nun aus (29),(31) und (33):

Satz 3. *Der reduzierte Bias (25) in Satz 2 beträgt*

$$(34) \qquad \overline{b(\alpha, K, L)} = \pm \overline{\mathrm{Det}\,\mathfrak{A}_K} \cdot \overline{\mathrm{Det}\,\mathfrak{A}_L}^{-1} \cdot \overline{(\,\mathrm{Det}\,\mathfrak{B})}^{r+s-1} \text{ in } \mathbb{Z}_{m'}.$$

Es sei darauf hingewiesen, daß in (34) $\mathfrak{B}$ *und* $\mathfrak{A}_L$ von α abhängen (um nicht durch den Vergleich mit (41) irritiert zu werden).

---

[11]Mit Hilfe dieses Quotienten von F nach dem 3. Term der Zentralreihe hat Reidemeister $\chi_{\min}(\overline{\pi}_1)$ ermittelt im Rahmen seiner Bestimmung der kommutativen Fundamentalgruppen von geschlossenen, orientierbaren 3-Mannigfaltigkeiten, siehe [19].

[12]Man beachte, daß im allgemeinen nicht alle freien Transformationen zu Automorphismen führen.

In praxi kann das Berechnungsverfahren oft abgekürzt werden, zum Beispiel, wenn die Komplexe sogleich mit π_1-Erzeugenden gegeben sind, welche beim Übergang zu $\overline{\pi}_1$ kanonisch liegen, und/oder, wenn einige Relatoren bei K respektive L eine H_2-Basis bestimmen.

5. Beispiele

A) Wir betrachten Standardkomplexe L zu Präsentationen

$$(35) \quad \mathfrak{Q} = \langle u_1, \dots, u_{t+1}, v_1, \dots, v_{r+t} | u_1 \cdot S_1, \dots, u_{t+1} \cdot S_{t+1},$$

$$v_1^{e_1} \cdot u_{t+1}, v_2^{e_2}, \dots, v_r^{e_r}, v_{r+1}^{e_{r+1}} \cdot u_1, \dots, v_{r+t}^{e_{r+t}} \cdot u_t, \{v_\mu, v_{\mu'}\}_{\mu < \mu' \le r} \rangle,$$

wobei die S_j in $[F(u_\lambda), F(u_\lambda)]$ liegen und $1 \ne e_1 | \dots | e_{r+t}$ mit $e_1 < e_2$ gilt.

$\mathfrak{Q}$ entsteht aus der Vereinigung einer Präsentation von $(\mathbb{Z}_{e_1} \times \dots \times \mathbb{Z}_{e_r}) *$ $\mathbb{Z}_{e_{r+1}} * \dots * \mathbb{Z}_{e_{r+t}}$ und einer balancierten Präsentation $\mathfrak{Q}' = \langle u_\lambda | u_\lambda \cdot S_\lambda \rangle$ einer perfekten Gruppe, indem man die Ordnungsrelationen für $v_1, v_{r+1}, \dots, v_{r+t}$ durch je ein u_λ "stört".

$([v_\mu, v_{\mu'}])$ sei das Basiselement von $H_2(L)$, welches durch $[v_\mu, v_{\mu'}]$ bestimmt ist. Den geschlossenen Weg $[v_1, v_2^{e_2}]$ kann man einmal als Folge von $v_2^{e_2}$, andererseits aber auch als Folge von $[v_1, v_2]$ ausdrücken und erhält so ein 2-Sphärenbild in L, welches ergibt, daß

$$(36) \quad e_2 \cdot ([v_1, v_2]) \in \Sigma_2(L)$$

gilt. Analog gilt insgesamt

$$(37) \quad e_\mu \cdot ([v_\mu, v_{\mu'}])_{\mu \ne 1} \in \Sigma_2(L) \quad \text{und} \quad e_{\mu'} \cdot ([v_\mu, v_{\mu'}]) \in \Sigma_2(L)$$

sowie, falls u_{t+1} endliche Ordnung hat,

$$(38) \quad e_1 \cdot \mathrm{ord}(u_{t+1}) \cdot ([v_1, v_{\mu'}]) \in \Sigma_2(L).$$

Es läßt sich ferner zeigen, daß die in (37) und (38) angegebenen Elemente $\Sigma_2(L)$ erzeugen:

Wenn in der durch $\mathfrak{Q}$ bestimmten Gruppe u_{t+1} endliche Ordnung hat, vergrößere man $\mathfrak{Q}$ um die definierende (Folge-)Relation $v_1^{e_1 \cdot \mathrm{ord}(u_{t+1})}$ zu $\mathfrak{Q}^*$. Dann erhält die 2. Homologiegruppe des zugehörigen Komplexes L^* einen neuen direkten $\mathbb{Z}$-Summanden, welcher sphärisch ist. T sei ein erzeugendes Element dieses neuen Summanden (T). Wenn u_{t+1} unendliche Ordnung hat, stimme $\mathfrak{Q}^*$ mit $\mathfrak{Q}$ und L^* mit L überein. L^* ist die Vereinigung aus dem Standardkomplex L_2, der zu $v_1, \dots, v_r$ und allen nur diese Erzeugenden enthaltenden definierenden Relationen von $\mathfrak{Q}^*$ gebildet wird und von L_1, der zu den $u_\lambda, v_1, v_{r+1}, \dots, v_{r+t}$ und den übrigen definierenden Relationen von $\mathfrak{Q}^*$ gehört. $\pi_1(L^*)$ ergibt sich als freies Produkt der $\pi_1(L_i)$ mit Amalgamierung über die von v_1 erzeugte Untergruppe. Die Betrachtung der Homologie von $\widetilde{L^*}$ oder die allgemeine Methode von Baik und Pride [1] (siehe [2], Thm. 3.1.) ergibt dann, daß $\Sigma_2(L^*)$ von $\Sigma_2(L_1), \Sigma_2(L_2)$ sowie gegebenenfalls (T) erzeugt wird. Dabei verschwindet $\Sigma_2(L_1)$ mit $H_2(L_1)$, und $\Sigma_2(L_2)$ läßt sich nach der im Beweis von Hilfssatz 3 genannten Methode von Hopf durch die in (37) und (38) genannten Elemente erzeugen. Im Fall $L^* = L$ ist damit gezeigt, daß diese Elemente $\Sigma_2(L)$ erzeugen. Für $L^* \neq L$ folgt die Behauptung, weil die Elemente von (37), (38) und T die Gruppe $\Sigma_2(L^*) = \Sigma_2(L) \oplus (T)$ erzeugen und T bei dieser Zerlegung nach (T) fällt und die Elemente von (37) und (38) nach $\Sigma_2(L)$, q.e.d.

(37) und (38) implizieren also, daß genau dann, *wenn u_{t+1} eine endliche Ordnung hat, welche zu e_2/e_1 relativ prim ist*, L den Biasmodul $m_1 = e_1$ besitzt. Im anderen Fall existiert m_1 ebenfalls und hat einen Wert $e_1 < m_1 \leq e_2$.

Das kursiv angegebene Problem, ob eine Erzeugende einer Präsentation in der zugehörigen Gruppe eine endliche, zu einer gegebenen Zahl relativ prime Ordnung hat, ist im allgemeinen unentscheidbar (siehe [18], Thm.2.10, Beweis). In unserem Fall ergibt sich die Ordnung von u_{t+1} aus der balancierten Präsentation $\mathfrak{Q}'$ einer perfekten Gruppe, welche injektiv in $\pi_1(L)$ liegt. Für Präsentationen vom Typ $\mathfrak{Q}'$ sind meines Wissens die (klassischen) Entscheidungsprobleme noch offen. Es könnte also sein, daß die Ermittlung des "wahren Wertes" von m_1 für L im allgemeinen nicht entscheidbar ist.

Andererseits läßt sich $\mathfrak{Q}$ durch Hinzunahme der restlichen Elementarkommutatoren der v_μ zu einer Präsentation der Abelschmachung $\overline{\pi_1} = \mathbb{Z}_{e_1} \times \ldots \times \mathbb{Z}_{e_{r+t}}$ auf Niveau $\chi_{\min}(\overline{\pi_1})$ ergänzen, denn, wenn alle v_μ miteinander vertauschbar sind, folgt das auch für die u_λ, und diese werden dann in $\overline{\pi_1}$ trivial.

Es gilt also

$$(39) \qquad m' = e_1 \text{ und } e_1 | m_1 | e_2.$$

K werde mit der Präsentation $\mathfrak{P}$ gebildet, welche mit $\mathfrak{Q}$ übereinstimmt, außer, daß $[v_1, v_2]$ durch die Relation $[v_1, v_2^x]$ für eine zu e_2 relativ prime Zahl x ersetzt wird.

K und L haben dann isomorphe Fundamentalgruppen und gleiche Eulersche Charakteristik. Auch ohne den genauen Wert von m_1 zu kennen, wollen wir K und L unter *einer geeigneten Kongruenzbedingung als nicht vom gleichen Homologietyp respektive Homotopietyp erweisen.*

Durch Vergleich von $[v_1, v_2]$ bei L und $[v_1, v_2^x]$ bei K ergibt sich:

$$(40) \qquad b(\mathrm{id}, K, L) \equiv \pm x \mod e_1.$$

Die folgende Abschätzung zeigt, daß dieser Bias durch die Komposition mit den $b(\alpha, L, L)$ für $\alpha \in \mathrm{Aut}(\pi_1(L))$ nicht immer trivialisiert werden kann:

Um eine diesbezügliche $H_2(K) \to H_2(L)$-Abbildungsmatrix zu ermitteln, zählen wir, wie oft $[v_\mu, v_{\mu'}]_{\mu < \mu' \leq r}$ in $[\alpha(v_\nu), \alpha(v_{\nu'})]_{\nu < \nu' \leq r}$ vorkommt. Hierfür darf der Quotient nach $F/[F, [F, F]]$ genommen werden, und es reicht, die Vielfachheiten modulo e_1 zu bestimmen. Von α benötigen wir daher auch nur die eingeschränkte und quotientierte Abbildung $\mathbb{Z}_{e_1} \times \ldots \times \mathbb{Z}_{e_1} \to \mathbb{Z}_{e_1} \times$

$\dots \times \mathbb{Z}_{e_1}$, wobei jeder Faktor für ein v_μ, $\mu \leq r$ steht. Sie werde durch $\mathfrak{B}$ gegeben. Durch Betrachtungen wie im Anschluß an (32) ergibt sich nun:

$$(41) \qquad b(\alpha, L, L) \equiv \pm \operatorname{Det} \mathfrak{B}^{r-1} \mod e_1, \text{ das heißt}$$
$$b(\alpha, K, L) \equiv \pm x \cdot \operatorname{Det} \mathfrak{B}^{r-1} \mod e_1.$$

Wenn x und $-x$ keine $(r-1)$-te Potenz in $\mathbb{Z}_{e_1}$ bestimmen, sind K und L also nicht homologie- respektive homotopieäquivalent. Dies ist zum Beispiel für $x = 2, e_1 = 5, r = 3$ der Fall.

Nach dem Muster von (35), nur ohne die u_λ-Anteile, lassen sich allgemein Präsentationen für *endliche freie Produkte aus endlich erzeugten abelschen Gruppen* bilden. Die Unterscheidung von $L^2 = [\mathfrak{Q}]$ und $K^2 = [\mathfrak{P}]$, bei welchem Erzeugende endlicher Ordnung (in π_1) in den Kommutatorrelatoren noch – wie im Anschluß an (39) – potenziert werden dürfen, gelingt dann in vielen Fällen. Man zerlegt dazu die π_1-Isomorphismen gemäß [6]. Hierdurch werden Ergebnisse aus [15] und [20] verallgemeinert. Im Gegensatz zu den von Sieradski betrachteten Situationen ist die *Klassifikation* von Homologie- und Homotopietypen im allgemeinen allerdings noch offen, insbesondere, wenn $\mathbb{Z}$-Faktoren beteiligt sind.

B) Daß eine genauere Analyse der π_1-Isomorphismen vonnöten sein kann, möge auch durch die Komplexe K respektive L zu

$$(42) \qquad \mathfrak{P} = \langle a, b, c \mid a^p, [a^2, b], aca^{-1}c^{-s} \rangle \text{ respektive}$$
$$\mathfrak{Q} = \langle a, b, c \mid a^p, [a, b], aca^{-1}c^{-s} \rangle$$

mit einer Primzahl $p \neq 2$ und $1 < s \equiv 1 \mod p$ verdeutlicht werden: $\mathfrak{P}$ und $\mathfrak{Q}$ präsentieren dieselbe Gruppe $\pi_1 = \mathbb{Z}_p \ltimes (\mathbb{Z} * \mathbb{Z}_q)$, wobei sich $q = s^p - 1$ als Ordnung von c ergibt. Eine Potenz c^y kommutiert in π_1 genau dann mit a, wenn y durch $1 + s + \dots + s^{p-1}$ teilbar ist. $\operatorname{Aut}(\pi_1)$ wird erzeugt von Automorphismen der Form

$$(43) \qquad \begin{array}{ll} (i) & \text{innere Automorphismen,} \\ (ii) & a \to a, \ b \to b, \ c \to c^v \text{ mit } (v, q) = 1, \\ (iii) & a \to a, \ b \to b \cdot a^i, \ c \to c \\ (iv) & a \to a, \ b \to c^{y_1} b^{\pm 1} c^{y_2}, \ c \to c \text{ mit } c^{y_i} \overset{\leftarrow}{\to} a. \end{array}$$

Dies läßt sich in Analogie zum Beweis von Hilfssatz 1 in [16] und [14], Lemma 2.7, siehe [8], zeigen. Lustig [14] hat für zu (42) verwandte 2-Komplexe den einfachen Homotopietyp vom Homotopietyp unterschieden. Wichtig ist, daß π_1 nur Automorphismen besitzt, welche in der Abelschmachung a in eine Potenz von sich abbilden, die modulo p kongruent zu 1 ist.

Analog zu den Biasberechnungen (40),(41) in A) erhalten wir dann

$$(44) \quad \begin{aligned} b(\operatorname{id}, K, L) &\equiv \pm 2 \quad \mod m'(= p) \text{ und vermöge (43)} \\ b(\alpha, L, L) &\equiv \pm 1 \quad \mod m'(= p), \end{aligned}$$

woraus sich ergibt, daß K und L homolog und homotop verschieden sind[13].
Da $\overline{\pi}_1$ den Automorphismus $a \to a^2, b \to b, c \to c$ besitzt, würde eine Abschätzung wie für (41) lediglich mit Hilfe der Automorphismen der Abelschmachung hier nicht ausreichen.

LITERATURVERZEICHNIS

[1] Y.G. Baik, S.J. Pride, Generators of the second homotopy module of presentations
     arising from group constructions, preprint, University of Glasgow (1992).
[2] W.A. Bogley, S.J. Pride, Calculating generators of π_2, in: C. Hog-Angeloni, W. Metz-
     ler, A.J. Sieradski (eds.): Two-dimensional Homotopy and Combinatorial Group The-
     ory, Lond. Math. Soc. Lecture Note Series 197, Cambridge University Press (1993),
     157-188
[3] K.S. Brown, The Geometry of Rewriting Systems: A Proof of the Anick-Groves-Squire
     Theorem, in: G. Baumslag, C.F. Miller III (eds.): Algorithms and Classification in
     Combinatorial Group Theory, Springer (1992), 137-164.
[4] M.N. Dyer, Invariants for distinguishing between stably isomorphic modules, J. Pure
     Appl. Algebra 37 (1985), 117-153.
[5] M.N. Dyer, A topological interpretation for the bias invariant, Proc. Amer. Math.
     Soc. 89 (1986), 513-523.
[6] D.I.Fouxe-Rabinovitch, Über die Automorphismengruppe der freien Produkte I, Math.
     Sb. 8 (1940), 265-276, II, Math. Sb. 9 (1941), 183-220.
[7] C.M. Gordon, Some embedding theorems and undecidability questions for groups, in
     A.J. Duncan, N.D. Gilbert, J. Howie (eds.): Combinatorial and Geometric Group The-
     ory, Lond. Math. Soc. Lecture Notes Series 204, Cambridge University Press (1995),
     105-110.
[8] C.Grabo, Ein verallgemeinerter Bias, Diplomarbeit, Frankfurt (1997).
[9] C. Hog-Angeloni, P. Latiolais, W. Metzler, Bias Ideals and Obstructions to simple-
     homotopy Equivalence, in: P. Latiolais (ed.): Topology and Combinatorial Group
     Theory, Springer Lecture Notes in Math. 1440 (1990), 109-121.
[10] C. Hog-Angeloni, W. Metzler, Geometric Aspects of Two-Dimensional complexes, in:
     C. Hog-Angeloni, W. Metzler, A.J. Sieradski (eds.): Two-dimensional Homotopy and
     Combinatorial Group Theory, Lond. Math. Soc. Lecture Note Series 197, Cambridge
     University Press (1993), 1-50.
[11] H. Hopf, Fundamentalgruppe und zweite Bettische Gruppe, Comment. Math. Helv.
     14 (1941), 257-309.
[12] P. Latiolais, When homology equivalence implies homotopy equivalence for 2-
     complexes, J. Pure Appl. Algebra 76 (1991), 155-165.
[13] P. Latiolais, Homotopy and homology classification of 2-complexes, in: C. Hog-
     Angeloni, W. Metzler, A.J. Sieradski (eds.): Two-dimensional Homotopy and Com-
     binatorial Group Theory, Lond. Math. Soc. Lecture Note Series 197, Cambridge Uni-
     versity Press (1993), 97-124.
[14] M. Lustig, Nielsen equivalence and simple homotopy type, Proc. London Math. Soc.
     62 (1991), 537-562.
[15] W. Metzler, Über den Homotopietyp zweidimensionaler CW-Komplexe und Ele-
     mentartransformationen bei Darstellungen von Gruppen durch Erzeugende und
     definierende Relationen, J. reine angew. Math. 285 (1976), 7-23.
[16] W. Metzler, Die Unterscheidung von Homotopietyp und einfachem Homotopietyp bei
     zweidimensionalen Komplexen, J. reine angew. Math. 403 (1990), 201-219.

---

[13]Es gilt in diesem Fall sogar $m' = \overline{m}_1 = m_1$.

[17] B. *Michalik*, Ein algebraisierter Biasbegriff, Diplomarbeit, Frankfurt (1993).

[18] *C.F. Miller III*, Decision problems for groups – survey and reflections, in: G. Baumslag, C.F. Miller III (eds.): Algorithms and Classification in Combinatorial Group Theory, Springer (1992), 1-59.

[19] *K. Reidemeister*, Kommutative Fundamentalgruppen, Monatshefte Math. Phys. 43 (1936), 20-28.

[20] *A.J. Sieradski*, A semigroup of simple homotopy types, Math. Z. 153 (1977), 135-148.

Wolfgang Metzler
Mathematisches Seminar der
Johann Wolfgang Goethe Universität
Robert-Mayer-Straße 6-10
D-60054 Frankfurt am Main
Germany

ON GROUPS WHICH ACT FREELY AND PROPERLY ON FINITE DIMENSIONAL HOMOTOPY SPHERES

GUIDO MISLIN AND OLYMPIA TALELLI

For Urs Stammbach on his 60th birthday

1. INTRODUCTION

In [28] C. T. C. Wall conjectured that if a countable group G of finite virtual cohomological dimension, $\text{vcd}\, G < \infty$, has periodic Farrell cohomology then G acts freely and properly on $\mathbb{R}^n \times S^m$ for some n and m. Obviously, if a group G acts freely and properly on some $\mathbb{R}^n \times S^m$ then G is countable since $\mathbb{R}^n \times S^m$ is a separable metric space. The Farrell cohomology generalizes the Tate cohomology theory for finite groups to the class of groups G with $\text{vcd}\, G < \infty$ (see for instance Chapter X of [2]). Wall's conjecture was proved by Johnson in some cases [12] and Connolly and Prassidis in general [4].

In [19] Prassidis showed that there are groups of infinite vcd which act freely and properly on some $\mathbb{R}^n \times S^m$. In particular, it follows from results of Prassidis [19] and Talelli [24] that if a countable group G has periodic cohomology after 1-step then G acts freely and properly on some $\mathbb{R}^n \times S^m$ [25].

A group G is said to have periodic cohomology after k-steps if there is a positive integer q such that the functors $H^i(G;\ )$ and $H^{i+q}(G;\ )$ are naturally equivalent for all $i > k$ (cf. [22], [26]). In [25] it was conjectured that the following statements are equivalent for a countable group G:

(1) G acts freely and properly on some $\mathbb{R}^n \times S^m$
(2) there is an integer q and an exact sequence

$$0 \to \mathbb{Z} \to A \to P_{q-2} \to \cdots \to P_0 \to \mathbb{Z} \to 0$$

with P_i projective $\mathbb{Z}G$-modules, $\mathbb{Z}$ with trivial G-action and proj. $\dim_{\mathbb{Z}G} A < \infty$
(3) G has periodic cohomology after some steps.

Note that if a group G has periodic cohomology with period q after k-steps and the isomorphisms are induced by cup product with an element $g \in H^q(G; \mathbb{Z})$, then g is represented by a q-extension

$$0 \to \mathbb{Z} \to A \to P_{q-2} \to \cdots \to P_0 \to \mathbb{Z} \to 0$$

1991 *Mathematics Subject Classification*. Primary 20J05, 18G20; Secondary 55J05.

with P_i projective $\mathbb{Z}G$-modules and proj. $\dim_{\mathbb{Z}G} A \leq k$; conversely, from (2) one can deduce that (3) holds, with the periodicity induced via a cup product (cf. [25]).

Now $(1) \Rightarrow (2) \Rightarrow (3)$ (see Corollary 5.2) and from the results mentioned above $(3) \Rightarrow (1)$ if vcd $G < \infty$ or if G has periodic cohomology after 1-step. Also, by a result of Talelli [22], the condition (3) is equivalent for an arbitrary group G to the condition

(3′) G admits a periodic (complete) resolution.

The definition of complete resolutions is recalled in Section 2, where we also review the definition of generalized Tate cohomology $\hat{H}^\bullet(G; M)$ for an arbitrary group G. It turns out (cf. Theorem 4.1) that (3′) is equivalent for an arbitrary group G to

(3″) $\hat{H}^\bullet(G; \mathbb{Z})$ contains a unit of non-zero degree.

The condition concerning units can be analyzed by considering suitable actions of G on finite dimensional contractible spaces and leads us to the following

Theorem A. *If G is a countable group in the class $\mathbf{H\mathfrak{F}}$ and there is a bound on the orders of the finite subgroups of G then* $(3) \Rightarrow (1)$.

The class $\mathbf{H\mathfrak{F}}$ of *hierarchically decomposable* groups was introduced by Kropholler [13] as follows. Let $\mathbf{H_0\mathfrak{F}}$ be the class of finite groups. Now define $\mathbf{H_\alpha\mathfrak{F}}$ for each ordinal α inductively: if α is a successor ordinal then $\mathbf{H_\alpha\mathfrak{F}}$ is the class of groups which admit a finite dimensional contractible G-CW-complex with cell stabilizers in $\mathbf{H_{\alpha-1}\mathfrak{F}}$, and if α is a limit ordinal then $\mathbf{H_\alpha\mathfrak{F}} = \cup_{\beta<\alpha}\mathbf{H_\beta\mathfrak{F}}$. A group belongs to $\mathbf{H\mathfrak{F}}$ if it belongs to $\mathbf{H_\alpha\mathfrak{F}}$ for some α.

Notation. If $\mathfrak{X}$ is a class of groups, we denote by $\mathfrak{X}_b$ the subclass consisting of those groups in $\mathfrak{X}$ for which there is a bound on the orders of their finite subgroups.

We show

Theorem B. *Let $G \in \mathbf{H\mathfrak{F}_b}$. Then the following statements are equivalent for G, and they all imply that $G \in \mathbf{H_1\mathfrak{F}_b}$:*

(I) *there is a finite dimensional free G-CW-complex homotopy equivalent to a sphere*

(II) *there is an integer q and an exact sequence*

$$0 \to \mathbb{Z} \to A \to P_{q-2} \to \cdots \to P_0 \to \mathbb{Z}$$

with P_i projective $\mathbb{Z}G$-modules, $\mathbb{Z}$ with trivial G-action and proj. $\dim_{\mathbb{Z}G} A < \infty$

(III) *G has periodic cohomology after some steps*

(IV) *there is an invertible element in the ring $\hat{H}^\bullet(G; \mathbb{Z})$ of non-zero degree.*

Moreover, for $G \in \mathbf{H_1\mathfrak{F}_b}$ the following is equivalent to (I), (II), (III) and (IV):

(V) *every finite subgroup of G has periodic cohomology.*

Note that in case the group G in Theorem B is countable, (I) gives rise to a free and proper G-action on $\mathbb{R}^n \times S^m$ for some n and m (Lemma 5.4) and therefore Theorem B implies Theorem A.

It follows from a theorem of Serre (e.g. Theorem 11.1, Chapter VIII in [2]) that the class of groups of finite vcd is contained in $\mathbf{H_1\mathfrak{F}_b}$. Connolly and Prassidis [4] proved essentially that (III) $\Rightarrow$ (I) if $\operatorname{vcd} G < \infty$, and Brown (Chapter X in [2]) that (V) $\Rightarrow$ (III) if $\operatorname{vcd} G < \infty$. Our proof of Theorem B is based on the methods developed in these papers.

Note that there are groups in $\mathbf{H_1\mathfrak{F}}$ such that (V) $\nRightarrow$ (III) and there are also groups in $\mathbf{H\mathfrak{F}_b}$ such that (V) $\nRightarrow$ (III); but there is also a family of groups in $\mathbf{H_1\mathfrak{F}} \setminus \mathbf{H_1\mathfrak{F}_b}$ such that (V) $\Rightarrow$ (III) $\Rightarrow$ (I) (for examples see Remark 4.11).

The class $\mathbf{H_1\mathfrak{F}_b}$ is a larger class than the class of groups of finite vcd. For example if $\operatorname{vcd} G_i < \infty$ $(i = 1, 2)$ and $G = G_1 *_S G_2$ then the group G need not be of finite vcd [20]. However, $G \in \mathbf{H_1\mathfrak{F}_b}$. Actually if a group G is the fundamental group of a finite graph of groups of finite vcd then $G \in \mathbf{H_1\mathfrak{F}_b}$; also $\mathbf{H_1\mathfrak{F}_b}$ is extension closed whereas the class of groups of finite vcd is not (see 3.9 and 3.10 for general results on groups in $\mathbf{H_1\mathfrak{F}_b}$). The Burnside group $B(d, e)$ of odd exponent $e > 665$ on d generators is another example of a group of infinite vcd in $\mathbf{H_1\mathfrak{F}_b}$. It turns out that it has periodic cohomology after 2-steps and it follows from Theorem A that it acts freely and properly on some $\mathbb{R}^n \times S^m$ (cf. Corollary 5.6).

The proof of Theorem B relies on a result of Kropholler and Mislin [15] which states that if $G \in \mathbf{H\mathfrak{F}_b}$ and $\operatorname{proj.dim}_{\mathbb{Z}G} B(G, \mathbb{Z}) < \infty$, where $B(G, \mathbb{Z})$ is the $\mathbb{Z}G$-module of bounded functions from G to $\mathbb{Z}$, then $G \in \mathbf{H_1\mathfrak{F}_b}$ and admits a finite dimensional $\underline{E}G$. (Recall that $\underline{E}G$, the classifying space for proper G-actions, is a G-CW-complex X characterized up to G-homotopy by the requirement that for every finite subgroup $H < G$ the fixed point space X^H is contractible, and for infinite $H < G$, X^H is empty).

We also show that if a group G contains a free abelian subgroup of infinite rank, then G does not act freely and properly on any $\mathbb{R}^n \times S^m$ (cf. Corollary 5.6).

For every group G there is a free G-CW-complex S_G homotopy equivalent to a sphere. For example, if Y is the universal cover of a $K(G, 1)$ complex then $Y \times S^n$, with diagonal G-action, trivial on S^n is such a complex S_G.

We believe that periodicity in cohomology after some steps is the algebraic characterization of those groups G which admit a *finite dimensional S_G*. We prove this for $G \in \mathbf{H\mathfrak{F}_b}$.

2. GENERALIZED TATE COHOMOLOGY

The classical Tate cohomology for finite groups was generalized by Farrell [7] to the case of groups of finite vcd and subsequently by Ikenaga [10] to the more general class of groups G admitting complete resolutions and having finite generalized cohomological dimension, $\underline{cd}\,G < \infty$ (for the definition of $\underline{cd}$ see below). In [17] generalized Tate cohomology groups $\hat{H}^i(G;M)$ are defined for arbitrary groups G and G-modules M, specializing to the ones defined by Farrell and Ikenaga, when the latter are defined. The definition of these generalized Tate groups is as follows:

$$\hat{H}^n(G;M) := \varinjlim_{j \geq 0} S^{-j}H^{n+j}(G;M), \quad n \in \mathbb{Z}$$

with $S^{-j}H^{n+j}(G;\ )$ denoting the jth left satellite of $H^{n+j}(G;\ )$ (for details, the reader is referred to [17]; different, but equivalent definitions of generalized Tate groups can be found in [1] and [9]). The following are three of their basic properties:

(T1) $\hat{H}^i(G;P) = 0$ for every projective P and $i \in \mathbb{Z}$

(T2) there is a canonical natural transformation

$$\tau : H^\bullet(G;\ ) \to \hat{H}^\bullet(G;\ )$$

such that every natural transformation from ordinary cohomology to a cohomological functor which vanishes on projectives, factors uniquely through τ

(T3) if there exists $n \in \mathbb{Z}$ such that $H^i(G;P) = 0$ for all projective P and all $i > n$ then

$$\tau : H^i(G;\ ) \cong \hat{H}^i(G;\ )$$

for all $i > n$.

Note that (T1) implies that generalized Tate cohomology is *effaceable*: there is *dimension-shifting upwards*

$$\hat{H}^i(G;M) \cong \hat{H}^{i+1}(G;\Omega M)$$

where ΩM denotes the kernel of a surjection of a projective module onto M.

Sometimes the generalized Tate cohomology groups can be computed using *complete resolutions*. For this we recall a few definitions.

Definition 2.1. A complete resolution for a group G is an acyclic complex $\mathcal{F} = \{F_i, \partial_i \mid i \in \mathbb{Z}\}$ of projective $\mathbb{Z}G$-modules, together with a projective resolution $\mathcal{P} = \{P_i, d_i \mid i \geq 0\}$ of G such that $\mathcal{F}$ and $\mathcal{P}$ coincide in sufficiently high dimensions:

$$F_{k-1} \to \cdots \to F_0 \to F_{-1} \to \cdots$$

$$\cdots \to F_{k+1} \to F_k \nearrow$$

$$\searrow$$

$$P_{k-1} \to \cdots \to P_0 \to \mathbb{Z} \to 0$$

The number $k \in \mathbb{N}$ is called the *coincidence index* of the complete resolution.

Clearly this definition generalizes the notion of complete resolution for finite groups and groups of finite vcd (e.g. Chapter X in [2]). In an analogous way one defines a complete resolution for a particular G-module M instead of $\mathbb{Z}$. It is easy to prove that G has a complete resolution if and only if every G-module M has a complete resolution.

We say that M has a complete resolution in the *strong sense* if it has a complete resolution $\mathcal{F}$ such that the complex $\mathrm{Hom}_{\mathbb{Z}G}(\mathcal{F}, P)$ is exact for all projective P (this is the way the term "complete resolution" is used in [5]). In case the trivial G-module $\mathbb{Z}$ has a complete resolution in the strong sense, we just say that *G has a complete resolution in the strong sense*.

Lemma 2.2. *If G admits a complete resolution and if every projective $\mathbb{Z}G$-module has finite injective dimension, then every $\mathbb{Z}G$-module admits a complete resolution in the strong sense.*

Proof. Let $\mathcal{F}$ be a complete resolution for G (cf. 2.1). We first show how to construct a complete resolution in the strong sense for a $\mathbb{Z}$-free $\mathbb{Z}G$-module M. Clearly $\mathcal{F} \otimes M$ with diagonal G-action yields a complete resolution for M. We need to show that for P projective, the complex $\mathrm{Hom}_{\mathbb{Z}G}(\mathcal{F} \otimes M, P)$ is exact. For this we choose an injective resolution

$$0 \to P \to I_0 \to I_1 \to \cdots \to I_n \to 0$$

and notice that for $0 \leq k \leq n$ the complexes $\mathrm{Hom}_{\mathbb{Z}G}(\mathcal{F} \otimes M, I_k)$ are exact, because I_k is injective. Thus $\mathrm{Hom}_{\mathbb{Z}G}(\mathcal{F} \otimes M, P)$ is exact too. To treat the case of a general M we choose a surjection $F \to M$, F a free $\mathbb{Z}G$-module and write ΩM for the kernel, which is $\mathbb{Z}$-free. Clearly, a complete resolution of ΩM in the strong sense yields one for M in an obvious way. $\quad\square$

Definition 2.3. A group G admits a periodic resolution, if it admits a complete resolution $\mathcal{F} = \{F_i, \partial_i \mid i \in \mathbb{Z}\}$ such that for some $k > 0$ and all $i \in \mathbb{Z}$ one has $F_{i+k} = F_i$ and $\partial_{i+k} = \partial_i$.

It was proved by Talelli [22] that a group G has period q after k-steps if and only if there is an exact sequence

$$0 \to R_{k+q} \to P_{k+q-1} \to \ldots \to P_0 \to \mathbb{Z} \to 0$$

with all P_j projective $\mathbb{Z}G$-modules and with R_{k+q} isomorphic to $R_k = \mathrm{Im}(P_k \to P_{k-1})$. Clearly by splicing together copies of

$$0 \to R_k \to P_{k+q-1} \to \ldots \to P_k \to R_k \to 0$$

one obtains a periodic resolution for G of coincidence index k.

Corollary 2.4. *A group G admits a periodic resolution if and only if G has periodic cohomology after some steps.*

If G has a complete resolution $\mathcal{F}$ in the strong sense, then by definition $H^\bullet(\mathrm{Hom}_{\mathbb{Z}G}(\mathcal{F}, P)) = 0$ for all projective P, and the universal property of the generalized Tate groups implies that one has a canonical equivalence of cohomological functors

$$\hat{H}^\bullet(G; \ ) \cong H^\bullet(\mathrm{Hom}_{\mathbb{Z}G}(\mathcal{F}, \ )).$$

We then say that "the generalized Tate cohomology can be computed using a complete resolution of G".

The following theorem characterizes groups for which the generalized Tate cohomology can be computed using a complete resolution of G (see also Theorem 3.10 of [5]). It involves the invariants $\mathrm{spli}\, G$, which is the supremum of the projective length of injective $\mathbb{Z}G$-modules

$$\mathrm{spli}\, G := \sup\{i : \mathrm{Ext}^i_{\mathbb{Z}G}(I, \ ) \neq 0 \,|\, I \ \mathbb{Z}G\text{-injective}\},$$

and Ikenaga's *generalized* cohomological dimension $\underline{\mathrm{cd}}\, G$, which is defined by

$$\underline{\mathrm{cd}}\, G := \sup\{i : \mathrm{Ext}^i_{\mathbb{Z}G}(M, F) \neq 0 \,|\, M \ \mathbb{Z}\text{-free}, \ F \ \mathbb{Z}G\text{-free}\}.$$

Occasionally we will also use the invariant $\mathrm{silp}\, G$, which is the supremum of the injective length of projective $\mathbb{Z}G$-modules

$$\mathrm{silp}\, G := \sup\{i : \mathrm{Ext}^i_{\mathbb{Z}G}(\ , P) \neq 0 \,|\, P \ \mathbb{Z}G\text{-projective}\}.$$

It is straightforward that $\underline{\mathrm{cd}}\, G$ and $\mathrm{silp}\, G$ are either both finite or both infinite, and more precisely

$$\underline{\mathrm{cd}}\, G \leq \mathrm{silp}\, G \leq 1 + \underline{\mathrm{cd}}\, G.$$

Theorem 2.5. *Let G be an arbitrary group. Then the following conditions (1) and (2) are equivalent and they imply (3):*

(1) $\mathrm{spli}\, G < \infty$

(2) *G admits a complete resolution and $\underline{\mathrm{cd}}\, G < \infty$*

(3) *G admits a complete resolution and the generalized Tate cohomology groups of G can be computed using any complete resolution of G.*

Proof. (1) $\Rightarrow$ (2): If $\mathrm{spli}\, G < \infty$ then G admits a complete resolution by Theorem 4.1 of [8]. In general $\mathrm{silp}\, G \leq \mathrm{spli}\, G$ (cf. [8]) and $\underline{\mathrm{cd}}\, G \leq \mathrm{silp}\, G$ as remarked above.

(2) $\Rightarrow$ (1): If $\mathcal{F}$ denotes a complete resolution for G and $\underline{\mathrm{cd}}\, G < \infty$, then $\mathrm{silp}\, G < 1 + \underline{\mathrm{cd}}\, G < \infty$. Therefore, by Lemma 2.2, every $\mathbb{Z}G$-module admits a complete resolution in the strong sense. By Theorem 3.10 of [5] this implies that $\mathrm{spli}\, G < \infty$.

(2) $\Rightarrow$ (3): If $\mathcal{F}$ denotes a complete resolution for G and $\underline{\mathrm{cd}}\, G < \infty$, then $H^i(G; P) = \mathrm{Ext}^i_{\mathbb{Z}G}(\mathbb{Z}, P) = 0$ for $i > \underline{\mathrm{cd}}\, G$ and P projective. As noted earlier, this implies that the generalized Tate cohomology groups can be computed using the complete resolution $\mathcal{F}$. Since $\underline{\mathrm{cd}}\, G < \infty$, complete resolutions of G are unique up to chain homotopy [10], thus any complete resolution of G can be used to compute the generalized Tate cohomology of G. $\square$

Ikenaga in [10] defined a class of groups $\mathfrak{C}_\infty$ via actions on finite dimensional acyclic complexes, and he proved that the groups in this class possess complete resolutions and have finite $\underline{\mathrm{cd}}$ [10, Theorem 2], hence by 2.5 these groups satisfy spli $G < \infty$.

The class $\mathfrak{C}_\infty$ is defined as follows. Let $\mathfrak{C}_0$ be the class of finite groups and for an integer $n > 0$ let $G \in \mathfrak{C}_n$ if and only if there is a finite dimensional acyclic G-*simplicial complex* X for which

- $G_\sigma \in \mathfrak{C}_{n-1}$ for all simplices σ of X
- $\sup_\sigma\{\underline{\mathrm{cd}}\,G_\sigma\} < \infty$ where σ runs over all simplices of X,

and $\mathfrak{C}_\infty = \cup_n \mathfrak{C}_n$.

Ikenaga's "G-simplicial complexes" are such that their barycentric subdivision are G-CW-complexes and they are therefore G-homeomorphic to G-CW-complexes. On the other hand it is an elementary fact that every G-CW-complex is G-homotopy equivalent to a "G-simplicial complex" of the same dimension. If X is an acyclic G-simplicial complex X of dimension k, then its join $X * G$ is a contractible G-simplicial complex of dimension $k + 1$, whose barycentric subdivision is a G-CW-complex, with point stabilizers being subgroups of the original stabilizers G_σ. As proved above, for any group G with spli $G < \infty$ the invariants $\underline{\mathrm{cd}}\,G$ and spli G differ at most by 1, and clearly $\underline{\mathrm{cd}}\,G = 0$ for finite G. Therefore we can record the following relationship with Kropholler's classes.

Corollary 2.6. *Ikenaga's class* $\mathfrak{C}_1$ *agrees with Kropholler's class* $\mathbf{H}_1\mathfrak{F}$, *and for every* $n \in \mathbb{N}$ *one has*

$$\mathfrak{C}_n = \{G \in \mathbf{H}_n\mathfrak{F} \mid \text{spli}\,G < \infty\}.$$

Moreover $\mathfrak{C}_\infty$ *consists of those groups in* $\mathbf{H}_\omega\mathfrak{F}$ *for which* spli $G < \infty$; *here* ω *denotes the first infinite ordinal.*

Remark 2.7. We do not know of an example of a group G with spli $G < \infty$ not belonging to $\mathbf{H}_1\mathfrak{F}$; it is conceivable that $\mathfrak{C}_\infty = \mathbf{H}_1\mathfrak{F}$.

In case the generalized Tate cohomology can be computed using complete resolutions, i.e. if spli $G < \infty$, the generalized Tate cohomology has many properties analogous to those of the Farrell theory, where the role of vcd G is played by $\underline{\mathrm{cd}}\,G$ (cf. [10]), namely:

(T4) the natural map $H^i(G; M) \to \hat{H}^i(G; M)$ is an isomorphism for $i > \underline{\mathrm{cd}}\,G$

(T5) Shapiro's Lemma holds: for any subgroup $H < G$ and $\mathbb{Z}H$-module M

$$\hat{H}^\bullet(H; M) \cong \hat{H}^\bullet(G; \text{Hom}_{\mathbb{Z}H}(\mathbb{Z}G, M))$$

(T6) $\hat{H}^\bullet(G; I) = 0$ for injective $\mathbb{Z}G$-modules I; hence generalized Tate cohomology is *coeffaceable* and one also has *dimension shifting downwards*

(T7) there is a cup product with the usual properties, compatible with that in ordinary cohomology:

$$H^p(G; M) \otimes H^q(G; N) \xrightarrow{\cup} H^{p+q}(G; M \otimes N)$$

$$\downarrow \qquad\qquad\qquad\qquad \downarrow$$

$$\hat{H}^p(G; M) \otimes \hat{H}^q(G; N) \xrightarrow{\cup} \hat{H}^{p+q}(G; M \otimes N)$$

Remark 2.8. Cup products as in (T7) exist for arbitrary G (cf. [14]). In particular, $R := \hat{H}^0(G; \mathbb{Z})$ is a commutative ring with 1, $\hat{H}^\bullet(G; \mathbb{Z})$ is an R-algebra and $\hat{H}^\bullet(G; M)$ is an R-module. In case of spli $G < \infty$ one can use a complete resolution $\mathcal{F}$ of G to define a cup product using a suitable diagonal (cf. [10])

$$\Delta : \mathcal{F} \to \mathcal{F} \hat{\otimes} \mathcal{F}.$$

The following result implies that not every group has a complete resolution.

Proposition 2.9. *If a group G has a complete resolution of coincidence index k, then $H^i(G; P) \neq 0$ for some projective $\mathbb{Z}G$-module P and some $i \leq k$.*

Proof. If G is finite, it admits a complete resolution of coincidence index 0 and $H^0(G; \mathbb{Z}G) \neq 0$. If G is infinite and

$$F_{k-1} \to \cdots \to F_0 \to F_{-1} \to \cdots$$

$$\nearrow$$

$$\cdots \to F_{k+1} \to F_k$$

$$\searrow$$

$$P_{k-1} \to \cdots \to P_0 \to \mathbb{Z} \to 0$$

is a complete resolution with coincidence index k, we define

$$\Lambda_j = \mathrm{Ker}(F_j \to F_{j-1}) \quad j \in \mathbb{Z},$$

and $\Omega_i = \mathrm{Ker}(P_i \to P_{i-1})$, $i > 0$. Since G is infinite $H^0(G; M) = 0$ for any submodule M of a projective module. Assume that $H^i(G; P) = 0$ for all projective P and all $i \leq k$. Then, by dimension shifting

$$H^0(G; \mathbb{Z}) \cong H^k(G; \Omega_k) = H^k(G; \Lambda_k) \cong H^0(G; \Lambda_{-1})$$

which is a contradiction, since $H^0(G; \mathbb{Z}) = \mathbb{Z}$ and Λ_{-1} is a submodule of the projective module F_{-1}. The result follows. $\square$

Corollary 2.10. (i) *If a group G contains a free abelian group of infinite rank then G does not admit a complete resolution.*
(ii) *The Thompson group*

$$T = \langle x_0, x_1, \ldots \mid x_i x_j x_i^{-1} = x_{j+1}, \quad i < j \rangle$$

is an example of a group of type FP_∞ which does not admit a complete resolution.

Proof. (i): A free abelian group A of countably infinite rank satisfies

$$H^i(A; P) = 0$$

for all $i \geq 0$ and all projective P. The result now follows from 2.9 since if a group G has a complete resolution then every subgroup of G has a complete resolution too.

(ii): Thompson's group is of type FP_∞ and contains the free abelian subgroup with basis $\{x_i x_{i+1}^{-1} \mid i \in \mathbb{N}\}$ (cf. [3]). $\square$

3. THE STABILIZER SPECTRAL SEQUENCE

The classical stabilizer spectral sequence of Farrell cohomology (cf. Chapter X in [2]) admits a generalization to the case of groups G with spli $G < \infty$.

Theorem 3.1. *Let X be a finite dimensional contractible G-CW-complex and write G_σ for the stabilizer of the cell σ of X. Assume that spli $G < \infty$. Then there is a finitely convergent spectral sequence*

$$E_1^{pq} = \prod_{\sigma \in \Sigma_p} \hat{H}^q(G_\sigma; M) \Rightarrow \hat{H}^{p+q}(G; M)$$

where Σ_p is a set of representatives for the p-cells of X mod G, and M is a $\mathbb{Z}G$-module.

The spectral sequence is obtained as in the case of Farrell cohomology from the double complex

$$\mathrm{Hom}_{\mathbb{Z}G}(\mathcal{F}, C^*(X; M)),$$

where $\mathcal{F}$ denotes a complete resolution of G and $C^*(X; M)$ the cellular cochain complex of X with coefficients in M and diagonal G-action. There is no need here to assume that the stabilizers G_σ be finite; however, we will mainly be interested in that case. The *finite* convergence results from the assumption that X be finite dimensional. This spectral sequence is discussed in [19]. An analysis of the first differential leads to the following useful result. For any cell $\sigma \subset X$ let

$$c(g^{-1})^* : \hat{H}^\bullet(G_\sigma; M) \to \hat{H}^\bullet(G_{g\sigma}; M)$$

be the isomorphism induced by conjugation with $g \in G$ and put $c(g^{-1})^*(u) = g \cdot u$. It follows that $E_2^{0,q}$ in 3.1 can be identified with the subgroup of *compatible families* in

$$\prod_{v \in X_0} \hat{H}^q(G_v; M), \quad X_0 \text{ the set of vertices of } X,$$

that is, the families (u_v) satisfying the following conditions:

- $g u_v = u_{gv}$ for all $g \in G$ and $v \in X_0$
- if v and w are vertices of a 1-cell σ of X, then u_v and u_w restrict to the same element of $\hat{H}^q(G_\sigma; M)$.

Since all groups in $\mathbf{H}_1\mathfrak{F}$ satisfy spli $< \infty$ we can argue as in the proof of Proposition 4.4, Chapter X of [2] to obtain the following.

Theorem 3.2. *Let X be a finite dimensional contractible G-CW-complex with finite stabilizers such that for every finite subgroup $H < G$ the fixed point set X^H is non-empty and connected. Let $\mathfrak{F}$ stand for the set of finite subgroups of G and write*

$$\mathcal{H}^q(G; M) \subset \prod_{H \in \mathfrak{F}} \hat{H}^q(H; M)$$

for those families $(u_H)_{H \in \mathfrak{F}}$ which are compatible with respect to the restriction maps $\hat{H}^q(H; M) \to \hat{H}^q(K; M)$ induced by embeddings $K \rightarrowtail H$ given by conjugation by elements of G. Then the E_2-term of the associated stabilizer spectral sequence 3.1 satisfies

$$E_2^{0,q} \cong \mathcal{H}^q(G; M).$$

Note that the theorem applies in particular to groups which admit a finite dimensional classifying space for proper actions $\underline{E}G$ (the definition of $\underline{E}G$ was recalled in the Introduction).

Corollary 3.3. *Suppose G admits a G-CW-complex X as in the previous theorem and let p be a prime number. Then the natural map induced by restricting to finite subgroups*

$$\rho : \hat{H}^\bullet(G; \mathbb{Z}/p\mathbb{Z}) \to \mathcal{H}^\bullet(G; \mathbb{Z}/p\mathbb{Z}) \subset \prod_{H \in \mathfrak{F}} \hat{H}^\bullet(H; \mathbb{Z}/p\mathbb{Z})$$

has the property that every element in the kernel of ρ is nilpotent, and that for any $u \in \mathcal{H}^\bullet(G; \mathbb{Z}/p\mathbb{Z})$ there is an integer k such that u^{p^k} lies in Im ρ (i.e. ρ is an F-isomorphism).

The proof is exactly the same as the one for Proposition 4.6 of Chapter X in [2].

We will also make use of the following consequence of 3.1.

Theorem 3.4. *Let G be a group in $\mathbf{H}_1\mathfrak{F}_b$ and let n be a positive integer such that the order of any torsion subgroup of G divides n. Then there is an integer k such that*

$$n^k \cdot \hat{H}^i(G; M) = 0$$

for all i and all $\mathbb{Z}G$-modules M. Moreover, if p is a torsion prime for G then $\hat{H}^0(G; \mathbb{Z})$ contains an element of order p.

Proof. Let n be a positive integer such that the order of every torsion subgroup of G divides n. Choose a contractible finite dimensional G-CW-complex X with finite stabilizers. Then the E_1-term of the associated stabilizer spectral sequence is annihilated by n. It follows that every $\hat{H}^i(G; M)$ is annihilated by

n^k, where $k = \dim X + 1$. It remains to show that the torsion group $\hat{H}^0(G; \mathbb{Z})$ contains an element of order p if p is a torsion prime for G. If $\mathbb{Z}/p\mathbb{Z} < G$ then the restriction map

$$\hat{H}^0(G; \mathbb{Z}) \to \hat{H}^0(\mathbb{Z}/p\mathbb{Z}; \mathbb{Z}) = \mathbb{Z}/p\mathbb{Z}$$

is surjective since it maps 1 to 1 and the claim follows. $\qquad\square$

Remark 3.5. If there is no bound on the order of finite subgroups of G, then $\hat{H}^0(G; \mathbb{Z})$ is not a torsion group, because $1 \in \hat{H}^0(G; \mathbb{Z})$ restricts to a generator of $\hat{H}^0(H; \mathbb{Z}) \cong \mathbb{Z}/|H|\mathbb{Z}$ for every finite $H < G$. On the other hand if $\hat{H}^0(G; \mathbb{Z})$ is torsion then all generalized Tate groups of G with coefficients in any $\mathbb{Z}G$-module are torsion, annihilated by the characteristic of the ring $\hat{H}^0(G; \mathbb{Z})$.

Note that for G as in 3.4, one has

$$\hat{H}^{\bullet}(G; \mathbb{Z}) \cong \prod_p \hat{H}^{\bullet}(G; \mathbb{Z})_{(p)} \cong \bigoplus_p \hat{H}^{\bullet}(G; \mathbb{Z})_{(p)},$$

where $\hat{H}^{\bullet}(G; \mathbb{Z})_{(p)}$ stands for the p-primary part. This p-primary part can sometimes be computed up to F-*isomorphism* by passing to Tate cohomology with coefficients in $\mathbb{Z}/p\mathbb{Z}$.

Lemma 3.6. *Let* $G \in \mathbf{H}_1\mathfrak{F}_b$ *and* p *a prime number. Then the natural map*

$$\alpha : \hat{H}^{\bullet}(G; \mathbb{Z})_{(p)} \to \hat{H}^{\bullet}(G; \mathbb{Z}/p\mathbb{Z})$$

has the property that every element in the kernel of α *is nilpotent, and for any* $u \in \hat{H}^{\bullet}(G; \mathbb{Z}/p\mathbb{Z})$ *there is an integer* k *such that* u^{p^k} *lies in* $\operatorname{Im}\alpha$ *(i.e. the map* α *is an* F-*isomorphism).*

The proof is the same as the one for Lemma 6.6, Chapter X in [2].

In view of our applications it is convenient to make use the following fact on groups in $\mathbf{H}\mathfrak{F}_b$, which is an easy consequence of the main theorem of [15].

Proposition 3.7. *For groups* $G \in \mathbf{H}\mathfrak{F}_b$ *the following are equivalent:*

(i) $\operatorname{spli} G < \infty$
(ii) G *admits a finite dimensional* $\underline{E}G$
(iii) $G \in \mathbf{H}_1\mathfrak{F}_b$.

Proof. (i) $\Rightarrow$ (ii): Let $\kappa(G)$ be the supremum over the projective dimension of those $\mathbb{Z}G$-modules which have finite projective dimension when restricted to any finite subgroup of G. It was shown in [6, Theorem C] that for G in $\mathbf{H}\mathfrak{F}$ one has $\kappa(G) = \operatorname{spli} G$. On the other hand, it is well known that for an arbitrary group G the module $B(G, \mathbb{Z})$ of bounded functions $G \to \mathbb{Z}$, is free over $\mathbb{Z}H$ for any finite subgroup $H < G$, (cf. [16]). Thus if $G \in \mathbf{H}\mathfrak{F}_b$ satisfies $\operatorname{spli} G < \infty$ then $\operatorname{proj.dim}_{\mathbb{Z}G} B(G, \mathbb{Z}) < \infty$ and therefore G admits by [15] a finite dimensional $\underline{E}G$.

(ii) $\Rightarrow$ (iii): This is clear since we assume that the torsion subgroups of G have bounded order.

(iii) $\Rightarrow$ (i): As observed in 2.6, all groups in $\mathbf{H}_1\mathfrak{F}$ satisfy spli $< \infty$. $\square$

By combining 3.3 with 3.6 and 3.7 we obtain the following

Corollary 3.8. *Let* $G \in \mathbf{H}\mathfrak{F}_b$ *with* spli $G < \infty$. *Then the natural map*

$$\hat{H}^\bullet(G; \mathbb{Z}) \to \prod_p \mathcal{H}^\bullet(G; \mathbb{Z}/p\mathbb{Z})$$

is an F-isomorphism.

The following lemma is useful for recognizing whether a group belongs to $\mathbf{H}_1\mathfrak{F}_b$.

Lemma 3.9. *Let X be a finite dimensional contractible G-CW-complex. Then the following holds:*

- *if X/G is compact and every cell stabilizer belongs to $\mathbf{H}_1\mathfrak{F}_b$ then $G \in \mathbf{H}_1\mathfrak{F}_b$*
- *if all cell stabilizers are finite of order dividing some fixed integer $n > 0$, then the order of every finite subgroup of G divides n and G belongs to $\mathbf{H}_1\mathfrak{F}_b$.*

Proof. If X/G is compact and $G_\sigma \in \mathbf{H}_1\mathfrak{F}_b$ is a cell stabilizer, then $\underline{\mathrm{cd}}\, G_\sigma < \infty$. Since

$$\underline{\mathrm{cd}}\, G \leq \dim X + \sup_\sigma \{\underline{\mathrm{cd}}\, G_\sigma\}$$

and the number of G orbits of cells is finite, we conclude that $\underline{\mathrm{cd}}\, G$ and thus spli G is finite and clearly $G \in \mathbf{H}\mathfrak{F}$. To check that the orders of the finite subgroups of G are bounded, it suffices to check that the orders of the p-subgroups are bounded by a bound independent of p. If $P < G$ is a finite p-subgroup of G then the fixed-point space X^P is not empty (cf. [2, Theorem 10.5, Chapter VIII]) which implies that $P < G_\sigma$ for some cell σ. But by assumption the order of the finite subgroups of each G_σ is bounded and, as X/G is compact, there are only finitely many G_σ's up to isomorphism. This implies that there is a universal bound independent of p for the order of the finite p-subgroups $P < G$. It follows that $G \in \mathbf{H}_1\mathfrak{F}_b$ by 3.7.

Next, if the order of every cell stabilizer divides n, then the order of every finite p-subgroup $P < G$ divides n, because P is a subgroup of some cell stabilizer. As a result the order of every finite subgroup of G divides n. $\square$

Corollary 3.10. *The class of groups belonging to $\mathbf{H}_1\mathfrak{F}_b$ is extension closed. Moreover, if G is the fundamental group of a finite graph of groups in $\mathbf{H}_1\mathfrak{F}_b$ then $G \in \mathbf{H}_1\mathfrak{F}_b$. In particular, $\mathbf{H}_1\mathfrak{F}_b$ is closed under amalgamated free products and HNN-extensions.*

Proof. Let $K \rightarrowtail G \twoheadrightarrow Q$ be an extension with K and Q in $\mathbf{H}_1 \mathfrak{F}_b$. Then $G \in \mathbf{H} \mathfrak{F}_b$ since $\mathbf{H} \mathfrak{F}$ is extension closed. But by a general fact $\mathrm{spli}\, G \leq \mathrm{spli}\, K + \mathrm{spli}\, Q$ (cf. [8, Theorem 5.5]) and therefore $G \in \mathbf{H}_1 \mathfrak{F}_b$ by 3.7.

Next, if G is the fundamental group of a finite graph of groups in $\mathbf{H}_1 \mathfrak{F}_b$ then G acts cocompactly on a tree T with stabilizers in $\mathbf{H}_1 \mathfrak{F}_b$ so that $G \in \mathbf{H}_1 \mathfrak{F}_b$ by 3.9. $\qquad\square$

Remark 3.11. An interesting group in $\mathbf{H} \mathfrak{F}_b$ which is not of finite vcd (because it is an infinite torsion group), is the Burnside group $B(d, e)$ of odd exponent $e > 665$ on d generators. It is known that there is a 2-dimensional contractible $B(d, e)$-CW-complex with all non-trivial stabilizers cyclic of order e (e.g. [18]). By 3.9, $B(d, e)$ belongs to $\mathbf{H}_1 \mathfrak{F}_b$. We will come back to this example in the next section.

4. Periodicity in Cohomology and Units

The existence of periodic resolutions is closely related to the existence of units of non-zero degree in generalized Tate cohomology. The precise relationship is as follows.

Theorem 4.1. *Let G be an arbitrary group. Then $\hat{H}^\bullet(G; \mathbb{Z})$ contains a unit of non-zero degree if and only if G admits a periodic resolution.*

Proof. Let

$$\ldots \to P_i \to P_{i-1} \to \ldots P_0 \to \mathbb{Z}$$

be a projective resolution and put $\Omega^i = \mathrm{Ker}(P_i \to P_{i-1})$. Let $u \in \hat{H}^k(G; \mathbb{Z})$ be a unit for some $k > 0$. According to [17] one has for all $n \in \mathbb{Z}$

$$\hat{H}^n(G; \mathbb{Z}) \cong \varinjlim_{j \geq |n|} [\Omega^{j+n}, \Omega^j]$$

where $[M, N]$ stands for $\mathrm{Hom}_{\mathbb{Z}G}(M, N)/S$, with S the group of $\mathbb{Z}G$-module homomorphisms $M \to N$ which factor through a projective module. By choosing j large enough we can thus represent u and u^{-1} by module maps $\Omega^{j+k} \to \Omega^j$ and $\Omega^j \to \Omega^{j+k}$ such that the composites of these two maps are equal to identity maps modulo maps which factor through projectives. But this implies that there is a projective module P such that

$$\Omega^j \oplus P \cong \Omega^{j+k} \oplus P.$$

Clearly, this implies that G admits a periodic resolution. For the converse we may assume that there are inverse isomorphisms

$$f : \Omega^{j+k} \to \Omega^j, \qquad g : \Omega^j \to \Omega^{j+k}$$

for some $j \geq 0$ and some $k > 0$. Then f and g represent inverse units in $\hat{H}^\bullet(G; \mathbb{Z})$ of non-zero degree. We used here the fact that the product xy of $x, y \in \hat{H}^\bullet(G; \mathbb{Z})$ is represented by $f \circ g$, if $g : \Omega^{m+r+s} \to \Omega^{m+r}$ resp. $f : \Omega^{m+r} \to \Omega^m$ represent y resp. x for some large m [14]. $\qquad\square$

Remark 4.2. It is conceivable that all groups G which admit complete resolutions actually satisfy spli $G < \infty$.

We will make use repeatedly of the following well-known property of F-isomorphisms (see for instance the proof of Proposition 6.1, Chapter X in [2]).

Lemma 4.3. *Let $\phi : R \to S$ be an F-isomorphism of rings with 1. If $u \in S$ is a unit, then there is a $k > 0$ such that $u^k = \phi(v)$ for some unit $v \in R$.*

The following result permits us, in some cases, to detect units in generalized Tate cohomology of G by looking at finite subgroups. If p is a prime we say that a finite group H has p-periodic cohomology, if the (trivial) $\mathbb{Z}H$-module $\mathbb{Z}/p\mathbb{Z}$ has a periodic resolution. This is equivalent to the existence of $q > 0$ such that the functors $H^i(H; - \otimes \mathbb{Z}/p\mathbb{Z})$ and $H^{i+q}(H; - \otimes \mathbb{Z}/p\mathbb{Z})$ are equivalent for all $i > 0$; it is well-known that in this situation the minimal such $q > 0$ divides $2(p-1)$ for p odd, or 4 for $p = 2$ (cf. [21]).

Theorem 4.4. *Let p be a prime and let X be a finite dimensional contractible G-CW-complex with finite stabilizers such that for all finite subgroups $H < G$ the fixed point spaces X^H are non-empty and connected. Then the following statements are equivalent:*

(i) *the ring $\hat{H}^\bullet(G; \mathbb{Z}/p\mathbb{Z})$ contains a unit of non-zero degree*

(ii) *every finite subgroup H of G has p-periodic cohomology.*

Proof. (i) $\Rightarrow$ (ii): Let x be an invertible element of $\hat{H}^\bullet(G; \mathbb{Z}/p\mathbb{Z})$ of some positive degree q. If H is a finite subgroup of G then since the restriction map

$$\hat{H}^\bullet(G; \mathbb{Z}/p\mathbb{Z}) \to \hat{H}^\bullet(H; \mathbb{Z}/p\mathbb{Z})$$

is a morphism of rings with 1, the image of x is a unit of $\hat{H}^\bullet(H; \mathbb{Z}/p\mathbb{Z})$ of degree q and this is equivalent to H having p-periodic cohomology with p-period q (cf. [2], Theorem 9.7, Chapter VI).

(ii) $\Rightarrow$ (i): Let $q = 2(p-1)$ for p odd or $q = 4$ for $p = 2$, which is a p-period for every finite subgroup H of G. Choose for each H a generator

$$v_H \in \hat{H}^q(H; \mathbb{Z}/p\mathbb{Z}) \cong \mathbb{Z}/p\mathbb{Z} \otimes \mathbb{Z}/|H|\mathbb{Z}.$$

Note that v_H is unique up to a unit in the ring $\mathbb{Z}/p\mathbb{Z}$. Therefore $u = (u_H)_{H \in \mathcal{F}}$ with $u_H = v_H^{p-1}$ for all H defines a compatible family and thus a unit in $\mathcal{H}^\bullet(G; \mathbb{Z}/p\mathbb{Z})$ of degree q^{p-1}. Since by 3.3

$$\rho : \hat{H}^\bullet(G; \mathbb{Z}/p\mathbb{Z}) \to \mathcal{H}^\bullet(G; \mathbb{Z}/p\mathbb{Z})$$

is an F-isomorphism the result follows from 4.3. $\qquad\square$

Definition 4.5. Let G be an arbitrary group. Then the *finitistic dimension* fin. dim G is the supremum of the projective dimension of all $\mathbb{Z}G$-modules of finite projective dimension.

Lemma 4.6. *Let G be a group such that either* $\operatorname{spli} G < \infty$ *or* $G \in \mathbf{H\mathfrak{F}}$. *Then* $\operatorname{fin.dim} G = \operatorname{spli} G$.

Proof. Without any assumtion on G one has $\operatorname{fin.dim} G \leq \operatorname{silp} G$. Indeed, if A is a module of finite projective dimension d, then $\operatorname{Ext}^d(A, P) \neq 0$ for some projective module P. It follows that the injective dimension of P is at least d. According to [8] for any G one has $\operatorname{silp} G \leq \operatorname{spli} G$. If $\operatorname{spli} G = k < \infty$, then there exists an injective module of projective dimension k, thus $\operatorname{fin.dim} G \geq \operatorname{spli} G$. It follows that for groups with $\operatorname{spli} G < \infty$ one has $\operatorname{fin.dim} G = \operatorname{spli} G$. If G is an arbitrary group in $\mathbf{H\mathfrak{F}}$ the $\operatorname{fin.dim} G = \operatorname{spli} G$ by [6, Theorem C]. $\qquad\square$

Lemma 4.7. *Let G be a group with periodic cohomology after k-steps. Then* $\operatorname{fin.dim} G \leq k + 1$.

Proof. By assumption there exist $k > 0$ and $q > 0$ such that the functors $H^i(G; \ )$ and $H^{i+q}(G; \ )$ are equivalent for all $i > k$. We claim that then $\operatorname{fin.dim} G \leq k+1$. Indeed if M is a $\mathbb{Z}G$-module of finite projective dimension $m + 1 > 0$, the ΩM, the kernel of a surjection $P \to M$ with P projective, has projective dimension m. Therefore there exists a $\mathbb{Z}G$-module A such that $\operatorname{Ext}^m(\Omega M, A) \neq 0$. But if we had $m > k$, then

$$\operatorname{Ext}^m(\Omega M, A) \cong \operatorname{Ext}^{m+q}(\Omega M, A) \neq 0$$

which is in contradiction with $\operatorname{proj.dim} \Omega M = m$. It follows that $\operatorname{fin.dim} G \leq k + 1$. $\qquad\square$

By combining the two previous results we conclude the following.

Corollary 4.8. *Let $G \in \mathbf{H\mathfrak{F}}$ and assume that G admits a periodic resolution. Then* $\operatorname{spli} G < \infty$.

By putting together some of our previous results we obtain

Theorem 4.9. *Suppose G has periodic cohomology after k-steps. Then the following holds:*

(i) *$H^i(G; P) \neq 0$ for some $i \leq k$ and some projective $\mathbb{Z}G$-module P; moreover $\operatorname{fin.dim} G \leq k+1$ and every subgroup $H < G$ of finite cohomological dimension satisfies $\operatorname{cd} H \leq k$*

(ii) *if $G \in \mathbf{H\mathfrak{F}}$ then $H^i(G; P) = 0$ for $i > k + 1$, $\operatorname{cd}_{\mathbb{Q}} G \leq k + 1$ and every torsion-free subgroup $H < G$ satisfies $\operatorname{cd} H \leq k$.*

Proof. (i): If G has periodic cohomology after k-steps then it admits a complete resolution of coincidence index k, thus $H^i(G; P) \neq 0$ for some $i \leq k$ and some projective P (cf. 2.9). That $\operatorname{fin.dim} G \leq k + 1$ follows from the proof of 4.7; moreover, since every subgroup $H < G$ has periodic cohomology after k steps too, it follows that if $\operatorname{cd} H < \infty$ then $\operatorname{cd} H \leq k$, because there is a complete resolution with coincidence index k.

(ii): If G is in $\mathbf{H}\mathfrak{F}$ then fin. dim $G = $ spli G (cf. 4.6), and obviously $H^i(G; P) = 0$ for $i > $ spli G and P projective, because $H^i(G; \ )$ vanishes on injectives for $i > 0$. Moreover, for any group G in $\mathbf{H}\mathfrak{F}$ one has

$$\text{cd}_{\mathbb{Q}} G = \text{spli}_{\mathbb{Q}} G \leq \text{spli} G = \text{fin. dim} G$$

and therfore $\text{cd}_{\mathbb{Q}} G \leq k + 1$ ($\text{spli}_{\mathbb{Q}} G$ is defined like spli G, but using $\mathbb{Q}G$-modules instead of $\mathbb{Z}G$-modules). Finally, if $H < G$ is a torsion-free subgroup, then H belongs to $\mathbf{H}_1\mathfrak{F}$, since spli $H \leq$ spli $G < \infty$. It then follows from (i) that $\text{cd}\, H \leq k$. $\qquad\square$

The following theorem corresponds to part of Theorem B of the Introduction.

Theorem 4.10. *Let* $G \in \mathbf{H}\mathfrak{F}_b$. *Then the following statements are equivalent and they all imply that* $G \in \mathbf{H}_1\mathfrak{F}_b$:

(i) *G has periodic cohomology after some steps*

(ii) *there exists an invertible element of non-zero degree in the ring* $\hat{H}^{\bullet}(G; \mathbb{Z})$.

Moreover, for $G \in \mathbf{H}_1\mathfrak{F}_b$ *the following is equivalent to* (i) *and* (ii):

(iii) *every finite subgroup of G has periodic cohomology.*

Proof. (i)⇔(ii): This holds for general G (cf. 4.1). Also, (i) and (ii) imply that G lies in $\mathbf{H}_1\mathfrak{F}_b$ because for groups in $\mathbf{H}\mathfrak{F}$ they imply that spli $G < \infty$ (cf. 4.8 and 3.7).

(i),(ii)⇒(iii): this is well-known ([2, Theorem 6.7, Chapter X]).

(iii)⇒(ii) (assuming that $G \in \mathbf{H}_1\mathfrak{F}_b$): in that case spli $G < \infty$ and it follows from 3.7 that G admits a finite dimensional $\underline{E}G$. We can thus apply 3.3, 3.6 and 4.4 to conclude (ii). $\qquad\square$

Remark 4.11. The following is an example of a group $G \in \mathbf{H}\mathfrak{F}_b$ satisfying (iii) but not (i) or (ii). Let $G = \oplus_{\mathbb{N}}\mathbb{Z}$. Clearly $G \in \mathbf{H}\mathfrak{F}_b$ and it satisfies (iii). But G does not satisfy (i) because of 2.10.

There is also an example of a group $K \in \mathbf{H}_1\mathfrak{F}$ such that (iii) does not imply (i) (see 2.2 in [23]). The group K is given as the fundamental group of a certain graph of finite cyclic p-groups for a fixed prime p. Note that it follows from 4.4 that $\hat{H}^{\bullet}(K; \mathbb{Z}/p\mathbb{Z})$ has an invertible element of non-zero degree.

On the other hand there are also groups in $\mathbf{H}_1\mathfrak{F} \setminus \mathbf{H}_1\mathfrak{F}_b$ such that (iii) does imply (i). For instance it was shown in [23] that a countable locally finite group has period q after 1-step if and only if all its finite subgroups have period q.

The following example illustrates 4.10

Lemma 4.12. *Let $B(d, e)$ be the Burnside group on d generators and of odd exponent $e > 665$. Then $B(d, e) \in \mathbf{H}_1\mathfrak{F}_b$ and*

(i) *$B(d, e)$ has periodic cohomology after 2-steps*

(ii) $H^i(B(d,e); \quad) \cong \prod_{\mathbb{N}} H^i(\mathbb{Z}/e\mathbb{Z}; \quad)$ *for all* $i > 2$

(iii) $\hat{H}^\bullet(B(d,e); \quad) \cong \prod_{\mathbb{N}} \hat{H}^\bullet(\mathbb{Z}/e\mathbb{Z}; \quad)$

(iv) *all finite subgroups of $B(d,e)$ are cyclic of order dividing e.*

Proof. We know already that $B(d,e) =: G$ belongs to $\mathbf{H}_1\mathfrak{F}_b$ (3.11). It was shown by Ivanov [11] that G has a presentation with associated relation module of the form $\oplus_{n\in\mathbb{N}}\mathbb{Z}[G/G_n]$ with each G_n cyclic of order e. It follows that

$$H^i(G; \quad) \cong \prod_{\mathbb{N}} H^i(\mathbb{Z}/e\mathbb{Z}; \quad), \qquad \text{for all} \quad i > 2.$$

This implies in particular that G has period 2 after 2-steps and that $H^i(G; \quad)$ vanishes on projectives for $i > 2$, implying (i), (ii) and (iii). It follows that $\hat{H}^2(G;\mathbb{Z})$ contains a unit and therfore, by restricting to finite subgroups, we see that every finite subgroup has periodic cohomology of period two. But finite groups of period two are known to be cyclic. Moreover, the order of the finite subgroups of G divides the exponent of $\hat{H}^0(G;\mathbb{Z})$, which is e and (iv) follows. $\qquad\square$

5. Free actions on finite dimensional homology spheres

Proposition 5.1. *Let G be a group which admits a finite dimensional free G-CW-complex X such that $H^\bullet(X;\mathbb{Z}) \cong H^\bullet(S^n;\mathbb{Z})$. Then*

(i) *there is a finite dimensional free G-CW-complex Y homotopy equivalent to S^{2n+1} such that G acts trivially on $H^{2n+1}(S^{2n+1};\mathbb{Z})$*

(ii) *there is an exact sequence*

$$0 \to \mathbb{Z} \to A \to P_{2n} \to \ldots \to P_0 \to \mathbb{Z} \to 0$$

with P_i projective $\mathbb{Z}G$-modules and proj. $\dim_{\mathbb{Z}G} A < \infty$.

Proof. (i): The join $Y = X * X$ with diagonal G-action has the homotopy type of S^{2n+1}, with homologically trivial G-action.

(ii): Take Y as in (i) and write $\{C_n, d_n \mid n \in \mathbb{N}\}$ for its cellular chain complex. Since $H^\bullet(C_*) \cong H^\bullet(S^{2n+1};\mathbb{Z})$ we obtain the following exact sequence of $\mathbb{Z}G$-modules

$$0 \to \operatorname{Ker} d_{2n+1} \to C_{2n+1} \to C_{2n} \to \cdots \to C_0 \to \mathbb{Z} \to 0$$

and a push-out square

$$
\begin{array}{ccc}
\operatorname{Ker} d_{2n+1} & \longrightarrow & C_{2n+1} \\
\tau \downarrow & & \sigma \downarrow \\
H^{2n+1}(Y;\mathbb{Z}) & \overset{\rho}{\longrightarrow} & A
\end{array}
$$

Then since $\operatorname{Ker}\sigma = \operatorname{Ker}\tau = \operatorname{Im} d_{2n+2}$, we obtain exact sequences

$$0 \to \mathbb{Z} \to A \to C_{2n} \to \cdots \to C_0 \to \mathbb{Z} \to 0$$

and
$$0 \to C_{\dim Y} \to \cdots \to C_{2n+1} \to A \to 0,$$
which proves (ii). □

Corollary 5.2. *Let G be an arbitrary group and consider the following statements:*

(1) *G acts freely and properly on $\mathbb{R}^n \times S^m$ for some $n \geq 0$ and some $m \geq 1$*

(2) *there is an exact sequence*
$$0 \to \mathbb{Z} \to A \to P_{q-2} \to \cdots P_0 \to \mathbb{Z} \to 0$$

with P_i projective and proj. $\dim_{\mathbb{Z}G} A < \infty$

(3) *G has periodic cohomology after some steps,* spli $G < \infty$ *and $H^i(G; P) = 0$ for P $\mathbb{Z}G$-projective and $i > 1 + $ proj. $\dim A$.*

Then (1)$\Rightarrow$(2)$\Rightarrow$(3).

Proof. (1) $\Rightarrow$ (2): Write X for $\mathbb{R}^n \times S^m$ with the given G-action. Since X/G is a topological manifold, it has the homotopy type of a CW-complex Z of dimension equal to the dimension of X/G. The covering space of Z associated with $\pi_1(X) < \pi_1(Z)$ is a finite dimensional G-CW-complex G-homotopy equivalent to X. Thus (2) follows from the previous proposition.

(2) $\Rightarrow$ (3): Clearly, (2) implies that
$$\operatorname{Ext}_{\mathbb{Z}G}^{i+q}(\mathbb{Z}, \ ) \cong \operatorname{Ext}_{\mathbb{Z}G}^i(\mathbb{Z}, \ )$$
for $i > $ proj. $\dim_{\mathbb{Z}G} A$ so that G has periodic cohomology with period q after (proj. $\dim_{\mathbb{Z}G} A$)-steps. It remains to prove that spli $G < \infty$. Let I be an injective $\mathbb{Z}G$-module. Note that the inclusion $\mathbb{Z} \to A$ in (2) is $\mathbb{Z}$-split. Therefore, upon tensoring with I, we obtain an injective map $I \to A \otimes I$, which is $\mathbb{Z}G$-split, because I is injective. Thus
$$\text{proj. } \dim I \leq \text{proj. } \dim A \otimes I \leq 1 + \text{proj. } \dim A$$
which shows that spli $G \leq 1 + $ proj. $\dim A$. Since for general G one has silp $G \leq$ spli G we conclude that for $i > 1 + $ proj. $\dim A$ and P projective the cohomology groups $H^i(G; P)$ vanish. □

Corollary 5.3. *Let G be a group of finite cohomological dimension and $\mathbb{Z} < G$ an infinite cyclic normal subgroup such that $G/\mathbb{Z}$ lies in $\mathbf{H}\mathfrak{F}$. Then* $\operatorname{cd}_{\mathbb{Q}} G/\mathbb{Z} < \infty$.

Proof. Let Y be the universal cover of a finite dimensional $K(G, 1)$. Then $Y/\mathbb{Z}$ is a finite dimensional free $(G/\mathbb{Z})$-CW-complex and, as Y is contractible, $Y/\mathbb{Z}$ is a $K(\mathbb{Z}, 1)$ thus homotopy equivalent to S^1. By (ii) of 5.1 and making use of the implication (2) $\Rightarrow$ (3) of 5.2 we infer that the group $G/\mathbb{Z}$ has periodic cohomology after some steps. Therefore (ii) of 4.9 implies that $\operatorname{cd}_{\mathbb{Q}} G/\mathbb{Z} < \infty$. □

Clearly, if G acts freely and properly on some $\mathbb{R}^n \times S^m$ then G must be countable. Conversely, the following holds.

Lemma 5.4. *Let X be a finite dimensional free G-CW-complex homotopy equivalent to S^m and suppose that G is countable. Then there exists for some $n \geq 0$ a free and proper G action on $\mathbb{R}^n \times S^m$.*

Proof. Since X/G has countable homotopy groups and is a finite dimensional CW-complex, it has the homotopy type of a finite dimensional locally finite countable simplicial complex Z. Choose a regular neighborhood $S \supset Z$ of a simplicial embedding of Z into some Euclidean space $\mathbb{R}^N$. It follows that Z is homotopy equivalent to the open submanifold $S \subset \mathbb{R}^N$. Using the h-cobordism theorem it follows that the universal cover of $S \times \mathbb{R}^q$ for q large enough is diffeomorphic to $\mathbb{R}^n \times S^m$ for some $n \geq 0$ (see also the proof of Theorem A in [4]).   □

To prove Theorem B of the introduction, we need the following.

Proposition 5.5. *Let $G \in \mathbf{H}\mathfrak{F}_b$ and assume that $\hat{H}^\bullet(G;\mathbb{Z})$ contains a unit of degree $k > 0$. Then for some $n > 0$ there exists a finite dimensional free G-CW-complex E homotopy equivalent to S^{nk-1} admitting an orientable spherical fibration*

$$\xi : S^{nk-1} \to E \to \underline{E}G$$

whose Euler class $e(\xi)$ induces isomorphisms

$$- \cup e(\xi) : H^i(G; \quad) \to H^{i+nk}(G; \quad)$$

for large i.

Proof. From 4.10 we infer that $G \in \mathbf{H}_1\mathfrak{F}_b$ and it admits therefore a finite dimensional $\underline{E}G$ (3.7). Let $u \in \hat{H}^k(G;\mathbb{Z})$ be a unit, $k > 0$. Since $G \in \mathbf{H}_1\mathfrak{F}_b$, the natural map

$$H^i(G;\mathbb{Z}) \to \hat{H}^i(G;\mathbb{Z})$$

is an isomorphism for large i and we can choose $m > 0$ such that u^m is the image of some $x \in H^{km}(G;\mathbb{Z})$. Choose a finite dimensional model for $\underline{E}G$. Since the finite subgroups of G have bounded order, G contains only finitely many isomorphism classes of finite subgroups and we can proceed as in the proof of Proposition 2.4 and Lemma 2.5 of [4] to construct an orientable spherical fibration

$$\xi : S^{kml-1} \to E \to \underline{E}G$$

for some large enough l, with E a finite dimensional free G-CW-complex. The construction is such that the Euler class $e(\xi) \in H^{kml}(G;\mathbb{Z})$ is of the form $x^l + \nu$, where ν is a nilpotent element in the ring $H^\bullet(G;\mathbb{Z})$. It follows that

$$- \cup e(\xi) : H^i(G; \quad) \to H^{i+kml}(G; \quad)$$

is an equivalence for i large, because $e(\xi)$ maps to a unit in $\hat{H}^\bullet(G;\mathbb{Z})$.   □

We are now ready to prove the theorems mentioned in the introduction.

Proof of Theorem A: Let $G \in \mathbf{H}\mathfrak{F}_b$ be a countable group with periodic cohomology after some steps. Then, by 4.10, $\hat{H}^\bullet(G;\mathbb{Z})$ contains an invertible element of non-zero degree. From 5.5 we infer that there is a finite dimensional G-CW-complex homotopy equivalent to a sphere, and 5.4 then implies that there is a proper and free G-action on some $\mathbb{R}^n \times S^m$. $\qquad\square$

Proof of Theorem B: Let $G \in \mathbf{H}\mathfrak{F}_b$. The assertion (I) $\Rightarrow$ (II) is a consequence of 5.1; (II) $\Rightarrow$ (III) according to 5.2 and (III) $\Rightarrow$ (IV) because of 4.10; next, (IV) $\Rightarrow$ (I) by 5.5, and 4.10 shows that (IV) implies that G must belong to $\mathbf{H}_1\mathfrak{F}_b$. Finally, with the assumption that G is in $\mathbf{H}_1\mathfrak{F}_b$ the assertion (IV) $\Leftrightarrow$ (V) follows from 4.10. $\qquad\square$

Corollary 5.6. *Let G be an arbitrary group.*

(i) *If G contains a free abelian subgroup of infinite rank then G does not act freely and properly on any $\mathbb{R}^n \times S^m$. In particular, the Thompson group (2.10) does not act freely and properly on any $\mathbb{R}^n \times S^m$.*

(ii) *The Burnside group $B(d, e)$ on d generators and of odd exponent $e > 665$ acts freely and properly on some $\mathbb{R}^n \times S^m$.*

Proof. (i): If G contains a free abelian subgroup of infinite rank then G does not admit a complete resolution (2.10), and therefore cannot have periodic cohomology after some steps. The result now follows from 5.2.

(ii): By 4.12 the group $B(d, e)$ belongs to $\mathbf{H}_1\mathfrak{F}_b$ and has periodic cohomology after 2-steps. Since $B(d, e)$ is countable, Theorem A implies the assertion. $\qquad\square$

REFERENCES

[1] D. J. Benson and J. Carlson, *Products in negative cohomology*, J. Pure Appl. Algebra **82** (1992), 107–130.

[2] K. S. Brown, 'Cohomology of Groups', Graduate Texts in Math. **87** (Springer, Berlin, 1982).

[3] K. S. Brown and R. Geoghegan, *An infinite dimensional FP$_\infty$-group*, Invent. Math. **77** (1984), 367–381.

[4] F. Connolly and S. Prassidis, *On groups which act freely on $\mathbb{R}^m \times S^{n-1}$*, Topology **28** (1989), 133–148.

[5] J. Cornick and P. H. Kropholler, *On complete resolutions*, Topology and its Applications **78** (1997), 235–250.

[6] J. Cornick and P. H. Kropholler, *Homological finiteness conditions for modules over group algebras*, J. London Math. Soc. **58** (1998), 49-62.

[7] F. T. Farrell, *An extension of Tate cohomology to a class of infinite groups*, J. Pure Appl. Algebra **10** (1977), 153–161.

[8] T. V. Gedrich and K. W. Gruenberg, *Complete cohomological functors on groups*, Topology and its Application **25** (1987), 203–223.

[9] F. Goichot, *Homologie de Tate-Vogel equivariant*, J. Pure Appl. Algebra **82** (1992), 39–64.

[10] B. M. Ikenaga, *Homological dimension and Farrell cohomology*, Journal of Algebra **87** (1984), 422–457.

[11] S. V. Ivanov, *Relation modules and relation bimodules of groups, semigroups and associative algebras*, Internat. J. Algebra Comput. **1** (1991), no.1, 89–114.

[12] F. E. A. Johnson, *On groups which act freely on $\mathbb{R}^m \times S^n$*, J. London Math. Soc. **32** (1985), 370–376.

[13] P. H. Kropholler, *On groups of type FP$_\infty$*, J. Pure Appl. Algebra **90** (1993), 55–67.

[14] P. H. Kropholler, *Complete cohomology*, Birkhäuser, to appear.

[15] P. H. Kropholler and G. Mislin, *Groups acting on finite dimensional spaces with finite stabilizers*, Comment. Math. Helv. **73** (1998), 122–136.

[16] P. H. Kropholler and O. Talelli, *On a property of fundamental groups of graphs of finite groups*, J. Pure Appl. Algebra **74** (1991), 57–59.

[17] G. Mislin, *Tate cohomology for arbitrary groups via satellites*, Topology and its applications **56** (1994), 293–300.

[18] A. Yu. Ol'shanskii, *Geometry of Defining Relations in Groups*, Kluver Academic Publishers.

[19] S. Prassidis, *Groups with infinite virtual cohomological dimension which act freely on* $\mathbb{R}^m \times S^{n-1}$, J. Pure Appl. Algebra **78** (1992), 85–100.

[20] H. R. Schneebeli, *On virtual properties and group extensions*, Math. Z. **159** (1978), 159–167.

[21] R. G. Swan, *The p-period of a finite group*, Illinois J. Math. **4** (1960), 341–346.

[22] O. Talelli, *On cohomological periodicity for infinite groups*, Comment. Math. Helv. **55** (1980), 178–192.

[23] O. Talelli, *On groups with periodic cohomology after 1-step*, J. Pure Appl. Algebra **30** (1983), 85–93.

[24] O. Talelli, *On cohomological periodicity isomorphisms*, Bull. London Math. Soc. **28** (1996), 600–602.

[25] O. Talelli, *Periodicity in cohomology and free and proper actions on* $\mathbb{R}^n \times S^m$, to appear in the *Proceedings of Groups at St. Andrews 1997: Vol. 2*, Cambridge Univ. Press.

[26] B. B. Venkov, *On homologies of groups of units in division algebras*, Trudy Mat. Inst. Steklov **80** (1965), 66–89 (English translation: Proc. Steklov Inst. Math. **80** (1965), 73–100.)

[27] C. T. C. Wall, *The topological space-form problems*, Topology of Manifolds, (eds. J. C. Cantrell and C. H. Edwards, Jr.), Proc. of the Univ. of Georgia Topology of Manifolds Institute, 1996, Markham, Chicago (1970), 319–331.

[28] C. T. C. Wall, *Periodic projective resolutions*, Proc. London Math. Soc. **39** (1979), 509–533.

Guido Mislin
Dept. of Mathematics
ETH Zürich
8092 Zürich
Switzerland
mislin@math.ethz.ch

Olympia Talelli
Dept. of Mathematics
University of Athens
Athens 15784
Greece
otalelli@cc.uoa.gr

ON CONFINAL DYNAMICS OF ROOTED TREE AUTOMORPHISMS

V V NEKRASHEVYCH AND V I SUSHCHANSKY

ABSTRACT. We study automorphism groups of spherically homogeneous rooted trees which are defined by properties of their actions on the tree boundaries. We introduce the notions of finitary, weakly finitary, confinal, and biconfinal automorphisms. We prove that every automorphism of a rooted tree is conjugate to a weakly finitary automorphism of width not greater than 2, any finite subgroup is conjugate to some subgroup of weakly finitary automorphisms and any automorphism of an n-regular rooted tree is a product of two automorphisms of finite order not greater than $n!$. If n is an odd prime then we construct some new examples of 2-generated confinally transitive periodic subgroups of the tree automorphism group.

1. INTRODUCTION

Rooted tree automorphism groups have proven to be an important topic in group theory. Any countable residually finite group can be faithfully represented by automorphisms of a rooted tree [Su1]. Moreover, many interesting groups are naturally introduced directly as automorphism groups of rooted trees. The most famous examples are the Grigorchuk group [G] and the Gupta-Sidki groups [GS], [Si].

Our article investigates the actions of rooted tree automorphisms on the boundary of the tree. If the tree admits a level transitive automorphism group (such trees are called spherically homogeneous) then the boundary of the tree is naturally represented as an infinite Cartesian product $B_T = \prod_{i=1}^{\infty} X_i$ of finite sets, or in other words as a space of infinite sequences $(x_1, x_2, \dots)$ where $x_i \in X_i$.

In this interpretation the set of permutations induced by the automorphisms of the tree can be characterized as the set of all *triangular* permutations. The space B_T has a natural metric on it (the Baire metric) and is homeomorphic as a metric space to the Cantor set. A permutation of B_T is triangular if and only if it is an isometry for this metric. Thus the theory of rooted tree automorphisms (in the case of a regular tree) becomes equivalent to the theory of strictly triangular permutations of the set B_T, the theory of bijective sequential functions [R] and the theory of the Baire space isometries.

It is natural to introduce on the set of sequences B_T a relation of confinality. Two sequences are said to be confinal if they differ only in a finite number of coordinates, or in other words, if they become equal beginning from some place. Confinality is an equivalence relation and its equivalence classes we call *confinality classes*.

Orbits on the tree boundary for periodic groups coincide with their confinality classes, by [Su5], [G], [GS] . Moreover, any level transitive automorphism group of a rooted tree is conjugate to a group whose orbits on the boundary are unions of the confinality classes. Therefore, the investigation of the confinal structure gives a deeper insight into the dynamics of groups acting on rooted trees.

The first section *"Rooted trees and their automorphisms"* defines the basic notions. We introduce the notions of triangular permutations and of Baire metric, and prove the necessary technical propositions.

In the second section *"Confinality and weakly finitary automorphisms"* we introduce the confinality relation and the main classes of tree automorphisms related to this relation: finitary, weakly finitary, confinal and biconfinal automorphisms. An automorphism is weakly finitary if it moves a point of the boundary to a confinal point, it is confinal if it preserves the relation of confinality and is biconfinal if it and its inverse are confinal. The sets of finitary, weakly finitary and biconfinal automorphisms are groups but the set of confinal automorphisms is a semigroup.

The main results of this section are the following statements

Theorem 4. *Any automorphism of a spherically homogeneous rooted tree T is conjugate in the automorphism group $\mathcal{A}(T)$ to a weakly finitary automorphism changing in every infinite sequence (viewed as an element of the boundary) not more than two coordinates.*

Theorem 4 has a natural generalization for finite automorphism groups of the rooted tree T.

Theorem 6. *Any finite automorphism group of a spherically homogeneous rooted tree T is conjugate in the group $\mathcal{A}(T)$ to some subgroup of the group of weakly finitary automorphisms.*

In the section *"Regular rooted trees"* we apply our results to the case of a regular rooted tree. A rooted tree is said to be n-regular if any vertex is connected exactly with n vertices of the next level.

Theorem 4 in this case implies

Theorem 9. *Any automorphism of an n-regular rooted tree T is a product of two automorphisms of finite order, not greater than $n!$. In particular, any automorphism of a 2-regular rooted tree T is contained in some (finite or infinite) dihedral subgroup of the group $\mathcal{A}(T)$.*

We define a natural embedding of the automorphism group of a regular tree into the group of the weakly finitary automorphisms.

In this section also the notion of a finitely automatic automorphism of a regular rooted tree and its connection with the notion of a weakly finitary automorphism are discussed.

In the section *"Confinally transitive groups"* we introduce the notion of confinally transitive groups and we finish our paper with a theorem showing that weakly finitary automorphisms may be used to construct confinally transitive infinite finitely generated periodic groups.

Theorem 12. *Let T be a p-regular tree, where p is an odd prime. Suppose that π is a finitary automorphism changing in every sequence not more than the first k coordinates and that the cyclic group $\langle \pi \rangle$ acts transitively on the k-th level of the tree T. Then one can find a weakly finitary automorphism ρ of T changing in every sequence not more than one coordinate such that the group $G = \langle \pi, \rho \rangle$ is an infinite p-group.*

If the group $\langle \pi \rangle$ acts quasi-regularly on the n-th level for all $n \leq k$ then the automorphism ρ can be chosen in such a way that the group G will be confinally transitive.

The results were announced in [NS].

Acknowledgement. This paper was written during the visit of the second author to Freiburg University. The author would like to thank the DFG for financial support and the Mathematisches Institut of Freiburg University for its hospitality.

2. ROOTED TREES AND THEIR AUTOMORPHISMS

Let $T = (T, x_0)$ be a rooted tree with a root x_0. We arrange the vertices of the tree into a countable set of levels (spheres) and number the levels by nonnegative integers so that the root is a single element of the zero level and any vertex of the i-th level is connected with a single vertex of the previous $(i-1)$-st level.

The rooted tree T is said to be *spherically homogeneous* [B] if any vertex of the i-th level is connected with n_i vertices of the $(i+1)$-st level, where n_i does not depend on the vertex of the level.

The vertices of the i-th level and only they are at the distance i from the root of the tree in the usual metric of a graph (i.e. in the maximal metric for which the distance between the adjacent vertices is equal to 1). In other words, the i-th level is the sphere of radius i with center x_0.

In the paper only locally finite trees are considered. A tree is said to be *locally finite* if the valence of every vertex is finite.

Any locally finite spherically homogeneous rooted tree T is uniquely defined up to isomorphism by a sequence of positive integers:

$$\tau(T) = \langle n_1, n_2, \ldots \rangle,$$

where n_i is the number of the vertices of the i-th level connected with a fixed vertex of the $(i-1)$-st level $(i \in \mathbb{N})$. Such a sequence will be called *the type* of the tree T and the numbers n_i will be called its *spherical indices*.

From the definition it follows that the number of vertices on the sphere of radius i and center the point x_0 is equal to the product of the spherical indices $n_1 n_2 \cdots n_i$.

A *ray* of the tree T is an infinite path beginning in the root x_0. Any ray is defined uniquely by an infinite sequence of vertices. The set of all the rays coincides with the boundary of the rooted tree T. This boundary may be identified with a Baire metric space over a sequence of sets $\{X_1, X_2, \dots , \}$ for which the equality $|X_i| = n_i$ holds for all $i \in \mathbb{N}$. In other words, the boundary B_T of the tree T with a natural metric on it is isometric to the space of sequences $\bar{t} = (t_1, t_2, \dots)$ where $t_i \in X_i$ for all $i \in \mathbb{N}$ and the metric is defined by the equality

$$\rho(\bar{x}, \bar{x}) = 0; \quad \rho(\bar{x}, \bar{y}) = \left(\frac{1}{2}\right)^{\nu(\bar{x}, \bar{y})} \quad \text{for} \quad \bar{x} \neq \bar{y},$$

where $\nu(\bar{x}, \bar{y})$ is the minimal number k for which $x_k \neq y_k$.

In this case the n-th level can be identified with the set

$$L_n = X_1 \times X_2 \times \cdots \times X_n.$$

(See the picture below.)

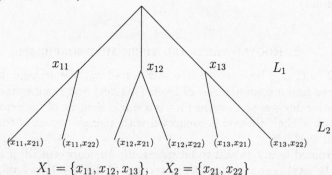

$$X_1 = \{x_{11}, x_{12}, x_{13}\}, \quad X_2 = \{x_{21}, x_{22}\}$$

The automorphism group of a spherically homogeneous rooted tree T of type $\langle n_1, n_2, \dots \rangle$ is isomorphic to the wreath product of an infinite sequence of symmetric groups $\langle S_{n_1}, S_{n_2}, \dots \rangle$ and acts on the boundary B_T as a full isometry group [Su1]. Some of the automorphism properties for the tree T are conveniently formulated in terms of transformations of the space B_T. Thereby a natural question arises: what transformations of this space are induced by tree automorphisms? An answer to this question is given in terms of triangular transformations.

Definition. A transformation π of the space B_T (of the set $L_n = \prod_{i=1}^{n} X_i$) is said to be triangular if for all $\bar{x} \in B_T$ (resp. $\bar{x} \in L_n$) and for every positive integer k (resp. every $1 \leq k \leq n$) the coordinate $(\overline{x}^\pi)_k$ depends only on the first k coordinates of $\overline{x}$.

Obviously, a transformation is triangular if and only if for any two sequences (finite or infinite) with a common beginning of length k their images also have a common beginning of length k. This is equivalent to the inequality $\rho(\overline{x}, \overline{y}) \geq \rho(\overline{x}^\pi, \overline{y}^\pi)$. We say that π is *strictly triangular* if it is an isometry of the space B_T. A transformation is strictly triangular if and only if for any $\overline{x}, \overline{y} \in B_T$ the length of the longest common beginning of $\overline{x}$ and $\overline{y}$ is equal to the length of the longest common beginning of $\overline{x}^\pi$ and $\overline{y}^\pi$.

Any permutation which is triangular on all the sets L_n induces in a natural way a triangular permutation of the space B_T.

A triangular transformation is not bijective if and only if it is not bijective on some L_n and thus not injective on L_n. Therefore a triangular transformation is bijective if and only if it is strictly triangular.

Let π be a triangular permutation. It is convenient to define the action of π on the sequences $(x_1, x_2, \dots)$ by a tableau [K]

$$[g_1, g_2(x_1), g_3(x_1, x_2), \dots],$$

where $g_1 \in Symm(X_1)$, and for $i = 2, 3, \dots$ $g_i \in Symm(X_i)^{L_{i-1}}$ are maps from L_{i-1} into the symmetric permutation groups of the set X_i. Then $(x_1, x_2, \dots)^\pi$ is equal to

$$\left(x_1^{g_1}, x_2^{g_2(x_1)}, x_3^{g_3(x_1, x_2)}, \dots \right).$$

Lemma 1. *A transformation of the space B_T (of the set L_n) is induced by an automorphism of the tree T if and only if it is strictly triangular.*

Proof. By construction, the vertex set of the tree is identified with the set of finite sequences $X^* = \cup_{n=1}^\infty L_n$, and a vertex $(x_1, x_2, \dots, x_n) \in L_n$ is connected with the vertex $(y_1, \dots, y_{n+1}) \in L_{n+1}$ if and only if $x_i = y_i$ for all $i = 1, \dots, n$.

Any automorphism of the rooted tree preserves the partition into the levels and the vertex adjacency. Thus it induces on the set X^* a triangular permutation and thus the permutation induced on the boundary is also triangular.

On the other hand, suppose π is a triangular permutation of the space B_T. Then the beginning of length n of the sequence $(x_1, \dots, x_n, x_{n+1}, x_{n+2}, \dots)^\pi$ does not depend on the sequence $(x_{n+1}, x_{n+2}, \dots)$. Thus we get a uniquely defined permutation of the set X^*. This permutation preserves the sets L_n, i.e., it preserves the levels of the tree and agrees with the relation of vertex adjacency. Thus it induces an automorphism of the tree T. $\square$

3. CONFINALITY AND WEAKLY FINITARY AUTOMORPHISMS

Definition. The sequences $\overline{x} = (x_1, x_2, \dots), \overline{y} = (y_1, y_2, \dots) \in B_T$ are said to be *s-confinal* if for all $k \geq s$ we have $x_k = y_k$.

The sequences $\overline{x}$ and $\overline{y}$ are said to be *confinal* if they are *s*-confinal for some number *s*.

The relation of s-confinality will be denoted by the symbol $\approx_s$ and that of confinality by $\approx$. The relations $\approx_s$ and $\approx$ are equivalence relations on the set B_T. The equivalence classes of the relation $\approx$ are called *confinality classes*.

Definition. An automorphism π of the tree T is called:

   a) *finitary*, if there exists a number s depending on π such that for any sequence $\overline{x} \in B_T$ we have $\overline{x}^\pi \approx_s \overline{x}$;

   b) *weakly finitary*, if for any $\overline{x} \in B_T$ we have $\overline{x}^\pi \approx x$;

   c) *confinal*, if it preserves the relation of confinality, i.e. for any $\overline{x}, \overline{y} \in B_T$ the relation $\overline{x} \approx \overline{y}$ implies $\overline{x}^\pi \approx \overline{y}^\pi$.

Denote by $\mathcal{AF}(T), \mathcal{AWF}(T), \mathcal{AC}(T)$ the sets of all finitary, weakly finitary and confinal automorphisms of the tree T respectively. Obviously, the following inclusions hold:

$$\mathcal{AF}(T) \subset \mathcal{AWF}(T) \subset \mathcal{AC}(T).$$

Lemma 2. *The set of finitary and the set of weakly finitary automorphisms are groups. The set $\mathcal{AC}(T)$ of confinal automorphisms is a semigroup but not a group.*

Proof. If g_1, g_2 are finitary automorphisms then for some positive integer s we have $\overline{x}^{g_1} \approx_s x$ and $\overline{x}^{g_2} \approx_s x$ for all $\overline{x} \in B_T$ (take the maximal s for the automorphisms g_i). Thus $\overline{x}^{g_1 g_2} \approx_s \overline{x}^{g_2} \approx_s \overline{x}$ for all $\overline{x} \in B_T$ and $g_1 g_2$ is finitary. If $\overline{x}^{g_1} \approx_s \overline{x}$ for all $\overline{x} \in B_T$ then $\overline{x}^{g_1^{-1}} \approx_s \overline{x}$. Therefore g_1^{-1} is also finitary.

The set of all weakly finitary automorphisms is a group by the same argument.

Let g_1, g_2 be confinal automorphisms. Then the relation $x \approx y$ implies $x^{g_i} \approx y^{g_i}$. Thus from the relation $x \approx y$ follows that $x^{g_1} \approx y^{g_1}$ and $x^{g_1 g_2} \approx y^{g_1 g_2}$, so the automorphism $g_1 g_2$ is confinal.

If an automorphism g is confinal then the automorphism g^{-1} is not necessary confinal. As an example, we may take the automorphism

$$g : (x_1, x_2, \dots) \mapsto (x_1, x_2 - x_1, x_3 - x_2, \dots), \quad ((x_1, x_2, \dots) \in B_T)$$

of a tree of the type $\langle n, n, \dots \rangle$, were the sets X_i used in the construction of B_T, are identified with the additive group $\mathbb{Z}_n$ of residues modulo n. Then the automorphism g will be confinal but its inverse:

$$g^{-1} : (x_1, x_2, \dots) \mapsto (x_1, x_1 + x_2, x_1 + x_2 + x_3, \dots)$$

is not confinal. (The transformations g and g^{-1} are obviously triangular and thus correspond to some automorphisms of the tree T.) $\square$

A tree automorphism $g \in \mathcal{AC}(T)$ is called *biconfinal* if $g^{-1} \in \mathcal{AC}(T)$. The set $\mathcal{AB}(T)$ of all biconfinal automorphisms is a subgroup of $\mathcal{A}(T)$ which can be characterized in the following way.

Theorem 3. *The group $\mathcal{AB}(T)$ of all biconfinal automorphisms of a spherically homogeneous tree T is the normalizer of the subgroup $\mathcal{AWF}(T)$ of all weakly finitary automorphisms in the full group of tree automorphisms $\mathcal{A}(T)$.*

Proof. Suppose $g \in \mathcal{AWF}(T)$ and let h be an arbitrary confinal automorphism. Then for any sequence $\overline{x} \in B_T$ we have

$$\overline{x}^{h^{-1}g} = \left(\overline{x}^{h^{-1}}\right)^g \approx \overline{x}^{h^{-1}}.$$

Thus

$$\overline{x}^{h^{-1}gh} = \left(\overline{x}^{h^{-1}g}\right)^h \approx \left(\overline{x}^{h^{-1}}\right)^h = \overline{x}$$

by the confinality of the automorphism h. Therefore $h^{-1}gh \in \mathcal{AWF}(T)$.

On the other hand, suppose h is an element of the normalizer of the subgroup $\mathcal{AWF}(T)$ in the group $\mathcal{A}(T)$. If $\overline{x} \approx \overline{y}$ then there exists an automorphism $g \in \mathcal{AWF}(T)$ such that $\overline{x}^g = \overline{y}$. Then $h^{-1}gh \in \mathcal{AWF}(T)$, so we have

$$\overline{y}^h = \overline{x}^{gh} = \overline{x}^{h(h^{-1}gh)} \approx \overline{x}^h.$$

So the relation $\overline{x} \approx \overline{y}$ implies $\overline{x}^h \approx \overline{y}^h$ and thus h is a confinal automorphism. In the same way we can prove that h^{-1} is confinal and thus any element of the normalizer is biconfinal. $\quad\square$

EXAMPLE. The image of $GL(n, \mathbb{Z})$ under the embedding into the automorphism group of a tree of the type $\langle 2^n, 2^n, \ldots \rangle$, constructed in [BS], is a subgroup of $\mathcal{AB}(T)$.

If a transformation g of the set B_T is weakly finitary then for any $\overline{x} \in B_T$ the number of coordinates where the sequences $\overline{x}$ and $\overline{x}^g$ differ is finite. We say that the automorphism g has *width k* if this number is less than or equal to k for any $\overline{x} \in B_T$ and equal to k for some sequence $\overline{x} \in B_T$.

The set of all weakly finitary automorphisms of finite width is a proper subgroup of the group $\mathcal{AWF}(T)$ containing the group $\mathcal{AF}(T)$.

Theorem 4. *Any automorphism of a spherically homogeneous rooted tree T is conjugate in the group $\mathcal{A}(T)$ to a weakly finitary automorphism of width ≤ 2.*

Proof. Let $\langle k_1, k_2, \ldots \rangle$ be the type of the tree T.

If $g \in \mathcal{A}(T)$ then *the orbit tree* of the automorphism g is the graph whose vertices are the vertex orbits of the cyclic group $\langle g \rangle$ and two orbits O_1, O_2 are connected by an edge if and only if there exist adjacent vertices $x_i \in O_i$. It is easy to see that this graph is a rooted tree. Let us label every orbit by a positive integer equal to the cardinality of the orbit. Such a labeled tree is called *the orbit type* of the automorphism. It is proved in [Su2] that two automorphisms are conjugate in $\mathcal{A}(T)$ if and only if they have isomorphic (as labeled graphs) orbit types.

One can prove that if two orbits O_1 and O_2 are adjacent and O_1 is of a smaller level then $|O_1|$ divides $|O_2|$. Thus a similar condition is fulfilled in the labelling of the orbit types of automorphisms.

Suppose $g \in \mathcal{A}(T)$. We shall construct the necessary automorphism $h \in \mathcal{A}(T)$ of width 2 by levels. This means that we are going to construct for every $n \geq 0$ a triangular permutation h_n of the set L_n such that h_n induces permutations of the levels $L_k, k \leq n$, equal to h_k and g_n is conjugate to h_n in the group of all triangular permutations of L_n. Here g_n is the permutation of the level L_n induced by g. If we construct such a sequence of permutations then we get a naturally defined automorphism h of the tree T which will be conjugate to g, since the orbit types of g and h will be isomorphic on every level, which implies that overall they are isomorphic.

We shall construct the sequence $\{h_n\}$ by induction on n. If we prove that the element x^{h_n} differs from x in not more than two coordinates for all $x \in L_n$ then the automorphism h will obviously be weakly finitary of width not more than 2. In order to proceed in our inductive constructions we will need additionally to require that in every orbit of h_n there exist at least two points $x \in L_n$ such that the elements x and x^{h_n} differ in at most one coordinate.

For $n = 1$ we get a standard situation in the symmetric group and the existence of the automorphism h_1 fulfilling all the necessary conditions can be easily checked (just take $h_1 = g_1$).

Suppose we have constructed h_n for some n and let us define h_{n+1}. Let O be any orbit of g on L_n. Let O' be the corresponding orbit in the orbit type of the automorphism h_n. It is easy to see that

$$\frac{|O_1| + \cdots + |O_l|}{|O|} = k_{n+1}$$

where $O_1, O_2, \ldots, O_l$ are all the orbits of the $(n + 1)$-st level adjacent to the orbit O. Let $X_{n+1} = U_1 \cup U_2 \cup \cdots \cup U_l$ be any partition such that $|U_i| = |O_i|/|O|$. If $|O_i| = |O|$ then we define the action of h_{n+1} on the elements $(\overline{x}, y)$ where $\overline{x} \in O'$ and $\{y\} = U_i$ by the rule

$$(\overline{x}, y)^{h_{n+1}} = \left(\overline{x}^{h_n}, y\right).$$

Then $\{(\overline{x}, y) : \overline{x} \in O_i, \{y\} = U_i\}$ will be an orbit of h_{n+1}. The automorphism h_{n+1} changes at most two coordinates in any element of the orbit and there exist at least two elements of the orbit such that h_{n+1} changes not more that one coordinate in each of them, because h_n has all of these properties.

If $|O_i| > |O|$ then let us define a cyclic permutation $\pi : y_i \mapsto y_{i+1}$ of the set $U_i = \{y_1 = y_{m_i+1}, y_2, \ldots, y_{m_i}\}$ and fix an element $\overline{x_0}$ of O' such that $(\overline{x_0})^{h_n}$ differs from $\overline{x_0}$ in not more than one coordinate. Then the action of h_{n+1} on the words $(\overline{x}, y)$ where $\overline{x} \in O'$ and $y \in U_i$ will be defined by the rule:

$$(\overline{x}, y)^{h_{n+1}} = \begin{cases} (\overline{x}^{h_n}, y) & \text{if } \overline{x} \neq \overline{x_0}; \\ (\overline{x}^{h_n}, y^{\pi}) & \text{if } \overline{x} = \overline{x_0}. \end{cases}$$

Then the sets $\{(\overline{x}, y) : \overline{x} \in O', y \in U_i\}$ are orbits of the defined automorphism h_{n+1} and the orbit types of the automorphisms h_{n+1} and g_{n+1} coincide. Let us prove that in any element $\overline{x} \in L_{n+1}$ the defined automorphism h_{n+1} changes at most two coordinates. If $\overline{x} = (\overline{x_0}, y)$ then $\overline{x}^{h_{n+1}} = (\overline{x_0}^{h_n}, y^{\pi})$, but h_n changes not more than one coordinate in $\overline{x_0}$. If $\overline{x} \neq (\overline{x_0}, y)$ then $\overline{x}^{h_{n+1}} = (\overline{x}^{h_n}, y)$ and thus h_{n+1} again changes at most two coordinates.

Recall that besides $\overline{x_0}$ there exists another element $\overline{x_1} \in O'$ such that h_n changes at most one coordinate in it. Then h_{n+1} changes not more than one coordinate in the elements $(\overline{x_1}, y_1)$ and $(\overline{x_1}, y_2)$ (we deal with the case $|U_i| > 1$ here, since the constructions for the other case was checked above). Thus all the conditions of the inductive construction are fulfilled. The theorem is proved. $\square$

Corollary 5. *Any conjugacy class of the group $\mathcal{A}(T)$ contains a weakly finitary element.*

Theorem 6. *Any finite automorphism group of a spherically homogeneous rooted tree T is conjugate in the group $\mathcal{A}(T)$ to some subgroup of $\mathcal{AWF}(T)$.*

Proof. Let G be a finite automorphism group of a spherically homogeneous rooted tree.

Let T' be the orbit tree of the group G (for the definition of the orbit tree see the proof of Theorem 4 or [S]). There exists a section $\phi : T' \to T$, i.e. such an embedding that $\pi(\phi(x)) = x$, where $\pi : T \to T'$ is the canonical surjective map. (This can be proved using Proposition 17 from the first chapter of [S].) Put $T_0 = \phi(T')$. Then every orbit of G on the vertex set of T has exactly one element from the subtree T_0.

For any sequence $\overline{x} = (x_1, x_2, \dots) \in B_T$ define the sequence of stabilizers

$$G = G_0(\overline{x}), G_1(\overline{x}), G_2(\overline{x}), \dots, G_n(\overline{x}), \dots$$

where $G_n(\overline{x})$ is the stabilizer of the sequence $(x_1, \dots, x_n)$ in the group G. Since the sequence $\{G_n(\overline{x})\}$ is descending (this follows from the triangularity condition) and the group G is finite, for some n we have $G_n(\overline{x}) = G_m(\overline{x})$ for all $m > n$. Let us fix the minimal number $n(\overline{x})$ with this property. Note that $n(\overline{x}) = 0$ if and only if the sequence $\overline{x}$ is a fixed point under the action of the group G on B_T.

If $g \in G$ then $n(\overline{x}^g) = n(\overline{x})$ since the stabilizers $G_k(\overline{x})$ and $G_k(\overline{x}^g)$ are conjugate in G.

Let us denote by $b(\overline{x})$ the beginning of $\overline{x} \in B_T$ of length $n(\overline{x})$ and put $\overline{x} = (b(\overline{x}), e(\overline{x}))$. If $n(\overline{x}) = 0$ then $b(\overline{x})$ is empty and $e(\overline{x}) = \overline{x}$. From triangularity and the identity $n(\overline{x}) = n(\overline{x}^g)$ follows that $b(\overline{x})^g = b(\overline{x}^g)$ for all $g \in G$.

The depth $d(\overline{x})$ of the sequence $\overline{x}$ is the number of all the beginnings of $b\,(\overline{x})$ which are equal to $b\,(\overline{y})$ for some $\overline{y} \in B_T$ and are shorter than $b\,(\overline{x})$. We shall define the action of the conjugating automorphism $h \in \mathcal{A}(T)$ on B_T by induction on depth.

If the depth of $\overline{x} \in B_T$ is equal to 0 then the action of h on $\overline{x}$ will be defined by the rule

$$\overline{x}^h = \left(b\,(\overline{x})\,, e\,(\overline{x}^g)\right),$$

where g is such an element of G that $b(\overline{x})^g \in T_0$.

If $g_1, g_2 \in G$ are such that both $b\,(\overline{x}^{g_1})$ and $b\,(\overline{x}^{g_2})$ belong to T_0 then $b\,(\overline{x}^{g_1}) = b\,(\overline{x}^{g_2})$ and $g_1 g_2^{-1}$ belongs to the stabilizer of $b\,(\overline{x})$. But this means that it stabilizes all the sequence $\overline{x}$. Thus $e\,(\overline{x}^{g_1}) = e\,(\overline{x}^{g_2})$ and the automorphism h is well defined.

Let us prove that the defined action of h on the set of the sequences $\overline{x} \in B_T$ of depth 0 is strictly triangular. Suppose $\overline{x} = (\overline{v}, \overline{w_1})$ and $\overline{y} = (\overline{v}, \overline{w_2})$ are two sequences having depth 0. Then either the common beginning $\overline{v}$ is shorter than both $b(\overline{x})$ and $b(\overline{y})$, or $b(\overline{x}) = b(\overline{y})$. Otherwise, one of the sequences $b(\overline{x})$ and $b(\overline{y})$ is a beginning of the other and this will contradict the definition of depth 0. But h acts trivially on $b(\overline{x})$ and $b(\overline{y})$, thus in the first case $(\overline{x})^h$ and $(\overline{y})^h$ have the longest common beginning $\overline{v}$ and thus h is triangular. In the second case triangularity follows from the fact that h is well defined.

If we have defined a triangular action of h on the set of all the elements of B_T having depth less than n then the action of h on the set of the sequences of depth n will be defined by the following rule. Suppose $d(\overline{x}) = n$. If $\overline{x}$ has a common beginning with a sequence $\overline{y}$ of depth less than n then we take among all such sequences $\overline{y}$ one having the longest common beginning $\overline{v}$ with $\overline{x}$. Then $b(\overline{x}) = (\overline{v}, \overline{v_1}, e\,(\overline{x}))$, where $(\overline{v}, \overline{v_1}) = b(\overline{x})$. Then we put

$$\overline{x}^h = \left((\overline{v})^h\,, \overline{v_1}, e\,(\overline{x}^g)\right),$$

where g is an element of G that $b(\overline{x})^g \in T_0$. It is well defined by the same arguments as for the case of depth 0. Note that the sequence $\overline{v_1}$ is nonempty, otherwise $\overline{v} = b(\overline{x})$. But in this case, if $\overline{v}$ is a beginning of $b(\overline{y})$ then $\overline{y}$ has depth greater than the depth of $b(\overline{x})$, but this is impossible. On the other hand, if $b(\overline{y})$ is a beginning of $\overline{v}$, then the stabilizer of $b(\overline{y})$ is equal to the stabilizer of $\overline{v}$ and thus equal to the stabilizer of $\overline{v}$: but this is a contradiction to the definition of $b(\overline{x})$ (to the minimality of $n(\overline{x})$). Thus h acts trivially on the last coordinate of $b(\overline{x})$.

If $\overline{x}$ has depth n and $\overline{y}$ has depth less than n and they have the longest common beginning of length k, then the images $\overline{x}^h$ and $\overline{y}^h$ also have a common beginning of length k by the definition of h and its triangularity for sequences of depth less than n. (Strict triangularity follows from the fact that h acts trivially on the last coordinates of $b(\overline{x})$.)

If both $\overline{x}$ and $\overline{y}$ have depth n and have a common beginning $\overline{v}$ of length k, then either $\overline{v}$ is shorter than both $b(\overline{x})$ and $b(\overline{y})$ or $b(\overline{x}) = b(\overline{y})$. (Otherwise, one of the sequences $b(\overline{x}), b(\overline{y})$ is a beginning of the other, but this will contradict the fact that they have equal depth.) In the first case $\overline{x}^h$ and $\overline{y}^h$ have a common beginning of length k, since h acts on the longest common beginning of $\overline{v}$ with sequences of depth less than n in the same way as on these sequences (and thus triangularly) and on the remaining part of $\overline{v}$ it acts trivially. In the second case we need additionally to use the fact that h is well defined. Thus the defined new action of h is strictly triangular.

Thus we get an inductive definition of a triangular permutation of the set B_T.

For any $\overline{x} \in B_T$ and $f \in G$ we have

$$\overline{x}^{fh} = \left(b\left(\overline{x}\right)^{fh}, e\left(\overline{x}^{fg_1}\right) \right)$$

where $g_1 \in G$ is such that $b(\overline{x})^{fg_1} = b\left(\overline{x}^f\right)^{g_1} \in T_0$. We have also

$$\overline{x}^h = \left(b\left(\overline{x}\right)^h, e\left(\overline{x}^{g_2}\right) \right)$$

where $b(\overline{x})^{g_2} \in T_0$. From the fact that h is well defined it now follows that $e\left(\overline{x}^{fg_1}\right) = e\left(\overline{x}^{g_2}\right)$, thus the sequences $\overline{x}^h$ and $\overline{x}^{fh}$ are $n\,(\overline{x})$-confinal. Thus for any $\overline{y}(= \overline{x}^h) \in B_T$ and $f \in G$ the sequences $\overline{y}$ and $\overline{y}^{h^{-1}fh}$ are confinal and $G^h \leq \mathcal{AWF}(T)$. $\square$

It is an interesting question as to which (finitely generated, periodic) subgroups of $\mathcal{A}(T)$ are conjugate to a subgroup of the group $\mathcal{AWF}(T)$.

4. Regular rooted trees

A spherically homogeneous rooted tree T is called n-regular if $\tau(T) = \langle n, n, \ldots \rangle$. In this case the boundary B_T of the tree T may be identified with the set of all infinite (to the right) words over an alphabet X, where $|X| = n$. The permutation group $(\mathcal{A}(T), B_T)$ coincides with the group of all automatic permutations acting on the set of infinite words over the alphabet X [Su3]. We consider only the case $|X| = n \geq 2$.

Proposition 7. *Let T be a regular rooted tree. Then there exists an embedding $\phi : \mathcal{A}(T) \hookrightarrow \mathcal{AWF}(T)$.*

Proof. Let $s, t \in X$ be different elements. If $\overline{x} \neq (s, s, \ldots)$ then put $\overline{x} = (\overline{v}, z, \overline{u_0}, \overline{u})$, where $\overline{v}$ is the longest beginning of $\overline{x}$ of the form $(s, s, \ldots, s)$, $z \in X$, the length of $\overline{u_0}$ is equal to the length of $\overline{v}$ and $\overline{u}$ is an infinite sequence. We define the embedding ϕ by the rule

$$(\overline{x})^{\phi(g)} = \begin{cases} \overline{x} & \text{if } \overline{x} = (s, s, \ldots) \\ \overline{x} & \text{if } \overline{x} \neq (s, s, \ldots) \text{ and } z \neq t \\ (\overline{v}, t, \overline{u_0}^g, \overline{u}) & \text{in other cases,} \end{cases}$$

(See the picture below.)

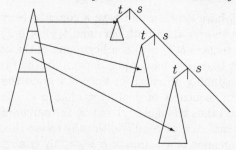

Obviously, the defined map ϕ is an embedding and $\phi\left(\mathcal{A}(T)\right) \leq \mathcal{AWF}(T)$.
□

Lemma 8. *Let T be an n-regular rooted tree. Then any weakly finitary automorphism $g \in \mathcal{A}(T)$ of width 1 has finite order not greater than $n!$.*

Proof. Let g be such an automorphism. For any $\overline{x} \in B_T$ we have $\overline{x} = (\overline{v}, y, \overline{w})$, where $\overline{v}$ has finite length and $\overline{x}^g = (\overline{v}, z, \overline{w})$. From the definition of triangular permutations we get that the map $\pi : y \mapsto z$ defines a permutation on the set X. Moreover, we get $(\overline{v}, t, \overline{w})^g = (\overline{v}, t^\pi, \overline{w})$. Any element of the symmetric group $Symm(X)$ has order dividing $n!$ (in fact dividing $lcm(1, 2, \ldots, n)$.) Thus for any $\overline{x} \in B_T$ we have $\overline{x}^{g^{n!}} = (\overline{v}, y^{\pi^{n!}}, \overline{w}) = (\overline{v}, y, \overline{w}) = \overline{x}$, and g has order not greater than $n!$.   □

Using Theorem 4 and Lemma 8 the following theorem is proved.

Theorem 9. *Any automorphism of an n-regular rooted tree T is a product of two automorphisms of finite order, not greater than $n!$. In particular, any automorphism of a 2-regular rooted tree T is contained in some (finite or infinite) dihedral subgroup of the group $\mathcal{A}(T)$.*

Proof. Let g be an automorphism of an n-regular tree T. By Theorem 4 it is conjugate to a weakly finitary automorphism $f^{-1}gf = h \in \mathcal{A}(T)$ of width not greater than 2.

Let us define an automorphism h_0 of width less than or equal to 1 by the rule:

$$(\overline{v}, x, \overline{w})^{h_0} = (\overline{v}, y, \overline{w}),$$

where $(\overline{v}, x, \overline{w})^h = (\overline{u}, y, \overline{w})$ and $\overline{v}, \overline{u}$ have equal length. Note that, since h has width not greater than 2, the sequences $\overline{v}$ and $\overline{u}$ differ in not more than one coordinate. Put $h_1 = h_0^{-1}h$. Then using the previous notations we have:

$$(\overline{u}, y, \overline{w}) = (\overline{v}, x, \overline{w})^h = \left((\overline{v}, x, \overline{w})^{h_0}\right)^{h_1} = (\overline{v}, y, \overline{w})^{h_1}.$$

Thus h_1 also has width not greater than 1. By Lemma 8 the automorphisms h_0 and h_1 are of finite order not greater than $n!$. But $g = (fh_0f^{-1})(fh_1f^{-1})$.   □

Remark. Lemma 8 remains true for the spherically homogeneous rooted trees whose spherical indices are bounded from above by n. Thus the first part of Theorem 9 also remains true for such trees.

An automorphism of an n-regular rooted tree T is said to be finitely automatic if it defines a finitely automatic transformation of the set B_T. A transformation of the set of infinite words over a given alphabet is said to be finitely automatic if it is defined by some automaton (in other terms: sequential machine) with a finite set of inner states [Su4]. A criterion for a triangular transformation of the set B_T to be finitely automatic can be easily formulated in the terms of remainders of transformations [R]. For any finite sequence $\overline{a} = (a_1, a_2, \ldots, a_k)$, any triangular transformation $\pi : B_T \to B_T$ induces a transformation $\pi_{\overline{a}} : B_T \to B_T$ called *the $\overline{a}$-remainder* of the transformation π, which is defined by the equality:

$$(\overline{a}^\pi, (x_1, x_2, \ldots)^{\pi_{\overline{a}}}) = (a_1, \ldots, a_k, x_1, x_2, \ldots)^\pi, \quad (x_1, x_2, \ldots) \in B_T.$$

Lemma 10. *An automorphism π of a regular rooted tree T is finitely automatic if and only if it induces on the set B_T a transformation having a finite number of different remainders.*

Using Lemma 10 one can prove

Theorem 11. *Any finitely automatic weakly finitary automorphism of a regular rooted tree T has finite width not greater than the number of its different remainders.*

Proof. Let π be a finitely automatic weakly finitary automorphism of a regular rooted tree T. Let π have n different remainders.

Suppose that $\overline{x} \in B_T$ is a sequence such that $\overline{x}^\pi$ and $\overline{x}$ differ in more than n coordinates. Let π change the coordinates numbered $k_1, k_2, \ldots k_m$. Since $m > n$, some of the remainders $\pi_{(x_1, \ldots, x_{k_i-1})}$ are equal. Let the remainder $\psi = \pi_{(x_1, \ldots, x_{k_i-1})}$ be equal to the remainder $\pi_{(x_1, \ldots, x_{k_j-1})}$ and $i < j$. Put $\overline{a} = (x_{k_i}, x_{k_i+1}, x_{k_i+2}, \ldots, x_{k_j-1})$ and $\overline{p} = (\overline{a}, \overline{a}, \ldots)$. The sequence $\overline{p}$ is periodic and it is easy to see that the automorphisms $\psi, \psi_{\overline{a}}, \psi_{(\overline{a}, \overline{a})}, \ldots$ are equal. But they are nontrivial and thus $\psi = \pi_{(x_1, \ldots, x_{k_i-1})}$ changes an infinite number of coordinates in $\overline{p}$, thus π changes infinitely many coordinates of the sequence $(x_1, \ldots, x_{k_i-1}, \overline{p})$. But this is impossible. $\square$

5. CONFINALLY TRANSITIVE GROUPS

Since the confinality classes are transitivity blocks for the subgroups of the group $\mathcal{AWF}(T)$, the following notion of transitivity for weakly finitary groups is natural.

Definition. A subgroup $G \leq \mathcal{AWF}(T)$ is said to be *confinally transitive* if the orbits of its action on B_T coincide with the confinality classes.

The class of confinally transitive groups is rather vast and contains many well known groups. The groups of Grigorchuk [G], the Gupta-Sidki groups [GS] and the groups from [Su5] are all confinally transitive.

In fact, all these groups satisfy a stronger transitivity condition. We say that a group $H \leq \mathcal{AWF}(T)$ acts quasi-regularly on level L_n if for any two sequences $\overline{t_1}, \overline{t_2} \in L_n$ there exists an automorphism $g \in G$ such that for any infinite sequence $\overline{x}$ we have $(\overline{t_1}, \overline{x})^g = (\overline{t_2}, \overline{x})$. Obviously, if a group $G \leq \mathcal{AWF}(T)$ acts quasi-regularly on every level then it is confinally transitive.

Using this the following theorem is proved.

Theorem 12. *Let T be a p-regular tree, where p is an odd prime. Suppose $\pi \in \mathcal{AF}(T)$ is an automorphism such that $\overline{x}^\pi \approx_{k+1} \overline{x}$ for every $\overline{x} \in B_T$ and the cyclic group $\langle \pi \rangle$ acts transitively on level L_k. Then one can find an automorphism $\rho \in \mathcal{AWF}(T)$ of width 1 such that the group $G = \langle \pi, \rho \rangle$ is an infinite p-group.*

If the group $\langle \pi \rangle$ acts quasi-regularly on levels L_n for all $n \leq k$, then the automorphism ρ can be chosen in such a way that the group G will be confinally transitive.

Proof. From triangularity it follows that the group $\langle \pi \rangle$ acts transitively on levels L_n for all $n \leq k$. Let $X = \{0, 1, 2, \ldots, p-1\}$. In this proof we use the tableau notation defined in the section 2.

The permutation π has a tableau of type:

$$\pi = [a_1, a_2(x_1), \ldots, a_k(x_1, \ldots x_{k-1}), \varepsilon, \varepsilon, \ldots],$$

where ε is a constant equal to the identity permutation. Let us define the permutation ρ by the tablaux:

$$\rho = [\underbrace{\varepsilon, \varepsilon, \ldots, \varepsilon}_{k}, b_{k+1}(x_1, \ldots x_k), b_{k+2}(x_1, \ldots, x_{k+1}), \ldots],$$

where the functions $b_l(x_1, \ldots, x_{l-1})$ are defined by the equalities

$$b_l(x_1, \ldots, x_{l-1}) = \begin{cases} c & \text{if } (x_1, \ldots, x_{l-1}) = (0, \ldots, 0, 1); \\ c^{-1} & \text{if } (x_1, \ldots, x_{l-1}) = (0, \ldots, 0, 2); \\ \varepsilon & \text{in the other cases,} \end{cases}$$

where c is the cyclic permutation $(0, 1, 2, \ldots p-1)$. It is easy to see that

$$(1) \qquad \rho^m = [\varepsilon, \ldots, \varepsilon, b_{k+1}^m(x_1, \ldots, x_k), b_{k+2}^m(x_1, \ldots, x_{k+1}), \ldots)]$$

and thus ρ is a permutation of order p.

We are going to prove that $G = \langle \pi, \rho \rangle$ is the necessary group.

An automorphism $u \in \mathcal{A}(T)$ has finite order if for any sequence $\overline{t} \in B_T = X^{\mathbb{N}}$ the orbit

$$O(\overline{t}, u) = \left\{ \overline{t}^{u^i} : i \in \mathbb{Z} \right\}$$

is finite and the lengths of all such orbits $O(\overline{t}, u)$ are bounded from above by a number not depending on $\overline{t}$.

Suppose $u = \pi^{\alpha_1} \rho^{\beta_1} \cdots \pi^{\alpha_s} \rho^{\beta_s}$ (where $0 \le \alpha_1 < |\pi|$, $0 \le \beta_s < p$, $0 < \alpha_i < |\pi|$ for $i = 2, 3, \ldots, s$ and $0 < \beta_i < p$ for $i = 1, 2, \ldots, s-1$) is an arbitrary element of the group G. Let $u = [g_1, g_2(x_1), g_3(x_1, x_2), \ldots]$ be the tableau of the automorphism u. Then from (1) follows that for $l > k$ we have

$$g_l(x_1, x_2, \ldots, x_{l-1}) =$$

$$b_l^{\beta_1}\left((x_1, \ldots, x_{l-1})^{w_1}\right) b_l^{\beta_2}\left((x_1, \ldots, x_{l-1})^{w_2}\right) \cdots b_l^{\beta_s}\left((x_1, \ldots, x_{l-1})^{w_s}\right),$$

where

$$w_i = \pi^{\alpha_1} \rho^{\beta_1} \cdots \pi^{\alpha_i}, \quad (0 \le i \le s).$$

For every $l \ge 1$ let $O(\bar{t}, u, l)$ be the orbit of the beginning $(t_1, \ldots, t_l) \in L_l$ of $\bar{t} = (t_1, t_2, \ldots)$ under the action of the cyclic group $\langle u \rangle$. The orbit $O(\bar{t}, u)$ will be finite if and only if the numbers $\left|O(\bar{t}, u, l)\right|$ are bounded from above.

If $l > k$ then $\left|O(\bar{t}, u, l)\right| = \left|O(\bar{t}, u, l+1)\right|$ if and only if the permutations $b_l^{\beta_1}\left((t_1, \ldots t_{l-1})^{u^j w_1}\right) \cdots b_l^{\beta_s}\left((t_1, \ldots t_{l-1})^{u^j w_s}\right)$ act trivially on t_l for every j such that $0 \le j \le |u_l|$, where $u_l = u|_{L_l}$. This will hold for example when these permutations are trivial.

Let $\tau_u(\bar{t}, l)$ be the number of elements $((\underbrace{0, \ldots, 0}_{l-2}), i)$ in the set

$$\left\{ \left((t_1, \ldots t_{l-1})^{u^j w_i}, i\right) : 0 \le j \le |u_{l-1}|, i = 1, 2, \ldots, s-1 \right\},$$

where $u_{l-1} = u|_{L_{l-1}}$.

Suppose

$$b_l^{\beta_i}\left((t_1, \ldots t_{l-1})^{u^j w_i}\right) \ne \varepsilon.$$

This is possible only for $(t_1, \ldots t_{l-1})^{u^j w_i} = (0, \ldots, 0, 1)$ or $(t_1, \ldots t_{l-1})^{u^j w_i} = (0, \ldots, 0, 2)$. Thus the sequence $(t_1, \ldots t_{l-2})^{u^j w_i}$ is equal to $(\underbrace{0, \ldots, 0}_{l-2})$ but the sequence $(t_1, \ldots t_{l-1})^{u^j w_i}$ is not equal to $(\underbrace{0, \ldots, 0}_{l-1})$.

Thus the inequality $\left|O(\bar{t}, u, l)\right| < \left|O(\bar{t}, u, l+1)\right|$ implies $\tau_u(\bar{t}, l+1) < \tau_u(\bar{t}, l)$. Note that $\left|O(\bar{t}, u, l)\right| / \left|O(\bar{t}, u, l-1)\right|$ is equal either to 1 or to p. So

$$\left|O(\bar{t}, u, l)\right| \le p^{\tau_u(\bar{t}, k+1)} M,$$

were M is the greatest cardinality of a u-orbit on the level L_{k+1}. Thus all the numbers $\left|O(\bar{t}, u, l)\right|$ do not exceed $p^{\tau_u} M$, were $\tau_u = \max\{\tau_u(\bar{t}, k+1) : \bar{t} \in B_T\}$. (Since $\tau_u(\bar{t}, k+1)$ depends only on the first $k+1$ coordinates of $\bar{t}$, the set $\{\tau_u(\bar{t}, k+1) : \bar{t} \in B_T\}$ is finite and τ_u is finite.) Thus the orbit lengths of $\langle u \rangle$ on the boundary are bounded from above and u has finite order.

Let us prove that for any sequences $(x_1, x_2, \ldots, x_{k+l})$ and $(y_1, y_2, \ldots, y_{k+l})$ there exists an element $u \in G$ such that for every $\overline{w} \in X^{\mathbb{N}}$ we have

$$(x_1, x_2, \ldots, x_{k+l}, \overline{w})^u = (y_1, y_2, \ldots, y_{k+l}, \overline{w}).$$

From this it will follow that the group G is infinite. In the case when $\langle \pi \rangle$ acts on levels L_n for $n \le k$ quasi-regularly this will imply that the group G acts an all the levels quasi-regularly and thus it is confinally transitive, since $G \le \mathcal{AWF}(T)$.

Let us prove this statement by induction on l. For $l = 0$ everything is trivial. Suppose we have proved the statement for all lengths $l < m$. Let us prove it for $l = m$. By the inductive assumptions, there exist elements $u_i \in G$ such that

$$(x_1, x_2, \dots, x_{k+m}, \overline{w})^{u_1} = (\underbrace{0, \dots, 0}_{k+m-2}, 1, x_{k+m}, \overline{w})$$

and

$$(y_1, y_2, \dots, y_{k+m}, \overline{w})^{u_2} = (\underbrace{0, \dots, 0}_{k+m-2}, 1, y_{k+m}, \overline{w}).$$

Then from the definition of ρ it follows that for some $0 \le r \le p - 1$

$$(\underbrace{0, \dots, 0}_{k+m-2}, 1, x_{k+m}, \overline{w})^{\rho^r} = (\underbrace{0, \dots, 0}_{k+m-2}, 1, y_{k+m}, \overline{w}).$$

Thus

$$(x_1, x_2, \dots, x_{k+m}, \overline{w})^{u_1 \rho^r u_2^{-1}} = (y_1, y_2, \dots, y_{k+m}, \overline{w}),$$

and the theorem is proved. $\square$

The theorem remains true also for an intransitive (but nontrivial) automorphism π. The intransitive case can be reduced to the transitive one by restricting the constructions to the $\langle \pi \rangle$-orbits on level L_k.

REFERENCES

[B]   Bass H., Otero-Espinar M.V., Rockmore D.N., Tresser C.P.L. *Cyclic renormalization and automorphism groups of rooted trees*, Berlin, Springer, 1996.

[BS]   Brunner A.M., Sidki S. *The generation of* GL$(n, \mathbb{Z})$ *by finite state automata*, International Journal of Algebra and Computations, v. 8, No.1, 1998, p.127–139.

[G]   Grigorchuk R.I. *Growth degrees of finitely generated groups and invariant mean theory*, Izv. AN SSSR Ser. Matem., 1984, No 5, p. 939–985, (in Russian).

[GS]   Gupta N., Sidki S. *On the Burnside problem for periodic groups*, Math. Z., 1983, 182, p. 385–388.

[K]   Kaluzhnin L. *On some generalization of Sylow p-subgroups of symmetric groups*, Acta Math. Hung., 1951, 2, No 3–4, p. 198–221.

[R]   Raney G.N. *Sequential functions*, J. Assoc. Comput. Math., 1958, v. 5, No 2, p. 177–180.

[NS]   Nekrashevych V.V., Sushchansky V.I. *Confinal Structure of Rooted Tree Automorphisms*, Int. Conference Dedicated to the 90th Anniversary of L.S.Pontryagin, Abstracts, Optimal Control and Appendices, 1998, p. 296–298.

[S]   Serre J.-P., *Trees*, Springer, New York, 1988.

[Si]   Sidki Said *Regular Trees and their Automorphisms*. Monografias de Matematica, Vol. 56, IMPA, Rio de Janeiro, 1998.

[Su1] Sushchansky V.I. *Representation of residually finite groups by isometries of homogeneous ultrametric spaces of a finite width*, Dop. AN URSR Ser. A., 1988, No 4, p. 19–22, (in Ukrainian).

[Su2] Sushchansky V.I. *Isometry groups of the Baire p-spaces*, Dop. AN URSR, 1984, No 8, p. 28–30, (in Ukrainian).

[Su3] Sushchansky V.I. *Automatic permutation groups* Dopovidi NAN Ukrainy, 1998, N6, p. 47–51 (in Ukrainian).

[Su4] Sushchansky V.I. *Groups of finitely automatic permutations* Dopovidi NAN Ukrainy, 1999, N2. p.29–32 (in Ukrainian).

[Su5] Sushchansky V.I. *Periodic p-groups permutation and the unrestricted Burnside problem*, Soviet. Math. Dokl., 1979, vol. 20, No 4, p. 766–770, (in English).

Volodymyr Nekrashevych
Dept. of Mathematics
Kyiv T.Shevchenko University
252033 Kyiv, Ukraine
nazaruk@ukrpack.net

Vitaly Sushchansky
Institute of Mathematics
Silesian Technical University
44-100 Gliwice, Poland
wsusz@zeus.polsl.gliwice.pl

AN ASYMPTOTIC INVARIANT OF SURFACE GROUPS

AMNON ROSENMANN

ABSTRACT. In this paper we compute an asymptotic invariant of the fundamental group G of a closed orientable surface of genus $n \geq 2$. We look at an increasing chain of balls in the Cayley graph of G with respect to the standard presentation, and compute the "average rank" defined by the quotient of the number of cycles by the number of vertices in each ball. This quotient is related to the growth of G and also to the ratio of the "areas" of the boundaries of the balls to their "volumes".

1. GROWTH FUNCTIONS OF SURFACE GROUPS

Let G be the fundamental group of a compact orientable closed surface of genus $n \geq 2$. Surface groups have been studied from the point of view of growth (see e.g. [2], [3], [6], [1]), Euler characteristic (see e.g. [4]), spectral radius of random walks (see e.g. [9], [1]), rewriting rules (e.g. [8]), and more. In the next section we examine an asymptotic invariant of surface groups. But first we look at some growth functions of these groups.

A standard presentation of G is

$$(1) \qquad G = \langle\, a_1, \ldots, a_n, b_1, \ldots, b_n \mid \prod_{i=1}^{n} [a_i, b_i] = 1 \,\rangle.$$

The Cayley graph Γ of G with respect to the above presentation corresponds to a tessellation of the hyperbolic plane, with the fundamental domain being a polygon with $4n$ sides. Thus, we will also refer to polygons in Γ - meaning their boundaries.

A vertex v of Γ is of *level* k if the length of a geodesic from the base vertex o (representing the identity element of G) to v is k. Similarly, a polygon P is of level k if it starts at level k, i.e. the distance from o to P is k. This means that at level k there is one vertex of P, at each of the levels $k+1, k+2, \ldots, k+2n-1$ there are two vertices of P, and one vertex of P is at level $k+2n$.

We may construct Γ inductively. Starting with the base point we first add the $4n$ polygons of level 0, then the $4n(4n-2)$ polygons of level 1, then the $4n(4n-1)(4n-2)$ polygons of level 2, etc. We distinguish between two types of vertices of Γ (for a more detailed partition see [2], [9], [1]). Suppose that v is a vertex of level $k > 0$ and Γ' is the subgraph of Γ formed by all polygons of level $0, \ldots, k-1$. Then either v has degree 3 or degree 4 in Γ'. In the first case we say that v is an *inner* vertex and in the second case - a *top* vertex. If

v is an inner vertex it is connected to one vertex of level $k-1$ (in Γ' and also in Γ) and to 2 vertices of Γ' of level $k+1$. It is a common vertex of 2 polygons of Γ', one of level $k-1$ and the other of level $k-j$, for some $1 \leq j \leq 2n-1$. If v is a top vertex it is connected to 2 vertices of level $k-1$ (in Γ' as well as in Γ) and to 2 vertices of level $k+1$ in Γ'. It is a common vertex of 3 polygons of Γ', 2 of level $k-1$ and one of level $k-2n$. It is because of this last polygon we call v a top vertex: it belongs to a polygon whose other vertices are of level less than the level of v. The base vertex o of Γ is a special vertex which is neither top nor inner.

We denote by t_k the number of top vertices of level k and by i_k the number of inner vertices of level k. There is a 1-1 correspondence between the top vertices of level k and polygons of level $k-2n$. Therefore, there are no top vertices at levels $0, \ldots, 2n-1$, and there are $4n$ top vertices at level $2n$. If v is an inner vertex of level $k-2n$ then it "contributes" $4n-2$ top vertices of level k through the $4n-2$ polygons that start at v. On the other hand, if v is a top vertex of level $k-2n$ then it contributes only $4n-3$ top vertices of level k. Thus, we get the recurrence formula

$$(2) \qquad t_k = (4n-3)t_{k-2n} + (4n-2)i_{k-2n}, \qquad k > 2n.$$

Let us examine now the inner vertices of level k. Each polygon of level $k-j$, $2 \leq j \leq 2n-1$, contributes 2 inner vertices of level k. Since the polygons that are of level less than $k-1$ become separated at the k-th level, these contributed vertices are distinct, and sum up to

$$2 \sum_{j=2}^{2n-1} [(4n-3)t_{k-j} + (4n-2)i_{k-j}].$$

As for the inner vertices of level k that come from polygons of level $k-1$, here the vertices are no longer distinct. If v is an inner vertex of level $k-1$ then it has degree 3 in Γ'. Thus, the $4n-2$ polygons of level $k-1$ that start at v contribute $4n-3$ inner vertices of level k. If v is a top vertex of level $k-1$ then it has degree 4 in Γ', and the $4n-3$ polygons that start at v contribute $4n-4$ inner vertices of level k. By the above, we get the following recurrence formula for the inner vertices

$$(3) \quad i_k = 2 \sum_{j=2}^{2n-1} [(4n-3)t_{k-j} + (4n-2)i_{k-j}] + (4n-4)t_{k-1} + (4n-3)i_{k-1},$$

where $k \geq 2n$. As for levels $0, \ldots, 2n-1$, there are no inner vertices at level 0, $4n$ inner vertices at level 1, and $4n(4n-1)^{k-1}$ inner vertices at levels $k = 2, \ldots, 2n-1$. (Each inner vertex of level $k = 1, \ldots, 2n-2$ is connected to $4n-1$ vertices of level $k+1$.)

Denoting by $T(z)$ and $I(z)$ the corresponding growth series

$$T(z) = \sum_{k=0}^{\infty} t_k z^k,$$

$$I(z) = \sum_{k=0}^{\infty} i_k z^k,$$

we get from (2) and (3) that

$$T(z) = (4n - 3)z^{2n}T(z) + (4n - 2)z^{2n}I(z) + 4nz^{2n},$$

and

$$I(z) = 2\sum_{j=2}^{2n-1}[(4n - 3)T(z) + (4n - 2)I(z)]z^j + (4n - 4)zT(z)$$

$$+(4n - 3)zI(z) + 4nz + 8n\sum_{j=2}^{2n-1}z^j.$$

These lead to the growth series of the top vertices

$$T(z) = \frac{4nz^{2n}}{1 - (4n - 2)\sum_{j=1}^{2n-1}z^j + z^{2n}},$$

and the growth series of the inner vertices

$$I(z) = \frac{4n(\sum_{j=1}^{2n-1}z^j - z^{2n})}{1 - (4n - 2)\sum_{j=1}^{2n-1}z^j + z^{2n}}.$$

We denote by a_k the number of vertices of Γ at level k, i.e. the number of elements of G of length k (with respect to the standard presentation), and by $A(z)$ the corresponding generating function

$$A(z) = \sum_{k=0}^{\infty} a_k z^k.$$

Then $a_0 = 1$,

(4) $a_k = t_k + i_k, \qquad k > 0,$

and

$$A(z) = T(z) + I(z) + 1,$$

and we obtain the growth function of G (see [2] [3]):

Theorem 1.1. (Cannon [2]). *The growth series of the standard presentation of the surface group G is*

$$A(z) = \frac{1 + 2\sum_{j=1}^{2n-1}z^j + z^{2n}}{1 - (4n - 2)\sum_{j=1}^{2n-1}z^j + z^{2n}}.$$

2. AN ASYMPTOTIC INVARIANT

In the spirit of Gromov ([7]) we compute here an asymptotic invariant of surface groups. In [10] the normalized cyclomatic quotient (NCQ) and the normalized balanced cyclomatic quotient (NBCQ) were introduced. These measure the asymptotic behaviour of a kind of an average rank: the ratio of the rank of the fundamental group of a finite subgraph of the Cayley graph to the size of the subgraph. In the NCQ we examine chains of increasing subgraphs which give the supremum of the cyclomatic quotient. It turns out that the NCQ characterizes amenable groups, it equals the Euler characteristic of the group in case the group is finite, amenable or free, and it sometimes behaves like the spectral radius of a symmetric random walk on the graph. In the NBCQ, which we compute here, we look at the same quotient but on the chain of increasing balls. From (5) below we see that it measures the asymptotic behaviour of the quotient of the "area" of the boundary of a ball by its "volume".

We use the following notation. If Γ' is a finite connected subgraph of the Cayley graph Γ of the surface group (with respect to the presentation (1)) then we denote by $\beta_0(\Gamma')$, $\beta_1(\Gamma')$, $\beta_2(\Gamma')$ the number of vertices, edges and cycles, respectively, in Γ'. Notice that $\beta_2(\Gamma')$ equals the number of 2-cells in an associated 2-dimensional complex or the rank of $\pi_1(\Gamma')$. Then we let

$$\xi(\Gamma') = \frac{\beta_2(\Gamma')}{\beta_0(\Gamma')}.$$

Let B_k be the ball of radius k around o, i.e. the subgraph of Γ containing all vertices of level $0, \ldots, k$. The *normalized balanced cyclomatic quotient* of Γ is defined to be

$$\hat{\Theta}(\Gamma) = 1 - 2n + \limsup_{k \to \infty} \xi(B_k),$$

where $2n$ is the number of generators. We know that the growth of Γ is exponential and this is equivalent to $\hat{\Theta}(\Gamma)$ being negative. Since

$$\beta_2(\Gamma') = \beta_1(\Gamma') - \beta_0(\Gamma') + 1,$$

we obtain

$$\hat{\Theta}(\Gamma) = -2n + \limsup_{k \to \infty} \frac{\beta_1(B_k) + 1}{\beta_0(B_k)}.$$

And since

$$\beta_2(\Gamma') = 1 + (2n - 1)\beta_0(\Gamma') - |E_{out}^+(\Gamma')|,$$

(see [11]) where $|E_{out}^+(\Gamma')|$ is the number of edges going out of Γ' in the directions of the generators $a_1, \ldots, a_n, b_1, \ldots, b_n$ (but not their inverses), we obtain

$$(5) \qquad \hat{\Theta}(\Gamma) = \limsup_{k \to \infty} \frac{1 - |E_{out}^+(B_k)|}{\beta_0(B_k)}.$$

We are going now to compute $\hat{\Theta}(\Gamma)$ via the quotients of $\beta_2(B_k)$ by $\beta_0(B_k)$. Let us first look at the asymptotic behaviour of the following difference sequences. The growth function of vertices

$$a_k = \beta_0(B_k) - \beta_0(B_{k-1}),$$

and the growth function of top vertices

$$t_k = \beta_2(B_k) - \beta_2(B_{k-1}).$$

Each vertex v of Γ of level $k > 0$ is connected either to $4n - 1$ vertices of level $k + 1$ (in case v is an inner vertex) or to $4n - 2$ vertices of level $k + 1$ (in case v is a top vertex). These vertices of level $k + 1$ are distinct, except for the case of inner vertices of level k belonging to the same polygon, which might share the top vertex of level $k + 1$ of this polygon. Thus, we see that a_k grows exponentially with

$$4n - 2 < \frac{a_{k+1}}{a_k} < 4n - 1,$$

for k large enough. In fact, a_{k+1}/a_k tends to a limit a, and moreover,

$$\lim_{k \to \infty} \frac{a_{k+1}}{a_k} = \lim_{k \to \infty} \frac{t_{k+1}}{t_k} = \lim_{k \to \infty} \frac{i_{k+1}}{i_k} = a.$$

This limit is a root of the polynomial

$$(6) \qquad\qquad P(z) = 1 - (4n - 2) \sum_{j=1}^{2n-1} z^j + z^{2n},$$

which is the same as the denominator appearing in the expressions for $A(z)$, $T(z)$ and $I(z)$. $P(z)$ is a *Salem polynomial*: monic, with integer coefficients, reciprocal $(P(z) = z^{2n}P(z^{-1}))$ and has exactly one root (of multiplicity 1) outside the unit circle (see [3]). The root of $P(z)$ which is outside the unit circle, called a *Salem number*, is the above limit a. This is proved in [1] based on the theorem expressing the coefficients of a rational function (see [5]).

Let us denote

$$\bar{a}_k = \beta_0(B_k)$$

and

$$\bar{A}(z) = \sum_{k=0}^{\infty} \bar{a}_k z^k.$$

Since

$$\bar{a}_k = \bar{a}_{k-1} + a_k,$$

we obtain

$$(7) \quad \bar{A}(z) = z\bar{A}(z) + A(z) = \frac{A(z)}{1 - z} = \frac{1 + 2\sum_{j=1}^{2n-1} z^j + z^{2n}}{1 - (4n - 1)z + (4n - 1)z^{2n} - z^{2n+1}}.$$

As before, since the denominator in the above rational function is $P(z)(1-z)$, where $P(z)$ is as in (6), we have

$$\lim_{k \to \infty} \frac{\bar{a}_{k+1}}{\bar{a}_k} = a$$

(a is the same limit as above).

In a similar way, let

$$\bar{t}_k = \beta_2(B_k)$$

and

$$\bar{T}(z) = \sum_{k=0}^{\infty} \bar{t}_k z^k.$$

By

$$\bar{t}_k = \bar{t}_{k-1} + t_k,$$

we obtain

$$(8) \quad \bar{T}(z) = z\bar{T}(z) + T(z) = \frac{T(z)}{1-z} = \frac{4nz^{2n}}{1 - (4n-1)z + (4n-1)z^{2n} - z^{2n+1}},$$

and

$$\lim_{k \to \infty} \frac{\bar{t}_{k+1}}{\bar{t}_k} = a.$$

We see that the same denominator appears in the expressions for $\bar{A}(z)$ and $\bar{T}(z)$. So let us denote

$$Q(z) = 1 - (4n-1)z + (4n-1)z^{2n} - z^{2n+1}$$

and let

$$S(z) = \frac{1}{Q(z)} = \sum_{k=0}^{\infty} s_k z^k.$$

Then

$$\lim_{k \to \infty} \frac{s_{k+1}}{s_k} = a.$$

By (7) and (8)

$$\bar{a}_k = s_{k-2n} + 2 \sum_{j=1}^{2n-1} s_{k-j} + s_k,$$

$$\bar{t}_k = 4n s_{k-2n},$$

for k large enough. Hence,

$$\lim_{k \to \infty} \frac{\bar{a}_{k+2n}}{\bar{t}_{k+2n}} = \lim_{k \to \infty} \frac{s_k + 2 \sum_{j=1}^{2n-1} s_{k+j} + s_{k+2n}}{4n s_k}$$

$$= \frac{1}{4n} \left(1 + 2 \sum_{j=1}^{2n-1} a^j + a^{2n} \right).$$

But a is a root of P, i.e.,

$$(9) \qquad 1 + (2 - 4n) \sum_{j=1}^{2n-1} a^j + a^{2n} = 0.$$

Thus,

$$\lim_{k \to \infty} \frac{\bar{a}_k}{\bar{t}_k} = \sum_{j=1}^{2n-1} a^j = \frac{a(a^{2n-1} - 1)}{a - 1},$$

and by (9)

$$\lim_{k \to \infty} \frac{\bar{a}_k}{\bar{t}_k} = \frac{a^{2n} + 1}{4n - 2}.$$

In fact, the same limit is obtained by looking at $\lim_{k \to \infty} a_k / t_k$ using

$$a_k = \frac{t_k + t_{k+2n}}{4n - 2}, \qquad k > 2n$$

(by (2) and (4)). We conclude that

Theorem 2.1. *The normalized balanced cyclomatic quotient (NBCQ) of the standard presentation Γ of the surface group G is*

$$\hat{\Theta}(\Gamma) = 1 - 2n + \frac{4n - 2}{a^{2n} + 1},$$

where a is the Salem number defined by the polynomial

$$1 - (4n - 2) \sum_{j=1}^{2n-1} z^j + z^{2n}.$$

Since a is a root of $Q(z)$ we get that

$$(10) \qquad 1 - (4n - 1)a + (4n - 1)a^{2n} - a^{2n+1} = 0,$$

$$4n(a^{2n} + 1) = (4n - 2)(1 + a) + (1 + a)(a^{2n} + 1),$$

$$\frac{4n}{1 + a} = 1 + \frac{4n - 2}{a^{2n} + 1},$$

and

$$\hat{\Theta}(\Gamma) = \frac{2n(1 - a)}{1 + a}.$$

We know that a is between $4n - 1$ and $4n - 2$, so let us write

$$a = (4n - 1) - \epsilon.$$

Then by (10)

$$[(4n - 1) - \epsilon]^{2n+1} - (4n - 1)[(4n - 1) - \epsilon]^{2n}$$

$$+ (4n - 1)[(4n - 1) - \epsilon] - 1 = 0,$$

and we obtain that

$$\epsilon[(4n - 1) - \epsilon]^{2n} - (4n - 1)[(4n - 1) - \epsilon] + 1 = 0,$$

$$\epsilon([(4n - 1) - \epsilon]^{2n} + (4n - 1)) = (4n - 1)^2 - 1.$$

So ϵ may be approximated by $(4n - 1)^{2-2n}$.

REFERENCES

[1] L. Bartholdi, T. Ceccherini-Silberstein (1998). Growth series and random walks on some hyperbolic graphs. Preprint.

[2] J. W. Cannon (1980). The growth of the closed surface groups and the compact hyperbolic Coxeter groups. Preprint.

[3] J. W. Cannon, Ph. Wagreich (1992). Growth functions of surface groups. *Math. Ann.* **293**, 239-257.

[4] W. J. Floyd, S. P. Plotnick (1987). Growth function on Fuchsian groups and the Euler characteristic. *Invent. Math.* **88**, 1-29.

[5] R. L. Graham, D. E. Knuth, O. Patashnik (1994). *Concrete Mathematics* (2nd edition). Addison-Wesley.

[6] R. I. Grigorchuk, T. Nagnibeda (1997). Complete growth functions of hyperbolic groups. *Invent. Math.* **130**, 159-188.

[7] M. Gromov (1993). Asymptotic invariants of infinite groups. In: (Niblo, G.A., Roller, M.A. eds) *Geometric Group Theory*, Vol. 2. London Math. Soc. Lecture Note Ser. **182**. Cambridge University Press.

[8] S. M. Hermiller (1994). Rewriting systems for Coxeter groups. *J. Pure Appl. Alg.* **92**, 137-148.

[9] T. Nagnibeda (1996). An estimate from above of spectral radii of surface groups. Preprint.

[10] A. Rosenmann (1997). The normalized cyclomatic quotient associated with presentations of finitely generated groups. *Israel J. Math.* **99**, 285-313.

[11] A. Rosenmann (1999). When Schreier transversals grow wild. In: (Campbell, C.M., Robertson, E.F., Ruskuč, N., Smith, G.C. eds) *Groups St. Andrews 1997 in Bath* II, London Math. Soc. Lecture Note Ser. **261**. Cambridge University Press.

Amnon Rosenmann
School of Mathematical Sciences
Tel Aviv University
Ramat Aviv
Tel Aviv 69978
Israel
aro@math.tau.ac.il

A CUTPOINT TREE FOR A CONTINUUM

ERIC L SWENSON

ABSTRACT. Given a Hausdorff continuum X and a set of cut points C of X, we construct a "tree" $T \supset C$ and a relation between X and T which preserves the separation properties of elements of C. We then give an application of this result which simplifies the proof of the cut point conjecture for negatively curved groups.

1. INTRODUCTION

The cut point conjecture for negatively curved groups has been proven. The pieces of the proof appear in [4], [5], [6], [7], [11], culminating in [8]. The steps in the proof are as follows:

1. Start with the convergence action of the negatively curved group G on the continuum $X = \partial G$, and assume that X has a cutpoint.
2. Construct an $\mathbb{R}$-tree R on which G acts in a non-nesting stable and virtually cyclic fashion.
3. Construct from R, an $\mathbb{R}$-tree T on which G acts by isometries.
4. Apply the Rips machine to this action to obtain a contradiction.

This paper provides a short easy version of step 2. The first written proof of step 2 appeared in [4]. The approach in this paper was first developed while the author was at Michigan Tech. in the winter of 94. The author was however unable to complete step 3, and so gave it up without ever writing up step 2. The present treatment was developed as part of a presentation of the cut point theorem in a graduate class at the university of Wisconsin, Milwaukee. The author was attempting to present the contents of [4] when it became apparent that it was simply too long to present in the allotted time. Thus this paper owes much to Bowditch's treatment in [4] in terms of methods and terminology, although the original idea predates it. Also I wish to thank G. A. Swarup for the suggestion that C could be taken to be a countable G-invariant set of cutpoints, G. Levitt for pointing out a topological difficulty with the original definition of the tree, and M. Bestvina for encouragement to write this up.

2. CONTINUA

Definition. A *continuum* is a compact connected Hausdorff space.

1991 *Mathematics Subject Classification*. Primary: 20F32; Secondary: 57N10.

Definition. In a continuum X, $c \in X$ is a *cutpoint* if $X = A \cup B$ where A and B are non-singleton continua and $A \cap B = \{c\}$. If in addition, $D \subset A - \{c\}$ and $E \subset B - \{c\}$ we say that c *separates* D from E.

Notation. For the remainder of the paper, let X be a continuum, G be a group (possibly trivial) of homeomorphisms of X, and $C \subset X$ be a G-equivariant $(GC = C)$ set of cut points of X. Notice that the set of all cutpoints will of course be G-equivariant.

For $a, b \in X$ and $c \in C$, we define $c \in (a, b)$ if there exist non-singleton continua $A \ni a$ and $B \ni b$ with $A \cup B = X$ and $A \cap B = \{c\}$. (a, b) will as one might imagine, be called an interval, and this relation an interval relation. We define the closed and half-open intervals in the obvious way. ie. $[a, b] = \{a, b\} \cup (a, b)$, and $[a, b) = \{a\} \cup (a, b)$ for $a \neq b$ ($[a, a) = \emptyset$).

Notice that if $c \in (a, b)$ then for any subcontinuum $Y \subset X$ from a to b $(a, b \in Y)$, $c \in Y$.

Definition. For $a, b \in X - C$ we say a is *equivalent* to b, $a \sim b$, if $(a, b) = \emptyset$.

For $c \in C$, c is equivalent only to itself. This is clearly an equivalence relation, so let P be the set of equivalence classes of X. We will abuse notation and say that $C \subset P$ since each element of C is its own equivalence class.

Notice that for $a, b, d \in X$, if $a \sim b$ then $(a, d) = (b, d)$. We can therefore translate the interval relation on X to P, and we also enlarge it as follows:

For $x, y, z \in P$, we say $y \in (x, z)$ if either

1. $y \in C$ and if $y \in (a, b)$ for $a, b \in X$ with $a \in x$ and $b \in z$.
2. $y \notin C$ and if $a, b, d \in X$ with $a \in x$, $b \in y$, and $d \in z$, then $[a, b) \cap (b, d] = \emptyset$.

Since C was chosen to be G-invariant, the action of G on X gives an action of G on P which preserves the interval structure (we have not given P a topology so it does not make sense to ask if the action is by homeomorphisms).

Example. Take X to be a locally finite tree of circles, ie. a union of circles so that any pair of them are either disjoint or tangent, and such that for any circles A, B there is a chain of tangent circles from A to B. (See Figure 1.) The points of tangency will be the set C of cut points. An equivalence class of $X - C$ will be a circle minus a finite number of points (those in C). To represent P, take all the points of C together with a point in the interior of each circle to represent the equivalence class of that circle minus the points of C. The interval relation is obtained by "connecting the dots", that is draw a dotted line segment from the point representing the equivalence class of each circle to the points of tangency (points of C) on that circle. (See Figure 1.) The interval relation is that which P inherits as a subset of the tree formed by the dotted line segments.

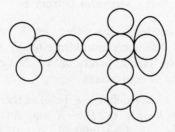

The continuum X

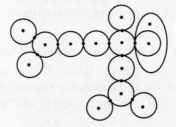

The pretree P and the continuum X

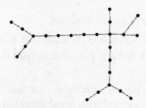

The interval structure on P

Figure 1

The following two Lemmas show that the interval relation on P satisfies the second and third axioms of a pretree. The first axiom (namely that $[x, y] = [y, x]$) is satisfied by definition in our case. The fact that the interval relation on X satisfies these properties is due to Bowditch [4].

Lemma 1. *For any* $x, y, z \in P$, $[x, z] \subset [x, y] \cup [y, z]$

Proof. Let $u \in (x, z)$. It suffices to show $u \in [x, y] \cup [y, z]$.

First consider when $u \in C$. There exist A, B non-singleton subcontinua of X with $A \cap B = u$, $A \cup B = X$, $x \subset A$ and $z \subset B$. If $y = u$ then we are done, and otherwise y is contained in A or in B. With no loss of generality $y \subset A$. It follows that $u \in (y, z)$.

Next consider when $u \notin C$. Suppose that $u \notin [x, y]$. Choose representatives $a, b, d, e \in X$ of u, x, y, z respectively. Since $u \notin [x, y]$, it follows that there exists $c \in [b, a) \cap (a, d]$. Since the case where $x = y$ is trivial, we may assume that $c \in C$. It can be shown that there are nonsingleton subcontinua $A, B \subset X$ such that $a \in A$, $A \cap B = \{c\}$, $A \cup B = X$, and $b, d \in B$. Since $u \in (x, z)$, it follows that $[b, a) \cup (a, e] = \emptyset$, and so $c \notin (a, e]$ implying that $e \in A - \{c\}$.

Suppose in addition that $u \notin [y, z]$. Arguing exactly as before, we have $c' \in C$ and nonsingleton subcontinua $A', B' \subset X$ with $A' \cap B' = \{c'\}$, $A' \cup B' = X$, $a, b \in A' - \{c'\}$, and $e, d \in B'$. Since $a, b \in A'$ it follows that $c \in A'$ (otherwise c could not separate a from b). However B' is a continuum from e to d missing c which is impossible since c separates e from d. Thus we must have $u \in [y, z]$ and the proof is complete.

Lemma 2. *For any $x, y \in P$, if $z \in (x, y)$ then $x \notin (y, z)$.*

Proof. Given $z \in (x, y)$ it follows that x, y, z are distinct. Take representatives $a, b, d \in X$ of x, y, z respectively.

If $d \in C$ (and therefore $z \in C$), then we have non-singleton subcontinua $A, B \subset X$ with $A \cap B = \{d\}$, $A \cup B = X$ and $a \in A$ and $b \in B$. Since $a \notin B$, it follows that if $a \in C$, then $a \notin (b, d)$ implying that $x \notin (y, z)$. Thus we are left with the case where $x \notin C$, but since $z \in (x, y] \cap (x, z]$ it follows by definition that $x \notin (y, z)$.

If $a \in C$ then we can argue as in the previous paragraph, and so we are left with the case where $a, d \notin C$. Since $a \not\sim d$, there is $c \in C$ and non-singleton subcontinua $A, B \subset X$ with $A \cap B = \{c\}$, $A \cup B = X$, $a \in A$ and $d \in B$. Since $z \in (x, y)$, it follows that $c \notin (x, z) \cap (z, y)$ and so $c \notin (z, y)$ implying that $b \in B$. Thus $c \in (b, a) \cap (a, d)$ and so by definition $x \notin (y, z)$.

The following two results are pretree results done first in [4]. We provide proofs for the sake of completeness.

Corollary. *If $x, y, z \in P$ with $y \in [x, z]$ then $[x, y] \subset [x, z]$.*

Proof. We may assume that $x \neq y \neq z$. Suppose $w \in [x, y] - [x, z]$. It follows $w \neq x, y, z$. By Lemma 1, $y \in (x, w) \cup (w, z)$ and $w \in (x, z) \cup (y, z)$ the latter of which implies that $w \in (y, z)$. However this in turn implies by Lemma 2 that $y \notin (w, z)$ and so by the first containment, $y \in (x, w)$. This however contradicts $w \in (x, y)$ by Lemma 2 and the proof is complete.

Lemma 3. *Let $[x, y]$ be an interval of P. The interval structure induces two linear orderings on $[x, y]$, one being the opposite of the other, with the* *property that if $<$ is one of the orderings, then for any $z, w \in [x, y]$ with $z < w$, $(z, w) = \{u \in [x, y] : z < u < w\}$. In other words the interval structure defined by one of the orderings is the same as our original interval structure.*

Proof. For $z, w \in [x, y]$, define $z < w$ if $z \in [x, w)$. We will show that this is a linear order and that the interval structure defined by it is the same as our original interval structure. The fact that the ordering we get by replacing x with y in the definition, is the opposite of this ordering follows from the fact that they define the same interval structure.

The fact that $z \not< z$ for any $z \in [x, y]$ follows from Lemma 2, and the fact that $[x, x) = \emptyset$.

We must next show that for any distinct $z, w \in [x, y]$ either $z < w$ or $w < z$. Suppose then that $w \notin [x, z)$. By Lemma 1 $w \in (z, y]$. By Lemma 2 $z \notin (w, y]$, and so by Lemma 1 $z \in [x, w)$. Hence $z < w$ as required.

Finally we must show transitivity. Let $z, u, w \in [x, y]$ with $z < u$ and $u < w$. In the language of intervals, $z \in [x, u)$ and $u \in [x, w)$. By the Corollary to Lemma 2, $z \in [x, w)$ and so $z < w$.

Now that we have shown that $<$ is a linear ordering on $[x, y]$ we will show that for $w, z \in [x, y]$ with $z < w$, $u \in (z, w)$ if and only if $z < u < w$.

Let $u \in (z, w)$. It follows from the Corollary to Lemma 2 that $u \in [x, w)$. Similarly $u \in [x, y]$, and so $u < w$. By Lemma 2 $z \notin (u, w)$, and so by Lemma 1 and $z < w$, $z \in [x, u)$. Thus $z < u$, and this implication is complete.

Let $u \in [x, y]$ with $z < u$ and $u < w$. Thus $z \in [x, u)$ and $u \in [x, w)$. By Lemma 1 $u \in [x, z) \cup (z, w)$. By Lemma 2 $u \notin [x, z)$, and so $u \in (z, w)$ as required.

Definition. We say distinct points $x, y \in P$ are *adjacent* if $(x, y) = \emptyset$.

Lemma 4. *If $x, y \in P$ are adjacent then exactly one of them is in C.*

Proof. If neither x nor y is in C, then an element c of C must separate them (otherwise $x = y$), and so $c \in (x, y)$ and x and y are not adjacent.

Suppose that both x and y are in C. Let $Y \subset X$ be an irreducible subcontinuum from x to y. Since x and y are adjacent, and Y is irreducible, it follows that no point of C is a cut point of Y. Let $C' = C \cap Y$. It suffices to show that C' is countable, for Y is uncountable, and so $Y - C'$ will be an equivalence class in (x, y) contradicting the adjacency of x and y.

For each $c \in C'$, choose non-singleton subcontinua $A_c, B_c \subset X$ with $A_c \cup B_c = X$, $A_c \cap B_c = \{c\}$ and $x \in A_c$. Notice that for any other $d \in C'$, $B_d \subset A_c$, and so $\{(B_c - \{c\}) : c \in C'\}$ is a collection of disjoint nonempty open sets of X, and therefore countable, since metric spaces are Lindeloef. Thus the set C' is countable and the proof is complete.

In fact this proves more about adjacent pairs, one of them will be a non-singleton equivalence class and the other will be an element of C in the closure (in X) of the non-singleton equivalence class.

Definition. We say $A \subset P$ is *collinear* if A is contained in some interval.

Lemma 5. *Any set A of three points of P two of which are adjacent is collinear. Also a union of two adjacent pairs is collinear.*

The proof is left to the reader.

Theorem 6. *A nested union of intervals of P is an interval of P*

Proof. We may assume that all the intervals are closed. We first reduce to the case where each of the intervals shares an endpoint y. If they do not, then pick an element y of the interior of one of them, and restrict to the intervals which contain y. Thus we can write each such interval $[x, z] = [x, y] \cup [y, z]$, and we deal with the two resulting nested unions separately.

Thus it suffices to show that if we have $\{[y, x_\alpha] : \alpha \in I\}$ where I is linearly ordered and $[y, x_\alpha] \subset [y, x_\beta]$ for $\alpha \leq \beta$, then $\cup [y, x_\alpha]$ is an interval of P. By taking a subnet, we may assume that all the x_α are distinct. If I has a last point the result is trivial, and otherwise we may assume that each $x_\alpha \in C$ (if not we replace x_α with an element of C in (x_α, x_β) where $\beta > \alpha$). Since X is compact, the net (or sequence if you prefer) $\{x_\alpha\}$ has a convergent subnet. Since the union is nested, it will not change if we switch to a subnet, and so we may assume that $x_\alpha \to x \in X$. We will abuse notation and use x both for the element of X and the corresponding equivalence class of P. Put the linear order on $[y, x]$ with $y < x$.

Case I: There is no last point of $[y, x)$. We will show that $\cup [y, x_\alpha] = [y, x)$. Notice in X, x_α separates y from x_β for all $\beta > \alpha$. Thus since $x_\alpha \to x$, it follows that $x_\alpha \in (y, x)$ for all α, and so by the Corollary to Lemma 2 $\cup [y, x_\alpha] \subset [y, x)$.

Now let $z \in [y, x)$. Since there is no last point of $[y, x)$, there is $c \in C$ such that $c \in (z, x)$. Since c cannot separate $\{x_\alpha\}$ from x, it follows that $c \in [y, x_\alpha)$ for all sufficiently large α. This implies $z \in [y, x_\alpha)$.

Case II: The last point of $[y, x)$ is z. Clearly $z \notin C$ (otherwise z would separate $\{x_\alpha\}$ from x). Similarly there can be no last point of $[y, z)$, for it would be in C. Arguing as in Case I, $\cup [y, x_\alpha] = [y, z)$.

Corollary. *Any interval of P has the supremum property with respect to either of the linear orderings derived from the interval structure.*

Proof. Let $[x, y]$ be the interval in question with the linear order $x \leq y$, and let $A \subset [x, y]$. Since A is linearly ordered, $\{[x, a] : a \in A\}$ is a set of nested intervals, and so by Theorem 6 its union will be an interval with one endpoint

x, and the other endpoint $\hat{a}$. The fact that we may choose $\hat{a} \in [x, y]$ is left as an exercise for the reader. It then follows that $\hat{a} = \sup A$.

For each pair $x, y \in P$ with x adjacent to y, let $\mathbb{R}_{x,y}$ be a copy of the real line. We will "sew in" $\mathbb{R}_{x,y}$ between x and y so that one of x, y, say x, is identified with $-\infty$ and the other, y, with ∞. The one identified with $-\infty$ will the first of the pair written, and so in $\mathbb{R}_{z,w}$, z is identified with $-\infty$ and w with ∞. Define

$$T = P \cup \bigcup_{x,y \text{ adjacent}} \mathbb{R}_{x,y}$$

We now extend the interval relation to T in the obvious way. Namely:

1. For $x, y, z \in P$, then $z \in (x, y)$ if this was so in P.
2. For $x, y \in P$, then $\mathbb{R}_{w,z} \subset (x, y)$ if $w, z \in [x, y]$.
3. For $z \in \mathbb{R}_{x,y}$ then $(x, z) = (-\infty, z) \subset \mathbb{R}_{x,y}$ and $(z, y) = (z, \infty)$.
4. For $z \in R_{x,y}$ and $w \in P$, by Lemma 5 $\{x, y, w\}$ is collinear, and so we may assume $x \in (y, w)$ in which case $(z, w) = (z, x) \cup [x, w)$.
5. For $z \in \mathbb{R}_{x,y}$ and $v \in \mathbb{R}_{u,w}$, by Lemma 5 $\{x, y, u, w\}$ is collinear and so we may assume $y, u \in (x, w)$ in which case $(z, v) = (z, y) \cup [y, u] \cup (u, v)$.

It is easily shown that T satisfies Lemmas 1, 2, 3, Theorem 6, and also the corresponding Corollaries. Notice also that there are no adjacent points in T. Also we can extend the action of G on P to an action of G on T which preserves the interval relation on T.

For $s \in T$ and $E \subset T$ finite, we define

$$U(s, E) = \{t \in T \,|\, [s, t] \cap E = \emptyset\}.$$

Remark. Notice that if $t \in U(s, E)$ then by Lemma 1 $U(t, E) = U(s, E)$. Observe that by definition, $U(s, E) \cap U(s, F) = U(s, E \cup F)$.

We now review some definitions from general topology.

Definition. A *big arc* is the homeomorphic image of a connected nonsingleton linearly ordered space. A separable big arc is an *arc*.

Definition. A *big tree* is a uniquely big-arcwise connected locally connected Hausdorff topological space. If the big arcs are all arcs, we say the space is a *real tree*. A metrizable real tree is called an $\mathbb{R}$-*tree*.

While this is not the standard definition of an $\mathbb{R}$-tree it is equivalent to the standard definition by [12].

Main Theorem. *The collection $\{U(s, E)\}$ is a basis for a topology on T such that G acts by homeomorphisms on T, and so that T is a regular big tree. If X is metrizable and every interval of P contains only countably many adjacent pairs, then T is a real tree.*

Proof. First we show that the $U(s, E)$ form a basis. Clearly they cover, for example $U(s, \emptyset) = T$. Let $r \in U(s, E) \cap U(t, F)$. By the remark,

$U(s, E) \cap U(t, F) = U(r, E) \cap U(r, F) = U(r, E \cup F)$. Thus the $U(s, E)$ form a basis for a topology on T. Furthermore since the topology was defined in terms of the interval structure which G preserves, it follows that G acts by homeomorphisms on T. The local connectivity will follow from the fact that the $U(s, E)$ are big-arcwise connected.

To see that T is Hausdorff, notice that for any $s, t \in T$, $(s, t) \neq \emptyset$. Thus let $F \subset (s, t)$ be nonempty and finite. We will show that $U(s, F) \cap U(t, F) = \emptyset$. Suppose that $r \in U(s, F) \cap U(t, F)$. Thus $F \cap [r, s] = \emptyset = F \cap [r, t]$. However by Lemma 1, $[s, t] \subset [r, s] \cup [r, t]$. This would mean $F \cap [s, t] = \emptyset$ contradicting the choice of F. Thus T is Hausdorff.

For regularity consider $U(s, \{x_1 \ldots x_n\})$ where $s \neq x_i$. By the Remark, such sets form a basis at the point s. Since no points of T are adjacent, we may choose $y_i \in (s, x_i)$. By the Corollary to Lemma 2,

$$U(s, \{y_1, \ldots y_n\}) \cup \{y_1, \ldots y_n\} \subset U(s, \{x_1, \ldots x_n\}).$$

The closure

$$\overline{U(s, \{y_1, \ldots y_n\})} = U(s, \{y_1, \ldots y_n\}) \cup \{y_1, \ldots y_n\}$$

and so the topology is regular.

Consider a closed interval $[x, y]$ ($x \neq y$) of T. Use Lemma 3 to put a linear order on $[x, y]$. The subspace topology on $[x, y]$ will be exactly the order topology on $[x, y]$. By the Corollary to Theorem 6, $[x, y]$ has the supremum property (that is every nonempty set has a supremum in the linear order), and since no two points of $[x, y]$ are adjacent, it follows by standard results in linear topology that $[x, y]$ is connected and therefore a big arc. For any $z \in (x, y)$, as in the proof that T is Hausdorff, $U(x, \{z\}) \cap U(y, \{z\}) = \emptyset$. Also $\{U(t, \{z\}) | t \neq z\}$ is a collection of nonempty disjoint open sets whose union is $T - \{z\}$. Thus any connected set of T which contains both x and y will also contain $[x, y]$, and so any big arc γ from x to y will contain $[x, y]$. However a big arc is an irreducible (maybe not metric) continuum between its endpoints, and therefore γ contains no sub-big arc from $[x, y]$. Thus T is uniquely big arcwise connected, and T is a big tree.

Now suppose that X is metrizable and every interval of P contains only countably many adjacent pairs. We must show that each closed interval $[s, t]$ of T is an arc. By standard results in linear topology, we need only find a countable dense set in $[s, t]$ (since $[s, t]$ is connected). We can easily reduce to the case where $s, t \in P$. Since X is a compact metric space, it follows that there exists a countable dense subset $\hat{D} \subset X$. Let $D \subset P \subset T$ be the set of equivalence classes of elements of $\hat{D}$. For any point $x \in T$, since T is uniquely big arcwise connected, there is a unique point $\pi(x) \in [s, t]$ such that $[x, \pi(x)] \cap [s, t] = \{\pi(x)\}$. This defines a continuous function $\pi : T \to [s, t]$.

Define

$$Q = \pi(D) \quad \cup \quad \bigcup_{\substack{x, y \in [s, t] \\ x, y \text{ adjacent in } P}} \mathbb{Q}_{x,y}$$

where $\mathbb{Q}_{x,y}$ is the set of rationals in $\mathbb{R}_{x,y}$. Since there are only countably many adjacent pairs in $[s, t] \cap P$, Q will be countable. We now will show that Q is dense in $[s, t]$. Let $u, v \in [s, t]$ with $u \in [s, v]$. If $(u, v) \not\subset P$ then by definition of T, $(u, v) \cap \mathbb{R}_{x,y} \neq \emptyset$ for some adjacent pair x, y of $[s, t] \cap P$ and so $(u, v) \cap \mathbb{Q}_{x,y} \neq \emptyset$. Thus we may assume $(u, v) \subset P$. Since there are no adjacent points in T, $(u, v) \neq \emptyset$ and in fact (u, v) is infinite since $[u, v]$ is connected. Every two nonsingleton equivalence class of X will be separated by at least one element of C, and so it follows that there are distinct $c, c' \in (u, v) \cap C$. Let $p \in (c, c')$. With no loss of generality $c \in (u, p) \subset (s, p)$ and $c' \in (p, v) \subset (p, t)$. This implies that there are non-singleton subcontinua $A, B, A', B' \subset X$ so that $A \cup B = X = A' \cup B'$, $A \cap B = \{c\}$, $A' \cap B' = \{c'\}$, $s, u \subset A$, $t, v \subset A'$, and $p \subset B \cap B'$. The set $U = (B - \{c\}) \cap (B' - \{c'\})$ is open and non-empty $(p \in U)$, and so there exist $\hat{d} \in U \cap (\hat{D})$. Let $d \in D$ the corresponding equivalence class.

It will suffice to show that $\pi(d) \in [c, c']$. Since $\hat{d} \in B - \{c\}$, it follows by definition that $c \in (s, d)$, and similarly $c' \in (d, t)$. Thus by the Corollary to Lemma 2, $[s, c] \subset [s, d]$ and $[c', t] \subset [d, t]$. It follows by definition that $\pi(d) \notin [s, c)$ and $\pi(d) \notin (c', t]$ Thus $\pi(d) \in [c, c'] \subset (u, v)$ as required, and so $[s, t]$ is separable and therefore an arc.

Corollary. *If C is countable then T is a real tree.*

Proof. It suffices to show that any interval of T with endpoints in P is an arc. Let I be an interval of T with endpoints in P. By Lemma 4 each adjacent pair of I will have exactly one point in C. Furthermore for any $c \in C \cap I$, there at most two points of I adjacent to c. Thus there is a function from the adjacent pairs of I to C which is at most 2 to 1. Thus the set of adjacent pairs of I is countable. Define

$$Q = (C \cap I) \quad \cup \quad \bigcup_{\substack{x, y \in I \\ x, y \text{ adjacent in } P}} \mathbb{Q}_{x,y}$$

where $\mathbb{Q}_{x,y}$ is the set of rationals in $\mathbb{R}_{x,y}$. Clearly Q will be countable. For $a, b \in I$ distinct, either $(a, b) \cap C \neq \emptyset$ or $(a, b) \cap \mathbb{Q}_{x,y} \neq \emptyset$ for some adjacent pair x, y of I. In either case $(a, b) \cap Q \neq \emptyset$. Thus Q is a countable dense set of I and I is an arc.

Conjecture. *If X is metrizable then every interval of P has only countably many adjacent points and so T is a real tree.*

Definition. The group G is called a convergence group if for each sequence of distinct elements of G, there exists a subsequence (g_i) and points $N, P \in X$ such that for any neighborhood U of P and any compact $K \not\ni N$, $g_i(K) \subset U$ for all $i \gg 0$.

We will not discuss the properties of convergence groups here, but refer the reader to [14].

Definition. An action on a real tree T is *non-nesting* if no arc of T is mapped to a proper subset of itself.

Definition. A non-nesting action is *stable* if for each arc γ of T, there is a non-trivial subarc α of γ so that $\text{Fix}(\alpha) = \text{Fix}(\beta)$ for all β nonsingleton subarcs of α.

Definition. A point of a big tree is *terminal* if it is not contained in the interior of any big arc.

Lemma 7. *Let G be a convergence group and $H < G$ a hyperbolic subgroup with $\Lambda H = \{a, b\} \subset X$. If a and b are separated by an element of C, then the corresponding equivalence classes $[a], [b] \subset T$ are terminal in T, and $[a]$ and $[b]$ are the only elements of T fixed by H. If on the other hand a and b are not separated by any element of C, then $|([a], [b]) \cap P| \leq 1$.*

Proof. First consider the case when a and b are separated by some $c \in C$. Thus we have $a \in A - \{c\}$, $b \in B - \{c\}$ where A and B are nonsingleton subcontinua with $A \cup B = X$ and $A \cap B = \{c\}$. By [14] we may choose a $h \in H$ non-torsion so that $h^n(B) \to b$ as $n \to \infty$ and $h^n(A) \to a$ as $n \to -\infty$. Replacing h with a power, we may assume that $h(B) \subsetneq B$ and $h^{-1}(A) \subsetneq A$.

We now show that $[b]$ is terminal in P and therefore in T; the argument for $[a]$ will be identical. Suppose not, then using Lemmas 1 and 2 there is $d \in X$ such that $[b] \in ([a], [d])$. Since h fixes a and b, it follows that $h^n(c) \in ([a], [b])$ and so by Lemma 3 applied to $([a], [d])$, $h^n(c) \notin ([b], [d])$ for all n. This implies that $d \in h^n(B)$ for all n, but $\cap h^n(B) = \{b\}$. Thus $b = d$ which implies $[b] \in ([a], [b])$ contradicting Lemma 2.

To see that only $[a]$ and $[b]$ are fixed by the action of H on T, notice that for any $x \in P - \{[a], [b]\}$ either $x \subset A$ or $x \subset B$. Without loss of generality $x \subset B$. Since the nested intersection $\cap g^n(B) = \{b\}$, it can be shown that $g^n(x) \to [b]$ as $n \to \infty$. Thus g cannot fix x.

In the case where no element of C separates a from b, it follows that $([a], [b]) \cap P \subset (P - C)$. Furthermore no two points of $([a], [b]) \cap P$ are separated by an element of C, and so $([a], [b]) \cap P$ is at most a single point of $P - C$.

Theorem 8. *If G is a countable convergence group, without any infinite torsion subgroup, acting on a non-singleton continuum $X = \Lambda G$ with cut points,* *then G has a non-nesting stable action on an $\mathbb{R}$-tree R. Furthermore the stabilizer of an arc will be an elementary subgroup of G (viewed as a convergence group), and no point of R is fixed by G.*

Proof. Choose C to be a non-empty countable G-invariant set of cutpoints of X (For example the orbit of a cutpoint). Construct the real-tree T as in the Corollary to the Main Theorem. Let $t \in P$, and consider $S = \bigcup_{g \in G} [t, g(t)]$. Clearly S is connected and therefore a real tree. It is easily seen that S is G invariant by using Lemma 1. We will show that S is metrizable by showing that it has a countable basis. For each $g \in G$ let $\mathbb{Q}_g$ be a countable dense subset of the arc $[t, g(t)]$. Let $\mathbb{Q} = \cup \mathbb{Q}_g$. $\mathbb{Q}$ is countable since G is and it is dense in any interval of S.

Consider $U = U(s, F) \cap S$ for some $s \in S$ with $s \notin F$ where of course F is finite. We may assume that $F \subset S$ since $U(s, (F \cap S)) \cap S = U$. For each $f \in F$, choose $q_f \in (s, f) \cap \mathbb{Q}$. Let $E = \{q_f : f \in F\}$. The point $s \in U(s, E) \cap S \subset U$. Clearly there is a $q \in U(s, E) \cap \mathbb{Q}$ implying by the remark that $U(s, E) = U(q, E)$. Thus the collection of all $U(q, E) \cap S$ for $q \in \mathbb{Q}$ and $E \subset \mathbb{Q}$ finite forms a basis for the subspace topology on S, and this collection is countable since $\mathbb{Q}$ is. Thus the subspace topology on S is Hausdorff and regular with a countable basis, and so by Urysohn's Metrization Theorem, S is metrizable.

Let $R = S - \{$ terminal points $\}$. Thus G acts on the $\mathbb{R}$-tree R. By Lemma 7 if $g \in G$ is hyperbolic then either some element of C separates the limit points, g^∞ and $g^{-\infty}$, of the hyperbolic subgroup $\langle g \rangle$, in which case g fixes no point of R, or $|[[g^\infty], [g^{-\infty}]] \cap P| \leq 3$. In the latter case either $[g^\infty] = [g^{-\infty}]$, $[g^\infty]$ is adjacent to $[g^{-\infty}]$ in P, or $g^\infty, g^{-\infty} \in C$ with $\{\Delta\} = (g^\infty, g^{-\infty}) \cap P$ where $g^{\pm\infty} \in \overline{\Delta}$ in X.

Suppose that the action is nesting. That is there exist $x, y \in R \cap P$ so that $g([x, y]) \subsetneq [x, y]$. By squaring g if need be, we may assume that g is order preserving ie. $g(y) \in (g(x), y)$. We may assume that $x \neq g(x)$ and it follows by the G invariance of C that there exists $c \in (x, g(x)] \cap C$. Thus there exist non-singleton subcontinua $A \supset x$, $B \supset y$, with $A \cap B = \{c\}$ and $A \cup B = X$. Since g preserves the order, it follows that $g(A) \subsetneq A - \{c\}$. By [14] g is hyperbolic with one limit point in $A - \{c\}$ and the other in $B - \{c\}$. By the preceding paragraph, g fixes no point of R which contradicts the Brouwer fixed point theorem since the arc $[x, y]$ is mapped inside itself by g. Thus G acts non-nestingly on R.

Suppose the action is not stable. That is there exist distinct $x, y \in P$ such that for any non-singleton interval $I \subset [x, y]$, there exists a non-singleton interval $J \subset I$ with the subgroups $\text{Fix}(J) \neq \text{Fix}(I)$.

If $[x, y]$ contains a pair of adjacent points p, q of P, then for any interval any $I \subset [p, q]$ in T, $\text{Fix}(I) = \text{Fix}([p, q])$ which is a contradiction. Thus $[x, y]$ contains no pair of adjacent elements of P. Let $I \subset [x, y]$ be a non-singleton interval. Since I will contain infinitely many elements of C, it will follow that Fix(I) contains no hyperbolic or parabolic elements. Thus Fix(I) is a finite group. Choose a sequence of intervals $[x, y] \supset I_1 \supset I_2 \ldots$ so that Fix(I_i) $\neq$ Fix(I_{i+1}) for all $i > 0$. It follows that the nested union $\cup$Fix(I_i) is an infinite torsion subgroup of G, contradicting the hypothesis.

We now show that arc stabilizers are elementary. Let $[x, y]$ be an arc of R. We may assume that $x, y \in P$, $x \neq y$. There exists $c \in [x, y] \cap C$. Fix($[x, y]$) < Fix($\{c\}$), but $c \in X$, and any subgroup of G which fixes a point of X is elementary by [14].

The fact that G fixes no point of R follows from the existence of a hyperbolic element $g \in G$ with g^∞ and $g^{-\infty}$ separated by an element of C. This in turn follows from [14] since $\Lambda G = X$.

REFERENCES

[1] J. Alonso, T. Brady, D. Cooper, T. Delzant, V. Ferlini, M. Lustig, M. Mihalik, M. Shapiro, and H. Short, *Notes on word hyperbolic groups*, Group Theory from a Geometrical Viewpoint (E. Ghys , A. Haefliger, and A. Verjovsky ed.) World Scientific, Singapore, 1992.

[2] M. Bestvina, *Local homology properties of boundaries of groups*, Michigan Math. J., vol. 43 (1996), 123-139.

[3] M. Bestvina and G. Mess, *The boundary of negatively curved groups*, J. Amer. Math. Soc., vol. 4 (1991), 469-481.

[4] B. Bowditch, *Treelike structures arising from continua and convergence groups*, Mem. Amer. Math. Soc. 139 (1999).

[5] B. Bowditch, *Group actions on trees and dendrons*, Topology, vol. 37 (1998), 1275-1298.

[6] B. Bowditch, *Boundaries of strongly accessible hyperbolic groups*, Geometry and Topology Monographs, vol 1: The Epstein Birthday Schrift (ed. I. Rivin, C. Rourke, C. Series.) 51-97.

[7] B. Bowditch, *Connectedness properties of limit sets* Trans. Amer. Math. Soc. 351 (1999), 3673-3686.

[8] B. Bowditch and G. A. Swarup, *Cut points in the boundary of hyperbolic groups* preprint 1998.

[9] M. Coornaert, T. Delzant, A. Papadopoulos, *Géométrie et théorie des groupes*, Springer Lecture Notes, vol. 1441 (1991).

[10] E. Freden, *Negatively curved groups have the convergence property I*, Ann. Acad. Sci. Fenn., vol 20 (1995), 157-187.

[11] G. Levitt, *Non-nesting actions on real trees*, Bull. London Math. Soc., vol. 30 (1998) 46-54.

[12] J. Mayer and L. Oversteegen, *A topological characterization of $\mathbb{R}$-trees*, Trans. Amer. Math. Soc., vol 320 (1990) 395-415.

[13] G. A. Swarup, *On the cut point conjecture*, Elec. Res. Anounc. Amer. Math. Soc., vol. 2, (1996) 98-100.

[14] P. Tukia, *Convergence groups and Gomov's metric hyperbolic spaces*, New Zealand Jour. of Math., vol. 23 (1994), 157-187.

Eric L. Swenson
Brigham Young University
Provo UT 84604, USA

GENERALISED TRIANGLE GROUPS OF TYPE $(2, m, 2)$

ALUN G T WILLIAMS

1. INTRODUCTION

A *generalised triangle group* is a group with presentation

(1.1) $$G = \langle a, b \mid a^l = b^m = w^n = 1 \rangle$$

where l, m, n are integers greater than 1, and w is a word of the form

$$a^{\alpha_1} b^{\beta_1} \ldots a^{\alpha_k} b^{\beta_k},$$

$k \geq 1$, $0 < \alpha_i < l$, $0 < \beta_i < m$ for all i, which is not a proper power. We say that two words w and v are *equivalent* if we can transform one to the other by a sequence of the following moves

1. cyclic permutation;
2. inversion;
3. automorphism of $\mathbb{Z}_l$ or $\mathbb{Z}_m$; or
4. interchanging the two free factors (if $l = m$);

and we write $w \sim v$. If in the presentation 1.1, we replace w by an equivalent word v, then we get an isomorphic copy of G. Thus it is enough to study generalised triangle groups up to equivalence of w.

It is well known that the ordinary triangle groups

$$T = \langle a, b \mid a^l = b^m = (ab)^n = 1 \rangle$$

satisfy a *Tits alternative*. That is, they either contain a soluble subgroup of finite index or have a non-abelian free subgroup. In [2], Rosenberger asks whether a Tits alternative holds for generalised triangle groups.

Work in [1, 2, 3, 4, 5] shows this conjecture to be true except possibly where $k > 4$ and $(l, m, n) = (3, 3, 2), (3, 4, 2), (3, 5, 2)$ or $(2, m, 2)$ $(m \geq 3)$. It is this last case we address here. Levin and Rosenberger have proved

Theorem 1.1. [3] *Let G be the generalised triangle group*

(1.2) $$G = \langle a, b \mid a^2 = b^m = w^2 = 1 \rangle$$

where $w = ab^{\alpha_1} \ldots ab^{\alpha_k}$. If k is even and $m \geq 7$, $m \neq 8, 10, 16$, then G contains a non-abelian free subgroup.

In this paper we prove two theorems, the first of which completes the omissions of, and extends Theorem 1.1.

Theorem A. *Let G be the group in 1.2. Suppose one of the following holds:*
1. $m = 2^r$ $(r \geq 3)$, $k > 1$

2. $m = 10$ *and* k *is even*

3. $m = 6$ *and* k *is divisible by 4*

Then G contains a non-abelian free subgroup.

Thus, if $m \geq 7$, G contains a non-abelian free subgroup whenever k is even. If k is odd, it is harder to show that G contains a non-abelian free subgroup, but we can show a slightly weaker property. Wilson and Zelmanov have proved

Theorem 1.2. [6] *Let G be a discrete group, and let d denote the minimum number of generators for G/G'. If $d \geq 2$ and*

$$(1.3) \qquad \operatorname{def}(G) + d^2/4 - d > 0$$

then, for some prime p, the pro-p completion of G contains a non-abelian free subgroup.

It is not hard to show that if a group G contains a non-abelian free subgroup H, then the pro-p completion of H also contains a non-abelian free subgroup (for some prime p). Also, if the pro-p completion of a group contains a non-abelian free subgroup, then the group is necessarily infinite. The converse is not true – a simple counterexample is provided by the integers. We can now address the group in 1.2 whether k is odd or even.

Theorem B. *Let G be the group in 1.2, where $k > 1$. If $m \geq 7$ then, for some prime p, G has a subgroup whose pro-p completion contains a non-abelian free subgroup.*

In [7], Wilson conjectures that a group satisfying the conditions of theorem 1.2 contains a non-abelian free subgroup. Howie [5] has proved this conjecture for groups of deficiency at least 1.

If Wilson's conjecture can be proved for groups of deficiency -1 then the proof of Theorem B shows that the groups in 1.2 contain non-abelian free subgroups for all $m \geq 7$. Conversely, if it can be shown that one of these groups does not contain a non-abelian free subgroup, then this will provide a counterexample to Wilson's conjecture.

As in [5], we shall say that a representation ρ of a generalised triangle group 1.1 is *essential* if $\rho(a), \rho(b), \rho(w)$ have orders l, m, n respectively. Baumslag, Morgan and Shalen [1] have shown that every generalised triangle group admits an essential representation to $PSL(2, \mathbb{C})$, which (for our purposes) occurs in the following manner. Let $f : \langle a, b \rangle \rightarrow SL(2, \mathbb{C})$ be given by $f(a) = A$, $f(b) = B$ where A and B are matrices with traces 0 and $2\cos(\pi/m)$ respectively, and let $\rho : G \rightarrow PSL(2, \mathbb{C})$ be given by $\rho(a) = A, \rho(b) = B$ where A and B are considered as elements of $PSL(2, \mathbb{C})$. Then $\operatorname{tr} f(w)$ is a polynomial of degree k in $\lambda = \operatorname{tr} f(ab)$, which we denote τ_w.

Let $W(A, B)$ be the word w in matrices A and B. Since $A^{-1} = -A$, we have that $\operatorname{tr}(A^{-1}B) = -\operatorname{tr}(AB) = -\lambda$ and $\operatorname{tr}(W(A^{-1}, B)) = \pm\operatorname{tr}(W(A, B))$.

This means

$$\tau_w(\lambda) = \text{tr}(W(A,B)) = \pm\text{tr}(W(A^{-1},B)) = \pm\tau_w(\text{tr}(A^{-1}B)) = \pm\tau_w(-\lambda)$$

and so λ and $-\lambda$ occur as roots of τ_w with equal multiplicity.

The representation ρ is essential if and only if $\tau_w(\lambda) = 0$. For the remainder of this paper, we will not mention the map f and consider λ to be $\text{tr}\rho(ab)$ and τ_w as $\text{tr}\rho(w)$.

Suppose $\rho(G)$ does not contain any free subgroup of rank 2, then by Corollary 2.5 of [2], $\rho(G)$ is elementary. By Corollary 2.4 of [2] there are then three possibilities: $\rho(G)$ is finite, $\text{tr}[A,B] = 2$, or $\lambda = 0$.

Since $\rho(G)$ contains an element of order $m \geq 6$, if it is to be a finite group then it is either cyclic or dihedral. A straightforward calculation using trace identities shows

$$\text{tr}[A,B] = (\text{tr}A)^2 + (\text{tr}B)^2 + (\text{tr}(AB))^2 - (\text{tr}A\ \text{tr}B\ \text{tr}(AB)) - 2$$

Thus if $\text{tr}[A,B] = 2$, then $\lambda = \pm 2\sin(\pi/m)$, which corresponds to an essential representation to a cyclic group. The value $\lambda = 0$ corresponds to an essential representation to a dihedral group.

Thus in proving Theorems A and B, it is enough to assume that the roots of τ_w correspond to essential cyclic or dihedral representations.

In Section 2, we will rule out the cyclic case. This will give us enough information about the trace polynomial to prove Theorem A. In Section 3, in proving Theorem B, we show that neither the cyclic nor dihedral cases can occur.

Acknowledgement. I would like to thank my supervisor Professor James Howie for his help in connection with this work.

2. The proof of Theorem A

We note the following lemma, which is a slight refinement of results by Howie [5].

Lemma 2.1. *Let G be the group in 1.2 where $m \geq 6$ and even, and suppose G has an essential cyclic representation. Then G contains a non-abelian free subgroup unless $m = 6$ and $w \sim abab^2$, in which case G is infinite soluble.*

Proof. By Theorems 4.7 and 4.8 of [5] we may assume that $m = 6$ and that 1 is not a repeated root of the trace polynomial $\tau_w(\lambda)$. Since $m = 6$ and G admits an essential cyclic representation, 1 occurs exactly once as a root of the trace polynomial, and hence so does -1.

If $k \geq 3$ then τ_w has other roots, and these correspond to dihedral representations. Thus the trace polynomial is given by $\tau_w(\lambda) = c\lambda^{k-2}(\lambda^2 - 1)$.

Consider an essential representation $\rho : G \to \mathbb{Z}_6$, $a \mapsto 3$, $b \mapsto 1$. Then $K = \ker\rho$ has a deficiency zero presentation with generators

$$x_1 = abab^{-1}, \quad x_2 = ab^2ab^{-2}, \quad x_3 = ab^3$$

and relators

$$v_i(x_1, x_2, x_3)v_i(x_1^{-1}, x_2^{-1}, x_3^{-1}) \qquad (1 \le i \le 3).$$

Thus K/K' is free abelian of rank 3. There exists an essential representation $\sigma : G \to D_{12} = \langle x, y \mid x^2 = y^6 = (xy)^2 = 1 \rangle$, $a \mapsto x$, $b \mapsto y$. Now $\sigma(K') = \langle y^2 \mid y^6 = 1 \rangle \le D_{12}$, so there exists $N \triangleleft K$ such that K'/N is abelian and nontrivial. In particular $K'/K'' \ne 1$. Furthermore, K admits an automorphism ϕ given by $\phi : x_i \mapsto x_i^{-1}$, which satisfies $\phi(K') = K'$, and induces the antipodal automorphism on K/K', so by Corollary 3.2 of [5], K' and hence G contain a free subgroup of rank 2.

Thus we are left with the cases $k = 1$ or 2. If $k = 1$, then there are no essential cyclic representations. If $k = 2$, then by proposition 8 of [2], G contains a free subgroup of rank 2 unless $w \sim abab^4$ or $abab^2$. In the first case, G does not admit any essential cyclic representation, and in the second case G is infinite soluble. $\qquad \square$

Lemmas 2.3-2.5 which will combine to prove Theorem A all start with the same argument, which we now describe.

Assume (for contradiction) that G contains no non-abelian free subgroup. By applying Lemma 2.1 we can assume that the roots of τ_w correspond to dihedral representations. Thus

$$\tau_w(\lambda) = c\lambda^k$$

where by [1]

$$c = \frac{1}{(\sin(\pi/m))^k} \prod_{i=1}^{k} \sin(\pi\alpha_i/m).$$

Let A and B be elements of $SL(2, \mathbb{C})$ with traces 0 and $2\cos(\pi/m)$ respectively. Let $\phi : \langle a, b \mid a^2 = b^m = 1 \rangle \to PSL(2, \mathbb{C})$ be given by $\phi : a \mapsto A$, $b \mapsto B$, where A and B are considered as elements of $PSL(2, \mathbb{C})$. If ϕ is a cyclic representation, then $\lambda = 2\sin(\pi/m)$. Clearly $\phi(a)$, $\phi(b)$ have orders 2 and m respectively; let δ be the order of $\phi(w)$. Then δ divides m, and $\delta \ne 2$, for otherwise ϕ induces an essential cyclic representation $G \to PSL(2, \mathbb{C})$, contrary to assumption. Hence

$$\text{tr}\phi(w) = 2\cos(q\pi/m)$$

for some $-m < q < m$, but

$$\begin{aligned} \text{tr}\phi(w) &= \tau_w(2\sin(\pi/m)) \\ &= c \cdot (2\sin(\pi/m))^k \end{aligned}$$

and so

$$\frac{1}{(\sin(\pi/m))^k} \prod_{i=1}^{k} \sin(\pi\alpha_i/m) \cdot (2\sin(\pi/m))^k \;=\; 2\cos(q\pi/m).$$

Hence

$$2^{k-1} \cdot \prod_{i=1}^{k} \sin(\pi\alpha_i/m) \;=\; \cos(q\pi/m)$$

but since $2^{k-1} \cdot \prod_{i=1}^{k} \sin(\pi\alpha_i/m) > 0$, we have that $0 \le q < m/2$ which gives

$$(2.1) \qquad\qquad 2^{k-1} \cdot \prod_{i=1}^{k} \sin(\pi\alpha_i/m) = \sin(q'\pi/m)$$

for some $0 < q' \le m/2$. We will use the following notation. For all $1 \le j \le m/2$, let $t_j = \sin(j\pi/m) = \sin(\pi\alpha_i/m)$ when $\alpha_i = j$ or $(m-j)$. Also, let k_j be the number of times α_i takes the value j or $(m-j)$ in the word w, so that $k = k_1 + k_2 + \ldots + k_{m/2}$.

For any word $w = ab^{\alpha_1} \ldots ab^{\alpha_k}$, we will denote the left hand side of equation 2.1 by $M(w)$, ie

$$M(w) = 2^{k-1} \cdot \prod_{i=1}^{k} \sin(\pi\alpha_i/m)$$

We aim to contradict equation 2.1, and thus prove the existence of a non-abelian free subgroup. That is, we want to show that there exists a word v, equivalent to w, such that $M(v) \ne t_j$ for any $1 \le j \le m/2$. Clearly two words v and w, equivalent under cyclic permutation or inversion have $M(v) = M(w)$. However, if w and v are equivalent under an automorphism θ of $\mathbb{Z}_m$, ie

$$w = ab^{\alpha_1} \ldots ab^{\alpha_k}$$
$$v = ab^{\theta(\alpha_1)} \ldots ab^{\theta(\alpha_k)}$$

then they (potentially) give different values for M:

$$M(w) = 2^{k-1} \cdot \prod_{i=1}^{k} \sin(\pi\alpha_i/m) = 2^{k-1} t_1^{k_1} \ldots t_{m/2-1}^{k_{m/2-1}}$$

$$M(v) = 2^{k-1} \cdot \prod_{i=1}^{k} \sin(\pi\theta(\alpha_i)/m) = 2^{k-1} t_1^{k(\theta(1))} \ldots t_{m/2-1}^{k(\theta(m/2-1))}$$

Let R be a largest set of equivalent words with different values of M, and let $P = \prod_{w \in R} M(w)$. We try to show $P > 1$, for then there is at least one v such that $M(v) \ne t_j$ for any $1 \le j \le m/2$. If we cannot find a contradiction by this method, we try to find a v such that $M(v) \ne t_j$ for any $1 \le j \le m/2$.

In the case $m = 2^r$, we have a formula for P given in the following proposition.

Proposition 2.2. *If $m = 2^r$ then $|R| = 2^{r-2}$, and $P = \prod_{w \in R} M(w)$ is given by*

$$P = 2^{(k-1) \cdot 2^{r-2}} \cdot 2^{-(2^{r-1}-1) \cdot \sum_{(i,2^{r-1})=1} k_i/2} \cdot 2^{-(2^{r-2}-1) \cdot \sum_{(i,2^{r-2})=1} k_{2i}/2} \cdots$$
$$\cdot 2^{-(2^2-1) \cdot 2^{r-3} \sum_{(i,2^2)=1} k_{2^{r-3}i}/2} \cdot 2^{-1 \cdot 2^{r-2} k_{2^{r-2}i}/2} \cdot 1$$

Proof. We rely on two claims about the cycle structure of permutations of $\{t_1, \ldots, t_{m/2-1}\}$.

Claim 1. *The subgroup of the automorphism group of $\mathbb{Z}_m$, generated by the map $i \mapsto 3i$, induces the following permutation on $\{t_1, \ldots t_{m/2-1}\}$*

$$(S_1)(S_2)(S_3) \ldots (S_r)$$

where (S_i) denotes some cyclic permutation of the elements of the set S_i, and

$$\begin{array}{ll}
S_1 = \{t_{2^{r-1}}\} & |S_1| = 1 \\
S_2 = \{t_{2^{r-2}}\} & |S_2| = 1 \\
S_3 = \{t_{2^{r-3}i} \mid (i, 2^2) = 1\} & |S_3| = 2^1 \\
\vdots & \vdots \\
S_{j+2} = \{t_{2^{r-(j+2)}i} \mid (i, 2^{(j+1)}) = 1\} & |S_{j+2}| = 2^j \\
\vdots & \vdots \\
S_r = \{t_{2^0 i} \mid (i, 2^{(r-1)}) = 1\} & |S_r| = 2^{r-2}
\end{array}$$

Proof of Claim 1. Assume that for $m = 2^r$, the t_i $(1 \le i \le 2^r/2)$ are permuted as stated in the claim. Let $m = 2^{r+1}$, and call the new values of t_i, s_i, so that

$$s_j = \sin(j\pi/2^{r+1}) \qquad \text{for all } 1 \le j \le 2^r.$$

Then $s_{2i} = t_i$, and these are permuted in the manner stated. This leaves the s_i where $(i, 2^r) = 1$, and there are 2^{r-1} of these. Now, the order of 3 modulo 2^{r+1} is 2^{r-1}, so the permutation of the remaining s_i, is a single 2^{r-1}-cycle, and the induction is complete.

Claim 2. $\prod_{t \in S_j} t = 2^{-(2^{j-1}-1)/2}$ *for each $1 \le j \le r$.*

Proof of Claim 2. Again, the proof is by induction on r. If $r = 3$ then the sets S_1, S_2, S_3 are given as above, and clearly

$$\begin{array}{lll}
\pi_1 &= t_4 = 1 = 2^{-(2^{1-1}-1)/2} \\
\pi_2 &= t_2 = 2^{-1/2} = 2^{-(2^{2-1}-1)/2} \\
\pi_3 &= t_1 t_3 = (\sqrt{(2-\sqrt{2})})/2 \cdot (\sqrt{(2+\sqrt{2})})/2 = 2^{-3/2} = 2^{-1(2^{3-1}-1)/2}.
\end{array}$$

where π_j denotes $\prod_{t \in S_j} t$.

The following result is identity 1.392 (1) of [8]. I would like to thank Alan Prince and Oliver Penrose for providing details of this result.

$$2^{n-1} \prod_{\alpha=0}^{n-1} \sin(x + \alpha\pi/n) = \sin(nx)$$

Differentiating this and setting $x = 0$, $n = 2^r$ provides

$$2^{2^r-1} \prod_{\alpha=1}^{2^r-1} \sin(\pi\alpha/2^r) = 2^r$$

Using this we get

$$\left(\prod_{i=1}^{2^r-1} t_i\right)^2 = 2^{r-2^r+1}$$

which implies

(2.2) $$\pi_1 \cdot \pi_2 \cdot \ldots \cdot \pi_r = 2^{(r-2^r+1)/2}.$$

Now assume inductively that

$$\pi_j = 2^{-(2^{j-1}-1)/2}$$

for all $j < r$. Then equation 2.2 implies

$$2^{-(2^{1-1}-1)/2} \cdot 2^{-(2^{2-1}-1)/2} \ldots 2^{-(2^{r-1-1}-1)/2} \cdot \pi_r = 2^{(r-2^r+1)/2}$$

and so

$$\begin{aligned} \pi_r &= 2^{((r-2^r+1)+(2^0+2^1+\ldots+2^{r-2}-(r-1)\cdot 1))/2} \\ &= 2^{-(2^{r-1}-1)/2} \end{aligned}$$

as required.

The order of each S_j divides 2^{r-2}, and this is the length of the longest cycle, so $|R| = 2^{r-2}$. Thus P is given by

$$\begin{aligned} P &= (2^{(k-1)})^{2^{r-2}} \prod_{j=1}^{r} (\pi_j)^{(2^{r-j} \sum_{(i,2^{r-j})=1} k_{2^{r-j}i}/2)} \\ &= 2^{(k-1)\cdot 2^{r-2}} \cdot 1 \cdot 2^{-1\cdot 2^{r-2}k_{2^{r-2}i}/2} \cdot 2^{-(2^2-1)\cdot 2^{r-3} \sum_{(i,2^2)=1} k_{2^{r-3}i}/2} \cdot \ldots \\ &\quad \cdot 2^{-(2^{r-2}-1)\cdot \sum_{(i,2^{r-2})=1} k_{2i}/2} \cdot 2^{-(2^{r-1}-1)\cdot \sum_{(i,2^{r-1})=1} k_i/2} \end{aligned}$$

as required. $\qquad\square$

Lemma 2.3. *Let G be the group in 1.2 where $m = 2^r$ $(r \geq 3)$, $k > 1$. Then G contains a non-abelian free subgroup.*

Proof. The proof is by induction on r. We first anchor at $r = 3$, so

$$G = \langle a, b \mid a^2 = b^8 = w^2 = 1 \rangle.$$

By proposition 2.2 , we have

$$P = 2^{(k_1 + 2k_2 + k_3 + 4k_4 - 4)/2}$$

If $k_1 + 2k_2 + k_3 + 4k_4 > 4$, then $P > 1$, and we have our contradiction. If $k_1 + 2k_2 + k_3 + 4k_4 \leq 4$, then $k_1 + k_2 + k_3 + k_4 \leq 4$, ie $k \leq 4$, and G contains a non-abelian free subgroup by [3], again a contradiction.

When $m = 2^r$, P is given by proposition 2.2 and it is easy to see that $P \leq 1$ if and only if

$$(2^{r-2}(k-1)) \ - \ \left(\frac{2^{r-1}-1}{2} \cdot 2^0 \sum_{(i,2^{r-1})=1} k_i \right) - \left(\frac{2^{r-2}-1}{2} \cdot 2^1 \sum_{(i,2^{r-2})=1} k_{2i} \right) - \ldots$$

$$- \ \left(\frac{2^2-1}{2} \cdot 2^{r-3} \sum_{(i,2^2)=1} k_{2^{r-3}i} \right) - \left(\frac{1}{2} \cdot 2^{r-2} k_{2^{r-2}} \right) \leq 0$$

equivalently

$$(2^{r-2}k) \ - \ \left((2^{r-2} - 2^{-1}) \cdot 2^0 \sum_{(i,2^{r-1})=1} k_i \right) - \left((2^{r-2}-1) \cdot 2^1 \sum_{(i,2^{r-2})=1} k_{2i} \right) - \ldots$$

$$- \ \left((2^{r-2} - 2^{r-4}) \cdot 2^{r-3} \sum_{(i,2^2)=1} k_{2^{r-3}i} \right) - \left((2^{r-2} - 2^{r-3}) \cdot 2^{r-2} k_{2^{r-2}} \right) \leq 2^{r-2}$$

where $k = k_1 + k_2 + \ldots + k_{2^{r-1}}$.

Noting that $k_{2^{r-1}}$ only occurs in the first term of the above equation, and that the coefficient of every k_i is positive, we can see that if $k_{2^{r-1}} \neq 0$, then $k_{2^{r-1}} = 1$, and $k_i = 0$ for all other i. This gives that $k = 1$ contrary to our hypothesis.

Hence we may assume $k_{2^{r-1}} = 0$, that is, in the word w, b never takes the power 2^{r-1}. Therefore in the factor group $H = G/\langle b^{2^{r-1}} \rangle^G$, with presentation

$$H = \langle a, b \mid a^2 = b^{2^{r-1}} = \overline{w}^2 = 1 \rangle$$

where $\overline{w} = ab^{\alpha_1 \bmod 2^{r-1}} \ldots ab^{\alpha_k \bmod 2^{r-1}}$, the length of $\overline{w}$ is equal to the length of w. Since $G = \langle a, b \mid a^2 = b^8 = w^2 = 1 \rangle$ has a free subgroup of rank 2, it follows by induction that

$$G = \langle a, b \mid a^2 = b^{2^r} = w^2 = 1 \rangle$$

has a free subgroup of rank 2 for all $r \geq 3$ – the required contradiction. $\quad\square$

Lemma 2.4. *Let G be the group in 1.2 where $m = 10$ and k is even. Then G contains a non-abelian free subgroup.*

Proof. Applying the automorphism $i \mapsto 3i$ of $\mathbb{Z}_{10}$ shows that $|R| = 2$, $R = \{v, w\}$ say, and

$$M(w) = t_1^{k_1} t_2^{k_2} t_3^{k_3} t_4^{k_4} \cdot 2^{k-1}$$
$$M(v) = t_1^{k_3} t_2^{k_4} t_3^{k_1} t_4^{k_2} \cdot 2^{k-1}$$

and so

$$P = M(w)M(v) = (t_1 t_3)^{k_1+k_3}(t_2 t_4)^{k_2+k_4} \cdot 2^{2(k-1)}$$

Now $t_1 t_3 = 1/4$, and $t_2 t_4 = \sqrt{5}/4 > 1/2$, so

$$P > 2^{2k-2} \cdot (1/4)^{k_1+k_3} \cdot (1/2)^{k_2+k_4} = 2^{k_2+k_4+2k_5-2}$$

If $k_2 + k_4 + 2k_5 > 2$, then $P > 1$, and we have a contradiction. Thus we may assume $k_2 + k_4 + 2k_5 \leq 2$. If $k_5 \neq 0$, then $k_5 = 1$, $k_2 = k_4 = 0$, which implies

$$M(w)M(v) = 2^{2k_1+2k_3+2-2}(t_1 t_3)^{k_1+k_3} = 1.$$

So, either one of $M(w)$, $M(v)$ is greater than 1, where we have a contradiction, or $M(w) = M(v) = 1$, in which case

$$(t_1/t_3)^{k_1-k_3} = 1 \Rightarrow k_1 = k_3$$

which cannot happen since $k = k_1 + k_3 + 1$ is even.

Hence we may assume $k_5 = 0$. In this case we consider the factor group $H = G/\langle b^5 \rangle^G$ of G, with presentation

$$H = \langle a, b \mid a^2 = b^5 = \overline{w}^2 = 1 \rangle$$

where $\overline{w} = ab^{\alpha_1 \bmod 5} \ldots ab^{\alpha_k \bmod 5}$. Since $k_5 = 0$, $\alpha_i \neq 5$ for any i, and so the length of $\overline{w}$ is equal to the length of w. Thus we may assume that H has no non-abelian free subgroup, for otherwise we can lift this to G and then G has a non-abelian free subgroup.

Now, ρ is an essential representation $G \to D_{20}$, given by $a \mapsto x$, $b \mapsto y$, $w \mapsto y^5$. Let $f_1 : G \to H$, and $f_2 : D_{20} \to D_{10}$ be natural maps. Then defining $\sigma_1 : H \to D_{10}$ by $a \mapsto x$, $b \mapsto y$ means that $\overline{w} \mapsto 1$ and the following diagram commutes.

$$G = \langle a, b \mid a^2 = b^{10} = w^2 = 1 \rangle \xrightarrow{\rho} D_{20} = \langle x, y \mid x^2 = y^{10} = (xy)^2 = 1 \rangle \leq PSL(2, \mathbb{C})$$

$$f_1 \downarrow \qquad\qquad\qquad\qquad\qquad\qquad\qquad \downarrow f_2$$

$$H = \langle a, b \mid a^2 = b^5 = \overline{w}^2 = 1 \rangle \xrightarrow{\sigma_1} D_{10} = \langle x, y \mid x^2 = y^5 = (xy)^2 = 1 \rangle \leq PSL(2, \mathbb{C})$$

Then

$$\text{tr}\sigma_1(ab) = \text{tr}(\sigma_1 f_1(ab))$$
$$= \text{tr}(f_2 \rho(ab))$$
$$= 0$$

and

$$\text{tr}\sigma_1(\overline{w}) = \text{tr}(\sigma_1 f_1(w))$$
$$= \text{tr}(f_2\rho(w))$$
$$= \pm 2.$$

Now, for any representation $\sigma : H \to PSL(2,\mathbb{C})$, $\text{tr}\sigma(\overline{w})$ is a polynomial $\psi_{\overline{w}}(\mu)$ (say) where $\mu = \text{tr}\sigma(ab)$. From the above, we have

$$\psi_{\overline{w}}(0) = \pm 2$$

Since k is even, there are no essential representations of H in $\mathbb{Z}_{10}$ or D_{10}, and so the only representations σ of H in $PSL(2,\mathbb{C})$ lie in A_5, which has elements of order 1,2,3,5. Now, if the product z of elements x and y of orders 2 and 5 respectively also has order 5, then z is conjugate to y^2. In terms of traces, this means that if $\text{tr}(y) = \pm 2\cos(\pi/5) = \pm(1 + \sqrt{5})/2$, then $\text{tr}(z) = \pm 2\cos(2\pi/5) = \pm(1 - \sqrt{5})/2$. Hence we may assume that the roots of the trace polynomial lie in $\{\pm 1, \pm \alpha\}$, where $\alpha = (1 - \sqrt{5})/2$. Thus

$$\psi_{\overline{w}}(\mu) = c(\mu^2 - 1)^{\eta_1}(\mu^2 - \alpha^2)^{\eta_2}$$

where $2\eta_1 + 2\eta_2 = k$, and

$$c = \frac{1}{(\sin(\pi/5))^k}(\sin(\pi/5))^{l_1}(\sin(2\pi/5))^{l_2}$$
$$= (2\cos(\pi/5))^{l_2} = ((1 + \sqrt{5})/2)^{l_2}$$

where l_1, l_2 denote the number of times α_i takes the values $\pm 1, \pm 2$ respectively in $\overline{w}$. This gives

$$\psi_{\overline{w}}(0) = \pm((1 + \sqrt{5})/2)^{l_2}((1 - \sqrt{5})/2)^{2\eta_2}$$
$$= \pm(((1 + \sqrt{5})/2)((1 - \sqrt{5})/2)))^{l_2}((1 - \sqrt{5})/2)^{2\eta_2 - l_2}$$
$$= \pm((1 - \sqrt{5})/2)^{2\eta_2 - l_2}$$

which contradicts $\psi_{\overline{w}}(0) = \pm 2$ and hence G has a non-abelian free subgroup. $\square$

Lemma 2.5. *Let G be the group in 1.2, where $m = 6$ and k is divisible by 4. Then G contains a non-abelian free subgroup.*

Proof. Since $m = 6$, $|R| = 1$ and

$$M(w) = 2^{k-1}t_1^{k_1}t_2^{k_2}t_3^{k_3}$$
$$= 2^{k-1}(1/2)^{k_1}(\sqrt{3}/2)^{k_2}(1)^{k_3}$$
$$= 2^{k_3-1}(\sqrt{3})^{k_2}$$
$$\leq 1 \iff (k_2, k_3) = (0,0), (0,1), (1,0).$$

Thus we may assume that

$$w = ab^{\alpha}ab^{\varepsilon_2}\dots ab^{\varepsilon_k}$$

where $\alpha = \pm 1, \pm 2$, or 3, $\varepsilon_i = \pm 1$. We are assuming that G admits only essential dihedral representations, and since k is even, we must have

$$-\alpha + \Sigma_{i=2}^{k}(-1)^i \varepsilon_i = 3 \bmod 6$$

and this can only happen if $\alpha = \pm 2$. Thus (up to equivalence), we may assume

$$w = ab^2 ab^{\varepsilon_2} \ldots ab^{\varepsilon_k}.$$

We now consider the factor group $H = G/\langle b^3 \rangle^G$ with presentation

$$H = \langle a, b \mid a^2 = b^3 = \overline{w}^2 = 1 \rangle$$

where $\overline{w} = ab^{-1}ab^{\varepsilon_2} \ldots ab^{\varepsilon_k}$. Since b never takes the power 3 in w, the length of $\overline{w}$ is equal to the length of w. As in the proof of Lemma 2.4, if H contains a non-abelian free subgroup, then so does G. Thus (for contradiction), we assume that H has no non-abelian free subgroup, and therefore the roots of its trace polynomial, $\psi_w(\lambda)$ correspond to representations to finite subgroups of $PSL(2, \mathbb{C})$. Now H has an essential representation to $\mathbb{Z}_6$ if and only if

$$3k + 2(-1 + \Sigma_{i=2}^{k}\varepsilon_i) = 3 \bmod 6$$

which cannot happen, since k is even. Also, since k is even, H does not admit any essential dihedral representation. Thus we need only consider essential representations onto A_4, S_4, or A_5 and hence the roots of $\psi_{\overline{w}}$ are $\pm 1, \pm\sqrt{2}, (\pm 1 \pm \sqrt{5})/2$. As we mentioned in Section 1, λ and $-\lambda$ occur with equal multiplicity. Now τ_w is a polynomial with integer coefficients so the roots $\pm(1 + \sqrt{5})/2$ and $\pm(1 - \sqrt{5})/2$ also occur with equal multiplicity, so

$$(2.3) \qquad \psi_{\overline{w}}(\mu) = (\mu^2 - 1)^{\eta_1}(\mu^2 - 2)^{\eta_2}(\mu^4 - 3\mu^2 + 1)^{\eta_3}$$

where $2\eta_1 + 2\eta_2 + 4\eta_3 = k$.

Let ρ be an essential representation $G \to D_{12}$, given by $a \mapsto x$, $b \mapsto y$, $w \mapsto y^3$. Let $f_1 : G \to H$, and $f_2 : D_{12} \to D_6$ be natural maps. Then defining $\sigma_1 : H \to D_6$ by $a \mapsto x$, $b \mapsto y$ means that $\overline{w} \mapsto 1$ and the following diagram commutes.

$$
\begin{array}{ccc}
G = \langle a, b \mid a^2 = b^6 = w^2 = 1 \rangle & \xrightarrow{\ \rho\ } & D_{12} = \langle x, y \mid x^2 = y^6 = (xy)^2 = 1 \rangle \leq PSL(2, \mathbb{C}) \\
f_1 \downarrow & & \downarrow f_2 \\
H = \langle a, b \mid a^2 = b^3 = \overline{w}^2 = 1 \rangle & \xrightarrow{\ \sigma_1\ } & D_6 = \langle x, y \mid x^2 = y^3 = (xy)^2 = 1 \rangle \leq PSL(2, \mathbb{C})
\end{array}
$$

This gives

$$\mathrm{tr}(\sigma_1(ab)) = 0$$
$$\mathrm{tr}(\sigma_1(\overline{w})) = \pm 2$$

and so $\tau_{\overline{w}}(0) = \pm 2$. But by equation 2.3 we have

$$\psi_{\overline{w}}(0) = \pm 2^{\eta_2}$$

so $\eta_2 = 1$.

Consider now an essential representation $\phi : \langle a, b \mid a^2 = b^3 = 1 \rangle \rightarrow \mathbb{Z}_6$ given by $a \mapsto 3$, $b \mapsto 2$, so $\overline{w} \mapsto 2(-1 + \sum_{i=2}^{k} \varepsilon_i)$ mod 6. Now $-1 + \sum_{i=2}^{k} \varepsilon_i \neq 3$ mod 6, for otherwise there exists an essential representation $G \rightarrow \mathbb{Z}_6$. Hence $\overline{w} \mapsto \pm 2$ mod 6. Since $ab \mapsto -1$ mod 6, we get

$$\begin{aligned} \mathrm{tr}\phi(ab) &= 2\cos(\pi/6) = \sqrt{3} \\ \mathrm{tr}\phi(\overline{w}) &= 2\cos(\pi/3) = 1 \end{aligned}$$

Thus

$$\psi_{\overline{w}}(\sqrt{3}) = 1.$$

But substituting $\mu = \sqrt{3}$ into 2.3 gives

$$\psi_{\overline{w}}(\sqrt{3}) = \pm 2^m$$

which gives $\eta_1 = 0$. Thus $k = 2\eta_1 + 2\eta_2 + 4\eta_3 = 2 + 4\eta_3$, which is not divisible by 4. This contradiction completes the proof. $\qquad\square$

3. THE PROOF OF THEOREM B

We need lemmas to deal with the cyclic and dihedral cases.

Lemma 3.1. *Let G be the group in 1.2, with $m \geq 7$. If G has an essential cyclic representation then, for some prime p, G has a subgroup whose pro-p completion contains a non-abelian free subgroup.*

Proof. If m is even, then Theorem 4.7 of [5] implies that G contains a non-abelian free subgroup, and hence so does the pro-p completion of that subgroup. We may therefore assume that m is odd.

The kernel of an essential representation of a generalised triangle group 1.1 to a finite group H has a presentation of deficiency $1 - \kappa \cdot |H|$ where $\kappa = (1/l + 1/m + 1/n) - 1$ (cf [5]). Thus the kernel K of an essential representation $\rho : G \rightarrow \mathbb{Z}_{2m}$ has a deficiency -1 presentation. Furthermore, if the generators of such a presentation are $x_1, \ldots, x_{m-1}$, then the relators have the form

$$v_i(x_1, \ldots, x_{m-1}) v_i(x_1^{-1}, \ldots, x_{m-1}^{-1}) \qquad (0 \leq i \leq m-1)$$

From this, it is clear that K/K' is free abelian of rank $m - 1$. That is, $d = m - 1$. Thus

$$\mathrm{def}(G) + d^2/4 - d = (m^2 - 6m + 1)/4 > 0$$

Now apply Theorem 1.2. $\qquad\square$

Proposition 3.2. *If matrices A and B over the ring $\Lambda = \mathbb{C}[\lambda]/((\lambda^2))$ are given by*

$$A = \begin{pmatrix} i & \lambda - i\alpha \\ 0 & -i \end{pmatrix}, \quad B = \begin{pmatrix} \alpha & -1 \\ 1 & 0 \end{pmatrix}$$

then

$$(AB)^2 = \lambda \begin{pmatrix} 0 & -i \\ -i & 0 \end{pmatrix} - I, \quad (AB^2)^2 = \lambda \begin{pmatrix} -i\alpha & 0 \\ -i\alpha^2 & i\alpha \end{pmatrix} - I$$

and

$$(AB^{j+3})^2 = \lambda \begin{pmatrix} -ir_{j+2}r_{j+1} & ir_{j+2}r_j \\ -ir_{j+2}^2 & ir_{j+2}r_{j+1} \end{pmatrix} - I$$

where $r_j(\alpha) = \sum_{i=0}^{j}(-1)^i \binom{j-i}{i}\alpha^{j-2i}$, for each $j \geq 0$.

The proof is by a straightforward inductive argument.

Lemma 3.3. *Let G be the group in 1.2, with $m \geq 7$ and $k > 1$. If 0 is a repeated root of τ_w then, for some prime p, G has a subgroup whose pro-p completion contains a non-abelian free subgroup.*

Proof. Since 0 is a root of the trace polynomial $\tau_w(\lambda)$, G admits an essential dihedral representation, ρ say. Let H be the preimage of $\rho(G)$ in $PSL(2, \Lambda)$ where $\Lambda = \mathbb{C}[\lambda]/((\lambda^2))$. Then there exists a homomorphism $\sigma : G \to H$ such that $\sigma \circ f = \rho$, where f is the natural map $PSL(2, \Lambda) \to PSL(2, \mathbb{C})$ induced by $\Lambda \to \mathbb{C}$. Let $K = \ker \rho$, then we have the following commuting diagram.

$$
\begin{array}{ccccc}
\ker \rho = K & \xrightarrow{\sigma} & \sigma(K) = L & \subseteq & M = \ker f \\
\downarrow & & \downarrow & & \downarrow \\
G & \xrightarrow{\sigma} & \sigma(G) \subseteq H & \subseteq & PSL(2, \Lambda) \\
& \searrow{\rho} & \downarrow{f} & & \downarrow{f} \\
& & D_{2m} & \subseteq & PSL(2, \mathbb{C})
\end{array}
$$

As in Lemma 3.1, we shall show that K satisfies inequality 1.3. Again K has a presentation of deficiency -1, so we must show that K^{ab} is generated by at least 5 elements.

Now the kernel M of f consists of elements $I + \lambda Y$ where Y is some 2×2 matrix with complex entries. For any two elements of M we have

$$(I + \lambda Y)(I + \lambda Z) = I + \lambda(Y + Z).$$

That is, M is abelian, so if L is at least 5-generated, then so is K^{ab}.

If we let σ map a and b to the matrices A and B in proposition 3.2 with $\alpha = 2\cos(\pi/m)$, then they are mapped to elements of order 2 and m respectively and $\operatorname{tr}\sigma(ab) = \lambda$. Since L is generated by $\sigma(ab^j ab^j)$ ($1 \leq j \leq m-1$), by proposition 3.2, L is generated by $X_j = I + \lambda N_j$ where

$$N_1 = \begin{pmatrix} 0 & i \\ i & 0 \end{pmatrix}, \quad N_2 = \begin{pmatrix} i\alpha & 0 \\ i\alpha^2 & -i\alpha \end{pmatrix}$$

$$N_{j+3} = \begin{pmatrix} ir_{j+2}r_{j+1} & -ir_{j+2}r_j \\ ir_{j+2}^2 & -ir_{j+2}r_{j+1} \end{pmatrix} \qquad (0 \le j \le m - 4)$$

So for L to be 5-generated, there must be 5 linearly independent matrices N_j. The top left entry in each of the N_j can be written $\alpha p_j(\alpha^2)$, where $p_j(\alpha^2)$ is a polynomial in α^2. Now if $j \ge 2$, p_j is a polynomial of degree $(j - 2)$, so if the degree t of the minimum polynomial $\chi(\alpha^2)$ of α^2 is greater than 3, the 4 matrices N_2, N_3, N_4, N_5 are linearly independent. Since $p_1 = 0$, and the other entries of N_1 are non-zero, this gives a 5th linearly independent matrix.

From Galois theory we have that $t = \phi(m)/2$, so if $\phi(m) > 6$, we have our result. This leaves the cases $\phi(m) \le 6$ namely $m = 7, 8, 9, 10, 12, 14, 18$. If $m = 8$, then Theorem A implies that G contains a non-abelian free subgroup. For the remainder, the minimum polynomial of α is found, then Maple is used to evaluate the polynomial entries in the matrices in their lowest form. A straightforward calculation then shows that $N_1, \ldots, N_5$ form a linearly independent set in each case. $\qquad\Box$

Proof of Theorem B.

If G admits an essential cyclic representation then Lemma 3.1 provides the result. We can therefore assume that G has only essential dihedral representations. Since $k \ge 2$, 0 is a repeated root of τ_w and Lemma 3.3 can be used.

REFERENCES

[1] G.Baumslag, J.Morgan and P.Shalen, Generalised triangle groups, *Math. Proc. Cambridge Phil. Soc.* **102** (1987) 25-31.

[2] G.Rosenberger, On free subgroups of generalised triangle groups, *Algebra i Logika* **28**:2 (1989) 227-240.

[3] F.Levin and G.Rosenberger, On free subgroups of generalised triangle groups, Part II, in : *Proceedings of the Ohio State-Denison Conference on Group Theory* (ed. S.Sehgal *et al*), World Scientific (1993) 206-222.

[4] B.Fine, F.Levin and G.Rosenberger, Free subgroups and decompositions of one-relator products of cyclics, Part 1: The Tits alternative, *Arch. Math.* **50** (1988) 97-109.

[5] J.Howie, Free subgroups in groups of small deficiency, *J. Group Theory* **1** (1998), 95-112.

[6] J.S.Wilson and E.I.Zelmanov, Identities for Lie algebras of pro-p groups, *J.Pure Appl. Algebra* **81** (1992), 103-109.

[7] J.S.Wilson, Finitely presented soluble groups, in P.H.Kropholler, G.A.Niblo, R.Stohr (eds.) *Geometry and Cohomology in Group Theory* London Math. Soc. Lecture Note series **252**, Cambridge University Press (1998).

[8] I.S.Gradsteyn and I.M.Ryzhik, Tables of Integrals, Series and Products (Fifth Edition), *Academic Press* (1994).

Alun G T Williams
Heriot-Watt University
Edinburgh EH14 4AS
gerald@ma.hw.ac.uk